S0-BAO-886

Over and Over Again

NEW MATHEMATICAL LIBRARY

PUBLISHED BY

The Mathematical Association of America

The New Mathematical Library (NML) was begun in 1961 by the School Mathematics Study Group to make available to high school students short expository books on various topics not usually covered in the high school syllabus. In a decade the NML matured into a steadily growing series of some twenty titles of interest not only to the originally intended audience, but to college students and teachers at all levels. Previously published by Random House and L. W. Singer, the NML became a publication of the Mathematical Association of America (MAA) in 1975. Under the auspices of the MAA the NML will continue to grow and will remain dedicated to its original and expanded purposes.

Over and Over Again

by
Gengzhe Chang
University of Science and Technology of China, Hefei, Anhui
and
Thomas W. Sederberg
Brigham Young University

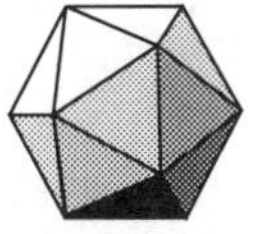

39

THE MATHEMATICAL ASSOCIATION OF AMERICA

ISBN 0-88385-641-7
Library of Congress Catalog Number 97-74344
Printed in the United States of America
Current Printing
10 9 8 7 6 5 4 3 2 1

NEW MATHEMATICAL LIBRARY

1. Numbers: Rational and Irrational *by Ivan Niven*
2. What is Calculus About? *by W. W. Sawyer*
3. An Introduction to Inequalities *by E. F. Beckenbach and R. Bellman*
4. Geometric Inequalities *by N. D. Kazarinoff*
5. The Contest Problem Book I Annual High School Mathematics Examinations 1950–1960. Compiled and with solutions *by Charles T. Salkind*
6. The Lore of Large Numbers *by P. J. Davis*
7. Uses of Infinity *by Leo Zippin*
8. Geometric Transformations I *by I. M. Yaglom, translated by A. Shields*
9. Continued Fractions *by Carl D. Olds*
10. Replaced by NML-34
11. } Hungarian Problem Books I and II, Based on the Eötvös Competitions
12. } 1894–1905 and 1906–1928, *translated by E. Rapaport*
13. Episodes from the Early History of Mathematics *by A. Aaboe*
14. Groups and Their Graphs *by E. Grossman and W. Magnus*
15. The Mathematics of Choice *by Ivan Niven*
16. From Pythagoras to Einstein *by K. O. Friedrichs*
17. The Contest Problem Book II Annual High School Mathematics Examinations 1961–1965. Compiled and with solutions *by Charles T. Salkind*
18. First Concepts of Topology *by W. G. Chinn and N. E. Steenrod*
19. Geometry Revisited *by H. S. M. Coxeter and S. L. Greitzer*
20. Invitation to Number Theory *by Oystein Ore*
21. Geometric Transformations II *by I. M. Yaglom, translated by A. Shields*
22. Elementary Cryptanalysis—A Mathematical Approach *by A. Sinkov*
23. Ingenuity in Mathematics *by Ross Honsberger*
24. Geometric Transformations III *by I. M. Yaglom, translated by A. Shenitzer*
25. The Contest Problem Book III Annual High School Mathematics Examinations 1966–1972. Compiled and with solutions *by C. T. Salkind and J. M. Earl*
26. Mathematical Methods in Science *by George Pólya*
27. International Mathematical Olympiads—1959–1977. Compiled and with solutions *by S. L. Greitzer*
28. The Mathematics of Games and Gambling *by Edward W. Packel*
29. The Contest Problem Book IV Annual High School Mathematics Examinations *1973–1982.* Compiled and with solutions *by R. A. Artino, A. M. Gaglione, and N. Shell*
30. The Role of Mathematics in Science *by M. M. Schiffer and L. Bowden*
31. International Mathematical Olympiads 1978–1985 and forty supplementary problems. Compiled and with solutions *by Murray S. Klamkin*
32. Riddles of the Sphinx *by Martin Gardner*
33. U.S.A. Mathematical Olympiads 1972–1986. Compiled and with solutions *by Murray S. Klamkin*
34. Graphs and Their Uses *by Oystein Ore.* Revised and updated *by Robin J. Wilson*
35. Exploring Mathematics with Your Computer *by Arthur Engel*
36. Game Theory and Strategy *by Philip D. Straffin, Jr.*

37. Episodes in Nineteenth and Twentieth Century Euclidean Geometry *by Ross Honsberger*
38. The Contest Problem Book V American High School Mathematics Examinations and American Invitational Mathematics Examinations 1983–1988. Compiled and augmented *by George Berzsenyi and Stephen B. Maurer*
39. Over and Over Again *by Gengzhe Chang and Thomas W. Sederberg*

Other titles in preparation.

MAA Service Center
P. O. Box 91112
Washington, DC 20090-1112
1-800-331-1MAA FAX: 301-206-9789

Note to the Reader

This book is one of a series written by professional mathematicians in order to make some important mathematical ideas interesting and understandable to a large audience of high school and beginning college students and laymen. Most of the volumes in the *New Mathematical Library* cover topics not usually included in the high school curriculum; they vary in difficulty, and, even within a single book, some parts require a greater degree of concentration than others. Thus, while you need little technical knowledge to understand most of these books, you will have to make an intellectual effort.

If you have so far encountered mathematics only in classroom work, you should keep in mind that a book on mathematics cannot be read quickly. Nor must you expect to understand all parts of the book on first reading. You should feel free to skip complicated parts and return to them later; often an argument will be clarified by a subsequent remark. On the other hand, sections containing thoroughly familiar material may be read very quickly.

The best way to learn mathematics is to *do* mathematics, and each book includes problems, some of which may require considerable thought. You are urged to acquire the habit of reading with paper and pencil in hand; in this way, mathematics will become increasingly meaningful to you.

The authors and editorial committee are interested in reactions to the books in this series and hope you will write to: Anneli Lax, New York University, The Courant Institute of Mathematical Sciences, 251 Mercer Street, New York, NY 10012.

The Editors

Preface

Transformation and iteration are two of the most basic notions in mathematics.

The three parts of this book discuss a variety of transformations and their iterations, arranged in order of sophistication. Chapters one through nineteen discuss iterations in elementary mathematics. Most problems in this part come from mathematical olympiads of different countries, many from China, drawn largely from the first author's extensive experience as coach of the Chinese delegation at the International Mathematical Olympiads (IMO).

We give special attention to transformations with a *smoothing* property. A variety of measures of smoothness occurs in our discussions. For example, for ordered n-tuples $(a_1, a_2, \ldots, a_n)$ we have occasion to consider the difference $\max_i\{a_i\} - \min_i\{a_i\}$ or the number of sign changes in the sequence; these can be regarded as measures of smoothness. Equilateral triangles can be considered the smoothest of all triangles. Similarly, the regular n-gon can be regarded as the smoothest of all n-sided polygons. In the set of all curve segments having given initial and terminal points, it is reasonable to identify the line segment joining these two points as the smoothest. Circles are considered the smoothest of all closed curves.

Two theorems contained in the first part should be spotlighted. The first (in Chapter 16) is the beautiful theorem discovered by Douglas and Neumann independently in the early 1940's; it gives a process for constructing a regular n-gon from an arbitrary n-gon by means of a sequence of transformations. Professor Neumann's simple and elementary proof is based on complex number computation, and prerequisite material is provided in Chapter 13. In contrast, the theorem in Chapter 18 is a classical result. It says that of all closed curves of a given length, the circle encloses the maximal area. The treatment of this theorem is based upon the Steiner transformation, which is smoothing.

Chapters 19–22 address functional iterations. Basic properties of continuous functions are briefly reviewed in Chapter 19. As a simple illustration Newton's method for finding roots is presented in Chapter 21. Chapter 23 discusses the main result of Li and Yorke's famous paper *Period Three Implies Chaos* and the beautiful theorem due to Sharkovskii is stated without proof. These chapters provide a basic knowledge about dynamical systems and chaos—a topic which has rapidly developed during the past two decades and which finds application in mathematics, physics, and other sciences.

The last part of the book involves Bézier curves and surfaces; they play an important role in computer aided geometric design (CAGD), a new branch of applied mathematics and computer science. They are a common research interest of the two authors. Bézier techniques are based on Bernstein polynomials, devised in the early 1900's, but which had no numerical application until the early 1960's. Bézier curves and surfaces enable a designer to produce a smooth and pleasing shape, and a program to direct a machine tool to actually create it, by adjusting the locations of control points. The designer does not need to know how this works, but the details of the process have connections with much interesting mathematics, and we explore this. For instance, we obtain estimates for the distances between the control points and the curve; this yields the Weierstrass approximation theorem. Some smoothing and convexity-preserving properties of the Bernstein transformation can be formulated as theorems. As a preparation we included a chapter on variation diminishing matrices.

Spline functions were discussed by I. J. Schoenberg in the 1940's and have become powerful tools in interpolation and approximation theory, as well as in CAGD. Chapter 27 discusses cubic spline functions for purposes of interpolation, and B-spline bases of higher degrees are presented in Chapter 28 by means of moving averages, a popular method for smoothing functions.

Chapters 1–18 require only high school mathematics, with a few exceptions. The level of the book becomes considerably more advanced in the later chapters. The additional background material required is mainly the basics of real analysis. This is not difficult for students who have an aptitude for mathematics and it is presented in an Appendix. It is the length and complexity of many of the discussions which makes the later chapters more difficult.

The first author conceived the idea for this book and began collecting material for it while serving as an associate member of the *International Center for Theoretical Physics*, Trieste, Italy. He greatly appreciated the hospitality of Professor Abdus Salam, the director of the Center. At the invitation of the second author, he first spent 20 months at Brigham Young University and greatly enjoyed the academic atmosphere at the Department of Civil Engineering and the hospitality of its faculty. In this setting both authors quickly developed a deep mutual respect, which led to the realization of this book.

Professor Qi Dongxu prepared the first draft of the chapter on moving averages. We received generous and insightful editorial assistance from Basil Gordon, Ross Honsberger, and Underwood Dudley. Peter Ungar added convergence proofs in the Bézier chapters, expanded the chapter on moving averages and wrote the chapter on variation diminishing matrices and the appendices. Anneli

Lax improved the presentation and the style throughout the book in addition to coordinating the work of the other editors, and took upon herself the labor of proofreading. We are greatly indebted to the editors for their help in this project.

Gengzhe Chang & Thomas W. Sederberg
Brigham Young University
April 1997

Contents

CHAPTER ONE

Transformations and their Iteration

The concept of *transformation* is of fundamental importance in mathematics. Examples of transformations occur even in elementary mathematics. Functions discussed in high school text books can be regarded as transformations. For example, the quadratic polynomial $f(x) = 2x^2 - 3x + 1$ represents a transformation which carries each real number x into $2x^2 - 3x + 1$, the latter being called the *image* of x under the transformation. Thus 6, 1, and 0 are the images of $-1, 0$, and 1, respectively. The trigonometric functions $f(x) = \sin x$ and $f(x) = \cos x$ can be regarded as transformations defined on the real axis. The logarithm function $f(x) = \log x$ is a transformation defined on the set of all positive numbers. We need to define "transformation" precisely.

Let M and S be two arbitrary sets. Suppose that a certain rule, f, associates with each element x in M a unique element y in S. Most commonly the rule is a formula for computing y from x, but y could also be obtained from x by an algorithm, a table or the observation of some quantity in the real world. The 'rule' f is called a *transformation* from M to S, denoted by $f : M \to S$. Furthermore we write $y = f(x)$, and y is called the *image* of x under the transformation f. The terms *map* or *function* can also be used instead of transformation.

In this book, a transformation is a more general concept than the high school textbook notion of a numerical function, as illustrated by the following examples:

EXAMPLE 1.1. Let M be any set; define $I : S \to S$ by $I(x) = x$ for every $x \in M$. This transformation is called the *identity map* of M.

EXAMPLE 1.2. Let M be the set of positive rational numbers and S the set of all ordered pairs (x, y) where x and y are positive integers. Given a rational number $r \in M$ we can write it as $r = m/n$, where m and n have no common factor and $n > 0$. Define $f : M \to S$ by $f(r) = (m, n)$.

EXAMPLE 1.3. An *n-dimensional vector* or an *ordered n-tuple* is a sequence $(a_1, a_2, \ldots, a_n)$ of n numbers; a_i is called the ith coordinate of the vector. Transformations can also be performed on vectors. For example, the vector above can be transformed into the vector $((a_1 + a_2)/2, (a_2 + a_3)/2, \ldots, (a_n + a_1)/2)$.

EXAMPLE 1.4. Transformations can also be defined on geometric figures. Let ABC be a given triangle. The midpoints of sides BC, CA, and AB are denoted by A_1, B_1, and C_1, respectively. We denote the transformation which maps $\triangle ABC$ to $\triangle A_1B_1C_1$ by T (Fig. 1.1).

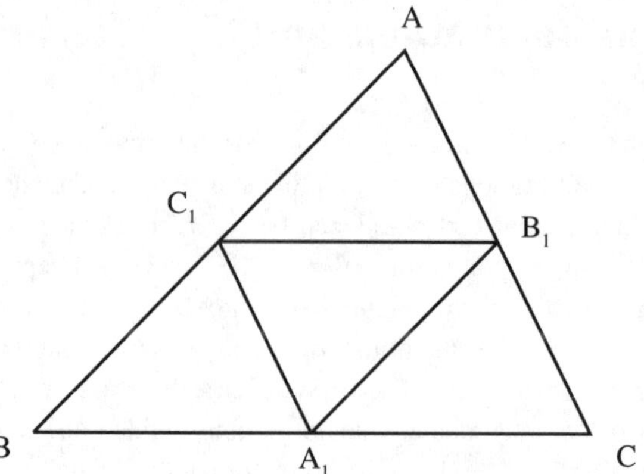

Figure 1.1. A geometric transformation.

Given a transformation $f : M \to S$; we define, for $s \in S$, the *inverse image* or *pre-image* of s with respect to f to be the set of all $m \in M$ such that $s = f(m)$. It may happen that for some s in S, its inverse image with respect to f is empty; that is, s is not the image under f of any element in M. In Example 1.2, (2, 4) is an element of S but it is not the image of any element in M under f.

DEFINITION. The transformation $f : M \to S$ is said to be *onto* S if every element of S is the image of at least one element of M.

DEFINITION. The transformation $f : M \to S$ is said to be *one-to-one* if each element of S has at most one pre-image; and if f is one-to-one and onto, it is often called a *one-to-one correspondence*.

DEFINITION. If $f : M \to S$ and $g : S \to U$, then the *composite* (or *composition*) of f and g is the transformation $g \circ f : M \to U$ defined by $g \circ f(m) = g(f(m))$ for every m in M. The word *product* is sometimes used in place of composite.

DEFINITION. Suppose that $f : M \to S$ is a one-to-one transformation of M onto S. In this case, we define a transformation $f^{-1} : S \to M$ by means of $f^{-1}(s) = m$, where $f(m) = s$; f^{-1} is called the *inverse* of f.

It is easy to see that $f^{-1} \circ f$ and $f \circ f^{-1}$ are the identity transformations on M and S, respectively.

The notation for the inverse, and the notation for iterates given are the same as those for powers. This double duty done by a limited number of available symbols is convenient, because the concepts are related and inverses and iterates obey analogues of the laws of exponents.

DEFINITION. If $M = S$, then $f : M \to M$ is often called an *operation* on M.

DEFINITION. Let $f : M \to M$. An element $m \in M$ is called a *fixed point* of the transformation f if $f(m) = m$.

We now survey the theme of this book, the *iteration* of transformations. When an operation f is repeated several times, each application of f is called an iteration. For example, suppose that a real number x is given and then we add 1 to x; this is an operation f which yields $x + 1$. If the same operation is applied to $x + 1$ then we have $(x + 1) + 1 = x + 2$. The expression $x + 2$ is called the *second iteration* of f on x. Similarly we can define the third iteration of f on x, which is $x + 3$. In general, the nth iteration of f on x yields $x + n$. If we set $x = 0$, we see that all the natural numbers are produced by this iteration.

Iterations of some functions can be easily implemented on a calculator. Let $f(x) = \sin x$. Take an arbitrarily chosen initial value x_0, for example $x_0 = 0.5$. Then start striking the function key SIN of the calculator over and over again. A few keystrokes will be enough to convince the reader that the sequence of iterates tends to zero.

DEFINITION. Let $f : M \to M$ be an operation on M. Define

$$
\begin{aligned}
f^0(x) &= x, \\
f^1(x) &= f(x), \\
f^2(x) &= f \circ f(x) = f(f(x)), \\
\dots &\quad \dots \\
f^{n+1}(x) &= f \circ f^n(x), \qquad n = 1, 2, 3, \dots.
\end{aligned}
$$

We call $f^n(x)$ the nth iteration of f, where $n = 0, 1, 2, \dots$. In this notation, f^0 always denotes the identity map. The following analogs of the laws of exponents are fairly obvious:

(i) $f^n \circ f^m(x) = f^m \circ f^n(x) = f^{n+m}(x)$;

(ii) $(f^m)^n = f^{nm}$.

We have already introduced the notation f^{-1} for the inverse function of f, in case it exists. If, for positive n, f^{-n} denotes the nth iterate of the inverse, then the above laws of indices hold without restrictions on the signs of the integers m, n.

We note that if c is a nonzero constant, the iterates of the function $f(x) = cx$ are $f^n(x) = c^n x$, where the n on the left side denotes iteration while on the right side it is an exponent. If c is positive, the formula could even be used to define fractional iterates of f.

The set M is quite general. Its elements can be real numbers, complex numbers, n-tuples, or even geometric figures like curves and surfaces.

PROBLEM 1.1. Let $f(x) = ax + b$, where a and b are constants. Find an explicit formula for $f^n(x)$.

SOLUTION. Applying our notation to this f, we get

$$f^2(x) = f \circ f(x) = f(ax + b) = a(ax + b) + b = a^2x + b(a + 1),$$
$$f^3(x) = a[a^2x + b(a + 1)] + b = a^3x + b(a^2 + a + 1),$$
$$\dots$$
$$f^n(x) = a^n x + b(a^{n-1} + a^{n-2} + \cdots + a + 1).$$

If $a = 1$, we have $f^n(x) = x + nb$; otherwise we sum the geometric series in the last line and get

$$f^n(x) = a^n x + b\frac{(a^n - 1)}{(a - 1)}. \tag{1.1}$$

We close this chapter with the following example of a common numerical computation.

PROBLEM 1.2. Find an approximate value of $\sqrt{a}$, where $a > 0$.

SOLUTION. Notice that if x is the actual square root of a, then $x = a/x$; that is, the two factors x and a/x are equal. If x is an underestimate of the square root, then a/x is an overestimate, and vice versa. The average of the underestimate and the overestimate is certainly a better estimate than at least one of them, and we hope that it is better than both. Hence we define the function

$$f(x) = \frac{1}{2}\left(x + \frac{a}{x}\right). \tag{1.2}$$

It turns out that for any given $x_0 > 0$, the sequence of iterates

$$f(x_0), f^2(x_0), \dots, f^n(x_0), \dots$$

converges quadratically to $\sqrt{a}$. Quadratic convergence means, roughly, that the number of correct digits doubles with each iteration. The number x_0 is called the *initial value* of the iteration.

To prove the above statements, let $x_n = f^n(x_0),\ n = 1, 2, \ldots$. Thus we have

$$x_{n+1} = \frac{1}{2}\left(x_n + \frac{a}{x_n}\right), \quad n = 0, 1, 2, \ldots. \tag{1.3}$$

By the well-known inequality between arithmetic and geometric means, discussed in Chapter 2, we have

$$x_{n+1} = \frac{1}{2}\left(x_n + \frac{a}{x_n}\right) \geq \sqrt{x_n \frac{a}{x_n}} = \sqrt{a}.$$

This tells us that $x_1, x_2, x_3, \ldots$ will never be less than $\sqrt{a}$, regardless of the initial value. It follows immediately that

$$x_{n+1} - x_n = \frac{1}{2}\left(x_n + \frac{a}{x_n}\right) - x_n = \frac{1}{2x_n}(a - x_n^2) \leq 0, \quad n = 1, 2, \ldots.$$

This implies that $x_1, x_2, x_3, \ldots$ is a non-increasing sequence and bounded by $\sqrt{a}$ below. Hence the sequence $\{x_n\}$ converges to a positive number, denoted by x^*. (Appendix B contains an introduction to convergence, limits and continuity.) Let n tend to infinity on both sides of equation (1.3). We get

$$x^* = \frac{1}{2}\left(x^* + \frac{a}{x^*}\right),$$

which is equivalent to $x^* = f(x^*)$, i.e., x^* is a fixed point of f. It is easy to see that $x^* = \sqrt{a}$.

Finally, let us examine the rate of convergence. Since

$$\begin{aligned} 0 \leq x_{n+1} - \sqrt{a} &= \frac{1}{2}\left(x_n + \frac{a}{x_n}\right) - \sqrt{a} \\ &= \frac{1}{2x_n}(x_n - \sqrt{a})^2 \leq \frac{1}{2\sqrt{a}}(x_n - \sqrt{a})^2, \end{aligned} \tag{1.4}$$

we see that if x_n is a good approximation to $\sqrt{a}$, then x_{n+1} will be a much better one.

For a numerical experiment, take $a = 2$. Start with $x_0 = 1$. Then successively, $x_1 = 1.5$, $x_2 = 1.41666\ldots$, $x_3 = 1.414215686\ldots$. The value of x_3 is already correct to 5 decimal digits.

Functional iterations and various applications of them will be discussed in detail in later chapters.

Exercises

1.1 An unknown concentration of a sugar and water solution completely fills a cup. Two-thirds of the solution is removed, and 10 grams of sugar are added to the remaining solution. Water is then added to fill the cup. This process is repeated 9 more times. After the cup is filled for the tenth time, g grams of sugar are determined to be in the cup. How much sugar was in the cup originally?

1.2 Define $f(x) = \frac{x}{1+x}$, $x > 0$. Find $f^n(x)$ for any positive integer n.

1.3 Let $f(x) = \frac{x}{\sqrt{1-x^2}}$. Find $f^n(x)$.

1.4 [Beijing Mathematical Olympiad (BMO) 1962] 1600 coconuts are given to 100 monkeys. Show that regardless of how the coconuts are distributed, there must be 4 monkeys with the same number of coconuts (that number may be as few as zero or as many as 400).

1.5 This problem involves five monkeys and a pile of apples on the beach. The first monkey privately divided those apples into five equal portions. There was one apple left, which he threw into the sea. He left with his portion of apples after combining the other four portions into a single pile. The second monkey came and did what the first monkey had done: 5 equal portions, 1 extra apple tossed into the sea, removed one portion. One by one, each monkey repeated this act.

What is the least possible number of apples initially on the beach? What is the least possible number of apples left after all five monkeys visited the beach?

1.6 [Chinese Mathematical Olympaid (CMO) 1988]. For any positive integer $n > 2$, let $f(n)$ denote the smallest positive integer which does not divide n; for example, $f(12) = 5$. If $f(n) > 2$, then we can consider the iterate $f^2(n)$, and so forth. Let L_n, the *length* of n, be the least j such that $f^j(n) = 2$. Determine L_n for all $n > 2$.

CHAPTER TWO

Arithmetic and Geometric Means

Let $a_1, a_2, \ldots, a_n$ be n nonnegative numbers. The inequality between their arithmetic and geometric means states that

$$A = \frac{a_1 + a_2 + \cdots + a_n}{n} \geq G = (a_1 a_2 \cdots a_n)^{1/n}. \tag{2.1}$$

with equality if and only if all the a_i's are equal. If one of the a_i's is 0, (2.1) obviously holds.

This inequality is very important in pure and applied mathematics. The following proof involves an elegant application of a smoothing transformation [Hardy et al. 1934].

It is obvious that if $a_1 = a_2 = \cdots = a_n$, then $A = G = a_1$; i.e., (2.1) holds with equality. Suppose that $a_1, a_2, \ldots, a_n$ are not all equal. Without loss of generality, we assume that a_1 is the smallest and a_2 is the largest of them. It is then clear that

$$na_1 < a_1 + a_2 + \cdots + a_n < na_2.$$

Dividing the above inequalities by n we obtain

$$a_1 < A < a_2,$$

from which it follows that $a_1 + a_2 - A > a_1 > 0$. Using the product of the a_i instead of their sum, we find similarly that

$$a_1 < G < a_2.$$

Thus we see that, if $a_1, a_2, \ldots, a_n$ are not all equal, then their arithmetic and geometric means lie between the smallest and the largest of them. When a_1 and a_2 are replaced by A and $a_1 + a_2 - A$ respectively, while the other numbers are left unchanged, we get a new set of n positive numbers:

$$A,\ a_1 + a_2 - A,\ a_3, \ldots, a_n. \tag{2.2}$$

We have defined a transformation on the set of all n-tuples with positive entries. Let A_1 and G_1 be the arithmetic mean and the geometric mean of the numbers

(2.2); we see immediately that

$$\begin{aligned} nA_1 &= A + (a_1 + a_2 - A) + a_3 + \cdots + a_n \\ &= a_1 + a_2 + \cdots + a_n = nA, \\ A &= A_1. \end{aligned}$$

From $a_1 < A < a_2$, we get

$$A(a_1 + a_2 - A) - a_1 a_2 = (A - a_1)(a_2 - A) > 0,$$

i.e., $A(a_1 + a_2 - A) > a_1 a_2$. Hence

$$A(a_1 + a_2 - A)a_3 \cdots a_n > a_1 a_2 a_3 \cdots a_n,$$

from which it follows that $G_1 > G$.

So far we have shown that the operation defined above does not change the arithmetic mean—in other words the arithmetic mean is an invariant under it—while the geometric mean is increased.

If the numbers in (2.2) are all equal then we have $A_1 = G_1$, and furthermore

$$A = A_1 = G_1 > G,$$

so inequality (2.1) is proved.

If the numbers in (2.2) are not all equal, we apply our transformation to (2.2) in order to obtain a new n-tuple; its arithmetic and geometric means are denoted by A_2 and G_2. More precisely, we rearrange the numbers (2.2) into $b_1, b_2, \ldots, b_n$, where b_1 and b_2 are the smallest and the largest numbers, respectively. We replace b_1 and b_2 by $A_1 = A$ and $b_1 + b_2 - A_1$, respectively; the other numbers remain unchanged. It is important to note that among our new numbers there are at least two A's. We also know that $A = A_1 = A_2$ and $G < G_1 < G_2$. If the latest n numbers are all equal, then we have

$$A = A_1 = A_2 = G_2 > G_1 > G,$$

and (2.1) is proved. Otherwise we continue to apply the transformation to the latest numbers. After applying it at most $n - 1$, say k times, we arrive at a vector with n equal coordinates. For this reason, we say that this is a *smoothing* transformation. Since $A_k = G_k$, we see that

$$A = A_1 = \cdots = A_k = G_k > \cdots > G_1 > G,$$

and the proof of arithmetic-geometric mean inequality is completed.

As an example of this process, consider the case $a_1 = 2$, $a_2 = 4$, $a_3 = 8$, $a_4 = 12$. The transformation described yields the following sequence of vectors:

$$\begin{gathered} (2, 4, 8, 12) \Rightarrow (4, 13/2, 15/2, 8) \\ \Rightarrow (11/2, 13/2, 13/2, 15/2) \Rightarrow (13/2, 13/2, 13/2, 13/2). \end{gathered}$$

It is obvious that any vector with equal coordinates is a fixed point of the transformation.

Other useful inequalities can be derived from (2.1). For example, suppose $a_1, a_2, \ldots, a_n$ are positive. The number

$$H = \frac{n}{\frac{1}{a_1} + \frac{1}{a_2} + \cdots + \frac{1}{a_n}}$$

is called the *harmonic mean* of $a_1, a_2, \ldots, a_n$; it is the reciprocal of the arithmetic mean of the $1/a_i$. By (2.1), we have

$$\frac{\frac{1}{a_1} + \frac{1}{a_2} + \cdots + \frac{1}{a_n}}{n} \geq \left(\frac{1}{a_1 a_2 \cdots a_n}\right)^{\frac{1}{n}} \quad \text{or} \quad \frac{1}{H} \geq \frac{1}{G}.$$

This implies

$$G \geq H \tag{2.3}$$

with equality if and only if $a_1 = a_2 = \cdots = a_n$. Combining (2.1) and (2.3), we obtain the *arithmetic-geometric-harmonic mean inequalities*

$$A \geq G \geq H, \tag{2.4}$$

with equality if and only if $a_1 = a_2 = \cdots = a_n$.

By using these inequalities we can solve some extreme value problems very easily without calculus. We illustrate the method by two examples.

PROBLEM 2.1. A rectangular box without a lid is to be constructed, so that it will contain a volume of 32 cubic feet. Find the dimension of the box that requires the least material.

SOLUTION. Let the box have length x, width y, and depth z. The amount of material used in constructing the box is proportional to the surface area, which is $xy + 2xz + 2yz$. We have the constraint $xyz = 32$. Inequality (2.1) yields

$$[(xy + 2xz + 2yz)/3]^3 \geq (xy)(2xz)(2yz) = 4(xyz)^2 = 16^3,$$

or equivalently

$$xy + 2xz + 2yz \geq 48.$$

Equality holds if and only if $xy = 2xz = 2yz$, i.e., if $x = y = 4$ and $z = 2$. The box should be 4 feet square and 2 feet deep.

PROBLEM 2.2. Find the maximum and the minimum of the function $y = x^2(1 - 3x)$ on the interval [0, 1/3].

SOLUTION. Since $1-3x \geq 0$ for x in $[0, 1/3]$, $y \geq 0$ on the interval. Hence the minimum is zero, assumed at $x = 0$ and $x = 1/3$. To find the maximum, rewrite y in the form

$$y = \frac{3}{2} x x \left(\frac{2}{3} - 2x\right).$$

The reason for taking 2/3 of the factor $1 - 3x$ in the rewritten expression is to make the sum of the three factors (and hence their arithmetic mean) independent of x. Note that x and $2/3 - 2x$ are both nonnegative. We obtain, by using the arithmetic-geometric mean inequality,

$$y \leq \frac{3}{2}\left(\frac{x + x + (\frac{2}{3} - 2x)}{3}\right)^3 = \frac{3}{2}\left(\frac{2}{9}\right)^3,$$

with equality if and only if $x = \frac{2}{3} - 2x$. We conclude that the maximum of y is $4/243$, attained at $x = 2/9$.

Countless problems in mathematical olympiads all over the world have been related to the arithmetic-geometric mean inequality. The reader can find many such problems in [Greitzer '78] and [Klamkin '86, 88].

We close this chapter by giving a construction based on the arithmetic-geometric mean inequality. It is due to Gauss who was only 14 years old at the time.

Let a and b be two distinct positive numbers. Let $a_0 = a$, $b_0 = b$, and define the successive arithmetic and geometric means:

$$(2.5) \qquad a_{n+1} = \frac{a_n + b_n}{2}, \qquad b_{n+1} = \sqrt{a_n b_n}, \quad n = 0, 1, 2, \ldots.$$

By inequality (2.1) we see that $a_n \geq b_n$ for $n = 1, 2, \ldots$. It follows that

$$a_{n+1} = (a_n + b_n)/2 \leq a_n, \qquad b_{n+1} = \sqrt{a_n b_n} \geq b_n.$$

These imply that a_n is non-increasing with a lower bound b_1 and b_n is non-decreasing with a upper bound a_1. Hence the sequences a_n and b_n both converge as n tends to infinity (see Appendix). Their limits are equal, denoted by $M(a, b)$ and called the *arithmetic-geometric mean* of Carl Friedrich Gauss (1777–1855).

A similar procedure with the arithmetic and harmonic means leads to a simple but useful result, see Exercise 2.8. Among many other ways of investigating $M(a, b)$, Gauss computed numerical values to great precision, among them $M(1, \sqrt{2}) = 1.1981402347\ldots$ He also studied integrals for which no formula was known, among them

$$(2.6) \qquad \tilde{\omega} = 2\int_0^1 \frac{dx}{\sqrt{1-x^4}} = 2.6220575542\ldots$$

On May 30th, 1799 he wrote in his diary: "We have confirmed that up to 11 digits

$$M(1, \sqrt{2}) = \pi/\tilde{\omega}. \tag{2.7}$$

When this equation is proved, a new field of analysis will surely open up." (The "we" does not mean that Gauss had an assistant. The "importance" of self-esteem for young people had not yet been recognized at the time, and saying "I did such and such" was discouraged as immodest.)

Apparently Gauss computed the right side of (2.7) and recognized it to be the same as the left side, which he had computed earlier. After about 6 months, he had a proof. Later the equality was recognized to be a consequence of an ingenious substitution discovered decades earlier by John Landen (1719–1790), and would probably have been found along that route sooner or later even without Gauss' unique combination of computing power, memory, and mathematical insight.

Gauss' discovery led to a convenient method for evaluating integrals similar to the one in (2.6), called elliptic integrals, when computations were done by hand. The algorithm defining $M(a, b)$ is similar to the square root algorithm we discussed in the first chapter, and converges at the same rapid rate.

Exercises

2.1 Prove inequality (2.1) by designing a smoothing transformation in which the geometric mean is invariant while the arithmetic mean is decreased.

2.2 Suppose that $a_1, a_2, \ldots, a_n$ are positive numbers and $b_1, b_2, \ldots, b_n$ form a rearrangement of $a_1, a_2, \ldots, a_n$. Show that

$$\sum_{i=1}^{n} \frac{a_i}{b_i} \geq n.$$

2.3 Let $a_1, a_2, \ldots, a_n$ be positive numbers. Show that

$$\left(\sum_{i=1}^{n} a_i\right)\left(\sum_{i=1}^{n} \frac{1}{a_i}\right) \geq n^2$$

with equality if and only if $a_1 = a_2 = \cdots = a_n$.

2.4 Each week Brenda and Tom shop together for sugar. Brenda always buys one pound of sugar while Tom always buys one dollar's worth of sugar. If the price of sugar is not constant from week to week, who gets the lowest average cost per pound of sugar?

2.5 [BMO 1964]. Let $a_1, a_2, a_3, \ldots$ be a sequence of positive numbers. Suppose that $a_n^2 \leq a_n - a_{n+1}$, $n = 1, 2, \ldots$. Show that

$$a_n < \frac{1}{n}$$

for all positive integers n.

2.6 [Chinese Mathematical Contest (CMC) 1984]. Let $a_1, a_2, \ldots, a_n$ be positive numbers.Show that

$$\frac{a_1^2}{a_2} + \frac{a_2^2}{a_3} + \cdots + \frac{a_n^2}{a_1} \geq a_1 + a_2 + \cdots + a_n.$$

2.7 Let $a_1, a_2, \ldots, a_n$ be positive numbers and $S = a_1 + a_2 + \cdots + a_n$. Show that

$$(1 + a_1)(1 + a_2) \cdots (1 + a_n) \leq 1 + S + \frac{1}{2!}S^2 + \cdots + \frac{1}{n!}S^n.$$

2.8 Start with two positive numbers a_0, b_0 and form the sequences

$$a_{i+1} = \frac{a_i + b_i}{2}, \qquad b_{i+1} = H(a_i, b_i).$$

where H denotes the harmonic mean we mentioned earlier. Show that the sequences $a_0, a_1, \ldots$ and $b_0, b_1, \ldots$ tend to a common limit and that this limit is the geometric mean of a_0 and b_0.

CHAPTER THREE

Isoperimetric Inequality for Triangles

Consider the set of all triangles. It is reasonable to regard equilateral triangles as the smoothest—their sides are all equal and their angles are all equal.

We consider the transformation which transforms a triangle with sides of lengths a, b, and c into the equilateral triangle with the same perimeter, i.e., into the triangle with sides

$$a' = b' = c' = \frac{a+b+c}{3}. \tag{3.1}$$

The equal perimeters of ABC and $A'B'C'$ are denoted by $2s$.

One might wonder which of the triangles ABC and $A'B'C'$ has greater area. By Heron's formula, the area of the original triangle is

$$\sqrt{s(s-a)(s-b)(s-c)}; \tag{3.2}$$

while the area of the transformed triangle is

$$\sqrt{s\frac{s}{3}\frac{s}{3}\frac{s}{3}} = \frac{\sqrt{3}}{9}s^2. \tag{3.3}$$

Note that the last 3 factors under the square root in (3.2) and in (3.3) are all positive, and that their sum is s. Hence, by the work done in the previous chapter, the area of the equilateral triangle is greater whenever the original triangle is not equilateral. If we denote the area of the original triangle by A and the perimeter of both triangles by p, so that $s = p/2$, our inequality becomes

$$A \leq \frac{p^2}{12\sqrt{3}}, \tag{3.4}$$

with equality only when the triangle is equilateral. The inequality in (3.4) is referred to as the *isoperimetric inequality for triangles*. It allows us to conclude: *the equilateral triangle is the triangle with maximum area for a given perimeter*, or equivalently, *the equilateral triangle is the triangle with minimum perimeter for a given area*.

There are transformations other than (3.1) which map triangles to triangles and keep their perimeters invariant. For example, consider the transformation

$$a_1 = \frac{b+c}{2}, \qquad b_1 = \frac{c+a}{2}, \qquad c_1 = \frac{a+b}{2}. \tag{3.5}$$

Since

$$b_1 + c_1 - a_1 = a > 0, \qquad c_1 + a_1 - b_1 = b > 0, \qquad a_1 + b_1 - c_1 = c > 0,$$

i.e., since the sum of any two of the numbers a_1, b_1, c_1 is greater than the third, there is a triangle $A_1B_1C_1$ with a_1, b_1, and c_1 as its three sides. The transformation in (3.5) maps $\triangle ABC$ to $\triangle A_1B_1C_1$ with the same perimeter, since $a_1 + b_1 + c_1 = a + b + c$; i.e., the perimeter $2s$ is an invariant of the transformation. We can also see that $a_1 = b_1 = c_1$ if and only if $a = b = c$. In other words, the image of a non-equilateral triangle under the transformation (3.5) can never be equilateral. This is a notable difference between the transformations (3.1) and (3.5). However, we will show that when (3.5) is applied over and over again, the iterated images approach an equilateral triangle.

To prove this smoothing property, denote our transformation by T; we have to show that the sequence of triangles $T^n(ABC)$ tends to an equilateral triangle as $n \to \infty$.

Let a_n, b_n, and c_n be the side lengths of $\triangle A_nB_nC_n$, $n = 1, 2, 3 \ldots$; then we have

$$a_n = \frac{b_{n-1} + c_{n-1}}{2} = s - \frac{a_{n-1}}{2},$$

or

$$a_n - 2s/3 = -\tfrac{1}{2}(a_{n-1} - 2s/3).$$

Thus the difference between a side and 1/3 of the perimeter is halved with each application of the transformation: $a_n - 2s/3 \to 0$, i.e., $a_n \to 2s/3 = (a + b + c)/3$ as $n \to \infty$. By symmetry b_n and c_n tend to the same limit. This means that the sequence of triangles $A_nB_nC_n$ tends to an equilateral triangle as n tends to infinity.

Inequality (3.4) suggests one way to decide which of two triangles is more nearly equilateral. The ratio A/p^2 depends only on the shape, not the size of a triangle. It is called the *isoperimetric quotient* of the triangle, jocularly abbreviated by Murray Klamkin as the I.Q. of a triangle. We saw that it is $\frac{1}{12\sqrt{3}}$ for equilateral triangles and smaller for all others. We have shown that if the transformation T is applied repeatedly, the limiting triangle is equilateral. From this, it does not necessarily follow that the I.Q. of $T(ABC)$ is $\geq$ the I.Q. of $\triangle ABC$, although intuition would suggest that this is true. Indeed it is, as we now show.

We want to prove that the ratio A/p^2 of $T(ABC)$ is $\geq$ that of ABC. Since the perimeter does not change when T is applied, it suffices to show that the area of $T(ABC)$ is $\geq$ that of ABC. The square of the area of the triangle ABC is

$$s(s-a)(s-b)(s-c) = \frac{s(b+c-a)(c+a-b)(a+b-c)}{8},$$

and the square of the area of $T(ABC)$ is

$$s\left(s-\frac{b+c}{2}\right)\left(s-\frac{c+a}{2}\right)\left(s-\frac{a+b}{2}\right)=\frac{sabc}{8}.$$

We want to show that the first is $\leq$ the second, i.e., that

$$(b+c-a)(c+a-b)(a+b-c) \leq abc \tag{3.6}$$

for the three sides a, b, and c of any triangle, with equality if and only if $a = b = c$. Since the sum of the lengths of any two sides of a triangle is greater than the length of the third, we have

$$0 < (b+c-a)(c+a-b) = c^2 - (a-b)^2 \leq c^2,$$

or

$$0 < (b+c-a)(c+a-b) \leq c^2,$$

and similarly

$$0 < (c+a-b)(a+b-c) \leq a^2,$$
$$0 < (a+b-c)(b+c-a) \leq b^2.$$

Multiplying the last three inequalities side by side and then taking the square root of each side, we obtain inequality (3.6), which is equivalent to the statement that the squares of the areas of triangles ABC and $T(ABC)$ satisfy the desired inequality, and hence that

$$\text{Area}(ABC) \leq \text{Area}(T(ABC)).$$

Another geometric inequality, also motivated by the isoperimetric inequality for triangles, will be discussed in the next chapter.

By expanding the left side of (3.6), we deduce a solution of an Olympiad problem.

PROBLEM 3.1 [IMO 1964/2]. Suppose a, b, c are the sides of a triangle. Prove that

$$a^2(b+c-a) + b^2(c+a-b) + c^2(a+b-c) \leq 3abc.$$

At this point, it is interesting to discuss another IMO problem, an inequality involving the three sides of a triangle.

PROBLEM 3.2 [IMO 1983/6]. Let a, b, and c be the lengths of the sides of a triangle. Prove that

$$a^2b(a-b) + b^2c(b-c) + c^2a(c-a) \geq 0. \tag{3.7}$$

SOLUTION. In establishing triangle inequalities, it is often helpful to use the substitution

$$a = (y+z)/2, \qquad b = (z+x)/2, \qquad c = (x+y)/2. \tag{3.8}$$

Its inverse is the transformation

$$x = b + c - a, \qquad y = c + a - b, \qquad z = a + b - c.$$

Thus x, y, and z are positive if and only if the triangle constraints $b + c > a$, $c + a > b$, $a + b > c$, are met. In the x, y, z formulation the problem inequality reduces to

$$[xy^3 + yz^3 + zx^3 - xyz(x+y+z)]/8 \geq 0. \tag{3.9}$$

Multiplying (3.9) by the positive number $8/xyz$, we obtain

$$\frac{x^2}{y} + \frac{y^2}{z} + \frac{z^2}{x} - x - y - z \geq 0.$$

Completing squares on the left side of the preceding inequality, we obtain

$$\left(\frac{x}{\sqrt{y}} - \sqrt{y}\right)^2 + \left(\frac{y}{\sqrt{z}} - \sqrt{z}\right)^2 + \left(\frac{z}{\sqrt{x}} - \sqrt{x}\right)^2 \geq 0,$$

and this inequality clearly holds for any positive numbers x, y, and z. Furthermore, equality holds if and only if each square is zero, and this implies that $x = y = z$. Reversing our steps we see that inequality (3.7) holds with equality if and only if $a = b = c$, or equivalently, if and only if the triangle is equilateral.

We mention here an interesting story of the IMO in 1983. Bernhard Leeb, a team member of the Federal Republic of Germany, was awarded the special prize of that year for his "one line proof" of (3.7). Leeb pointed out that (3.7) has an equivalent form

$$a(b-c)^2(b+c-a) + b(a-b)(a-c)(a+b-c) \geq 0. \tag{3.10}$$

Note that the first term in (3.10) is always nonnegative and that a cyclic permutation of (a, b, c) leaves the given inequality (3.7) unchanged, so we can assume without loss of generality that a is the maximal side length. Thus the second term in (3.10) is nonnegative too. Equality holds only if $a = b = c$.

One wonders how Leeb arrived at the form (3.10) of (3.7). The following consideration seems plausible. As we pointed out before, in dealing with an inequality involving the sides of a triangle, it is routine to use the substitution (3.8). Hence we may assume that Leeb had arrived at the inequality (3.9). Expanding the left-hand side of (3.9), we get

$$\tfrac{1}{8}(xy^3 - xy^2z - xyz^2 + yz^3 + x^3z - x^2yz) \geq 0. \tag{3.11}$$

The first three terms in (3.11) can be factored as $x(y^3 - y^2z - yz^2)$. After adding xz^3 to this expression, and subtracting the same quantity from the last three terms, we can write the entire expression as a sum of two products:

$$\frac{1}{8}(x(y-z)^2(y+z) + z(y-x)(z-x)(z+x)) \geq 0. \tag{3.12}$$

Assume that x is the minimum of the three positive numbers x, y, and z; then (3.12) holds, with equality if and only if $x = y = z$. The proof is now complete, but we can make it shorter by expressing x, y, z in (3.12) in terms of a, b, c. If we do that, we obtain Leeb's formula.[1] For other solutions, see [Klamkin '86].

Exercises

3.1 For any triangle with sides a, b, and c, show that

$$a(b-c)^2 + b(c-a)^2 + c(a-b)^2 + 4abc > a^3 + b^3 + c^3.$$

3.2 For any triangle with sides of lengths a, b, and c, show that

$$3(ab + bc + ca) \leq (a+b+c)^2 < 4(ab + bc + ca).$$

3.3 Prove that of all triangles with a given base and height, the isosceles triangle has the minimal perimeter.

3.4 Suppose a convex quadrilateral has area 1. Show that the perimeter plus the sum of the two diagonals is $\geq 4 + 2\sqrt{2}$.

[1] We contacted Bernhard Leeb and he confirmed that he had indeed proceeded via (3.8), (3.9) and (3.12).

CHAPTER FOUR

Isoperimetric Quotient

We saw in Chapter 3 that the I.Q. of a triangle is greatest for equilateral triangles. Having this in mind, Professor Murray Klamkin proposed the following problem for the eleventh USA Mathematical Olympiad (1982).

PROBLEM 4.1. If A_1 is a point in the interior of an equilateral triangle ABC and A_2 is in the interior of triangle A_1BC, prove that

$$\text{I.Q.}(A_1BC) > \text{I.Q.}(A_2BC).$$

A solution to this problem may be found in [Klamkin '88]. A somewhat different solution is as follows.

We first obtain a trigonometric formula for the I.Q. of a triangle ABC. It will save some writing to denote by $\alpha,\ \beta,\ \gamma$ the half-angles at the vertices instead of the whole angles, as one normally does. We see from Fig. 4.1 that $\triangle ABC$ can be decomposed into three triangles of height r and bases $AB,\ BC,\ CA$. Adding their areas we get

(4.1) $$\text{Area}(ABC) = rp/2 = rs.$$

We also have $AF = AE = r\cot\alpha$ with similar formulas for the other two

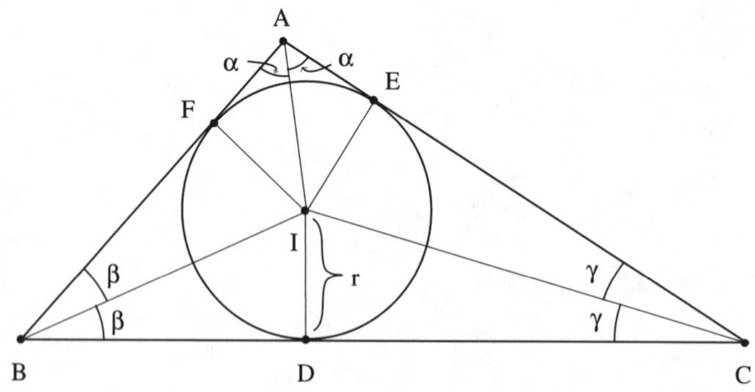

Figure 4.1. Perimeter, inradius and area.

pairs of segments. Adding these formulas, we get

$$p = 2r(\cot\alpha + \cot\beta + \cot\gamma) \tag{4.2}$$

Thus we get the trigonometric formula

$$\frac{1}{\text{I.Q.}(ABC)} = \frac{p^2}{\text{area}} = \frac{2p}{r} = 4(\cot\alpha + \cot\beta + \cot\gamma). \tag{4.3}$$

Klamkin's problem is essentially the following: A triangle A_1BC has angles $\leq 60^\circ$ at B and at C. We construct another triangle A_2BC by decreasing one or both of the angles at B and C. Show that the isoperimetric quotient decreases.

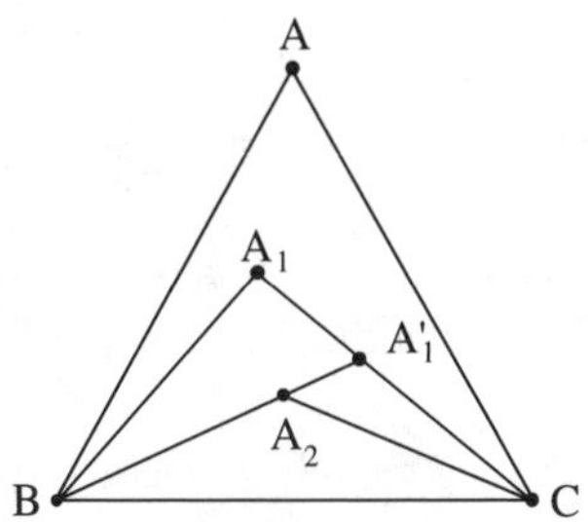

Figure 4.2. Changing one angle at a time.

We can make the change from $\triangle A_1BC$ to $\triangle A_2BC$ in two steps. First we decrease only the angle at B to obtain $\triangle A_1'BC$, and then we decrease only the angle at C to obtain $\triangle A_2BC$. (See Fig. 4.2). So it suffices to prove that the I.Q. decreases when the angle at B is decreased and the angle at C is unchanged, as in Fig. 4.3. By Formula (4.3) this amounts to showing that

$$\cot\alpha_2 + \cot\beta_2 > \cot\alpha_1 + \cot\beta_1, \tag{4.4}$$

provided that

$$\beta_2 < \beta_1 \leq 30^\circ \leq \alpha_1 < \alpha_2 < 90^\circ \quad \text{and} \quad \alpha_1 + \beta_1 = \alpha_2 + \beta_2. \tag{4.5}$$

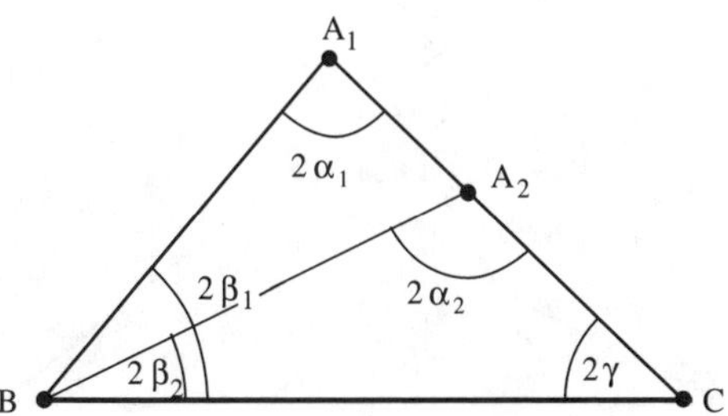

Figure 4.3. The two triangles to be compared.

From the last equation we get $0 < \alpha_2 - \alpha_1 = \beta_1 - \beta_2 = \delta$, say, so that $\alpha_1 = \alpha_2 - \delta$ and $\beta_2 = \beta_1 - \delta$. Hence Inequality (4.4) is equivalent to

$$\cot(\beta_1 - \delta) - \cot\beta_1 > \cot(\alpha_2 - \delta) - \cot\alpha_2. \tag{4.6}$$

Thus we have to show that $\cot(x - \delta) - \cot x$ is a decreasing function of x in $\delta < x < 90°$.

The addition formula for $\cot x$ is

$$\cot(x - \delta) = \frac{\cot x \cot\delta + 1}{\cot\delta - \cot x},$$

and hence

$$\cot(x - \delta) - \cot x = \frac{\cot^2 x + 1}{\cot\delta - \cot x}.$$

The numerator is positive and decreasing, and the denominator is positive and increasing in $\delta < x < 90°$, so that the function is indeed decreasing. Thus the proof is completed.

The isoperimetric quotient and the related inequality can be extended to tetrahedra, see [Klamkin '84]. In reviewing the first draft of this book, Peter Ungar suggested the following extension and geometric proof of Klamkin's inequality.

PROBLEM 4.2. If the greatest angle in triangle ABC is at A and E is any point on side AC, then

$$\text{I.Q.}(ABC) > \text{I.Q.}(EBC).$$

PROOF. Similar triangles have the same I.Q. Instead of triangle ABC we will work with triangle A_1B_1C where $A_1B_1 \parallel AB$ and passes through the midpoint H of BE (see Fig. 4.4). We show that triangle A_1B_1C has greater area and smaller perimeter than triangle EBC, which makes its I.Q. greater.

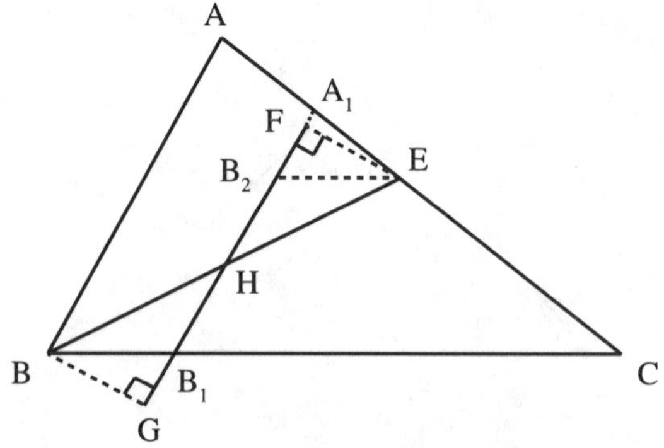

Figure 4.4. A geometric proof.

Let us first compare the areas. It is clear that

$$[A_1B_1C] - [EBC] = [HEA_1] - [HBB_1], \tag{4.7}$$

where $[K]$ denotes the area of figure K. If in triangle HEA_1 we draw a line segment parallel to BC, starting at E and ending at B_2 as shown in Fig. 4.4, we form a triangle HEB_2 congruent to HBB_1. Therefore the quantity in (4.7) is indeed positive.

Next we compare the perimeters.

$$\text{perimeter}(EBC) - \text{perimeter}(A_1B_1C) = (BB_1 - EA_1) + (EB - B_1A_1).$$

We show that both terms on the right side are positive.

The first term is equal to $B_2E - EA_1$. Since $\triangle A_1B_2E \sim \triangle ABC$ and the largest angle in $\triangle ABC$ is at A, then the largest angle of $\triangle A_1B_2E$ is at A_l, the first term is indeed positive.

The points F and G are the orthogonal projections of E and B onto the line A_1B_1, hence $FG < EB$. We show that $FG > A_1B_1$. We have

$$FG = A_1B_1 + B_1G - A_1F.$$

We note that $\angle ABC$ is acute, so G is always outside $\triangle ABC$, as shown in Fig. 4.4. The point F is inside $\triangle ABC$, as shown, if the angle at A is acute; otherwise it is not inside. In the second case, $FG > A_1B_1$, so we need to consider only the case drawn.

Now the right-angled triangles B_1GB and B_2FE are congruent because $BG = EF$ (remember H bisects EB), and the angles at B_1 and B_2 are the same. Hence $B_2F = B_1G$. That $B_2F > FA_1$ follows from the fact that the largest angle of $\triangle A_1B_2E$ is at A_1.

This completes the proof.

CHAPTER FIVE

Colored Marbles

A useful strategy in dealing with smoothing transformations is to pay special attention to the maximum and the minimum of the set of numbers being transformed. For example, in Chapter 2, the arithmetic–geometric mean inequality was proved by means of a carefully designed transformation which replaces the maximum and the minimum of a set of positive numbers by two numbers lying between them, while the rest of the numbers remain unchanged. The following example, based on a problem from the Invitational Mathematical Competition of three provinces in Northeastern China, 1987, further illustrates the usefulness of this strategy.

Given a box containing an unlimited supply of red, yellow, and blue marbles, select at random a set of 1,987 marbles from the box. Define a trading operation in which any two differently colored marbles from the selected set can be traded for two marbles of the third color from the box.

- Show that no matter how the colors are initially distributed, it is always possible to perform a finite number of trades which will result in 1,987 marbles of the same color.
- Based on the initial distribution of colors, predict what the final color must be.

To prove the first statement, let b, r, and y be the number of blue, red, and yellow marbles in the set. Note that $b + r + y = 1{,}987$. Note that the case for which $b = r$ (or $b = y$ or $r = y$) has the following simple solution. Take one blue and one red marble at a time, and then replace them by two yellow marbles. After such an operation is applied b times, we get 1,987 yellow marbles.

We treat the general case by showing that regardless of the initial color distribution, we can always apply a finite number of trade operations which will result in either $b = r$, $b = y$, or $r = y$. Without loss of generality, assume that $b < r < y$ initially. In this case, the following trade is made: one red and one yellow marble are replaced by two blue ones. After this trade the numbers of blue, red, and yellow become $b + 2$, $r - 1$, and $y - 1$, respectively. It is clear that $r - 1 < y - 1$. Note that we must have $b + 2 \le y - 1$, for if $b + 2 > y - 1$, then $2 \le y - b < 3$, i.e., $y - b = 2$. It follows that $b = r - 1$

and $y = r + 1$, thus $b + r + y = 3r = 1987$, a contradiction since 1987 is not a multiple of 3.

This shows that after the first trade, $y - 1$ is the maximum. If the three new numbers are not distinct, we have reached the special case. Otherwise, a similar trade can be made once more, and the maximum number again decreases by 1. Since y is a finite integer, the special case must occur after a finite number of such trades. Hence the first statement is proved.

This type of proof is *constructive*, which means that it gives instructions which allow us to solve specific numerical examples. We can see from the above proof that the number 1,987 is not essential and can be replaced by any positive integer which is not a multiple of 3. Suppose that we are given 14 blue, 2 red, and 1 yellow marble. The following table illustrates the desired sequence of trades:

Blue	Red	Yellow
14	2	1
13	1	3
12	3	2
11	2	4
10	4	3
9	3	5
8	5	4
7	4	6
6	6	5

We now show how to predict the final color. Imagine that the different colored marbles also have different weights:

Color	Weight
Blue	1
Red	2
Yellow	3

Denote the total weight of the set by W, and consider the variation of W after one trade. If a blue and red marble are replaced with two yellow marbles, W increases by 3. If a blue and yellow marble are replaced with two red marbles, W is not changed. If a red and yellow marble are replaced with two blue marbles, W decreases by 3. This means that the remainder of the simple division $W/3$ does not change during any trade. In mathematical jargon, $W \pmod 3$ is an invariant under any trade.

If all the marbles are blue, the total weight $W = 1987$ and $1987 \equiv 1 \pmod 3$. If all the marbles are red, the total weight $W = 2 \times 1987$ and $2 \times 1987 \equiv 2 \pmod 3$. If all the marbles are yellow, the total weight $W = 3 \times 1987$ and $3 \times 1987 \equiv 0 \pmod 3$. Therefore, if initially $W \equiv 1 \pmod 3$, finally all the marbles must be blue. If initially $W \equiv 2 \pmod 3$, then finally all the marbles must be red, and if initially $W \equiv 0 \pmod 3$, finally all the marbles must be yellow.

Without using weights, we now give a slightly simpler way of predicting the final color. Suppose that we have at a certain stage the color distribution (b, r, y) and that after some trade the color distribution becomes (b', r', y'). It is easy to verify that $b' - r' \equiv b - r \pmod 3$. This means that $b - r \bmod 3$ never changes. Analogous statements are true for $b - y$ and $r - y$. The possible final color distributions are $(1987, 0, 0)$, $(0, 1987, 0)$ and $(0, 0, 1987)$, for which $b - r \bmod 3$ is 1, 2, and 0 respectively. This implies that the final color distribution is entirely determined by the initial color distribution.

For any initial distribution of 1987 marbles, there is exactly one pair of the three numbers b, r, and y whose difference is a multiple of 3. This observation enables us to predict the final color distribution at the beginning: let (b, r, y) with $b + r + y \equiv 1 \pmod 3$ be the initial color distribution; if, say, $b \equiv r \pmod 3$, then eventually we will have all yellow marbles. For example, if initially there are 1,900 blue marbles, 80 red marbles, and 7 yellow marbles, then $b = 1{,}900$, $r = 80$, and $y = 7$. Since $b \equiv y \pmod 3$, the only single color that we can arrive at through the trading operation is red.

We have mentioned that the number of marbles can be any positive integer not divisible by 3. If the number *is* a multiple of 3, then we may not be able to achieve a single color in a finite number of trades; and even if this is possible, the final color may depend on what trades we make.

For example, take three marbles, two red and one blue. It is clear that no sequence of trades can produce three marbles of the same color, since after each trade, there will always be two marbles of one color and one of a different color.

To see that we cannot predict the final color from only the initial color distribution, consider 3 marbles colored blue, red and yellow. If one blue and one red are replaced by two yellows, we get three yellows. If one blue and one yellow are replaced by two reds, we get three reds. This means that the final result depends on the trades we make.

Exercises

5.1 1,001 cups sit right side up on a table. The operation of turning over any 100 of them at a time may be applied over and over again. Prove that no matter how many times the operation is applied, not all cups will be upside down.

5.2 Let $\{a_1, a_2, \ldots, a_{1995}\}$ be any permutation of $\{1, 2, \ldots, 1995\}$. Show that the maximum of $\{a_1, 2a_2, 3a_3, , \ldots, 1995a_{1995}\}$ must be greater than or equal to $(998)^2$.

5.3 We start a list of numbers with 1 and 2. The operation $\mathcal{O}$ consists of taking all pairs a, b of different numbers already listed and adding the numbers $a + b + ab$ to the list. Give a simple description of the numbers which will be in the list if we repeat $\mathcal{O}$ infinitely often.

CHAPTER SIX

Candy for School Children

The first problem in this Chapter illustrates again how powerful the strategy of maximum and minimum is in solving problems related to smoothing transformations.

PROBLEM 6.1 [BMO 1962]. A number of students sit in a circle while their teacher gives them candy. Each student initially has an even number of pieces of candy. When the teacher blows a whistle, each student simultaneously gives half of his or her own candy to the neighbor on the right. Any student who ends up with an odd number of pieces of candy gets one more piece from the teacher. Show that no matter how many pieces of candy each student has at the beginning, after a finite number of iterations of this transformation all students have the same number of pieces of candy.

SOLUTION. Each distribution of candy is described by an ordered set of non-negative integers. The transformation just defined carries one ordered set into another. We are required to prove that the transformation possesses a smoothing property.

Initially there is a student who has the greatest number of candies, denoted by $2m$, and there is a student who has the smallest number of candies, denoted by $2n$. If $m = n$ then all students have the same number of candies, and the transformation preserves this status quo.

Suppose $m > n$. Three observations can be made about the distribution of candy after the transformation:

1. The number of candies each student has is still between $2n$ and $2m$, inclusive.

   Suppose that a student initially has $2k$ candies and the left neighbor has $2h$ candies. After the transformation, the student has $k + h$ candies if $k + h$ is even, and $k + h + 1$ candies if $k + h$ is odd. Since $n \leq k$ and $h \leq m$, we have $2n \leq k + h \leq 2m$ in general, and $2n < k + h + 1 \leq 2m$ for odd $k + h$.

2. All students who had more than $2n$ pieces of candy still have more than $2n$ after the transformation.

   To see this, note that if $k > n$, then $h + k > n + n = 2n$.

3. At least one of the students who had $2n$ pieces of candy before the transformation, has more than $2n$ after the transformation.

There must be a student who had $2n$ candies, but whose left neighbor had more than $2n$, otherwise all students had $2n$ candies. After the transformation, this student has more than $2n$ pieces.

We conclude from the above observations that the maximum number of pieces of candy any student has cannot increase, whereas the minimal number eventually does increase unless all children have the same number of candies. Thus, after a finite number of iterations, the minimum will equal the maximum, and all students have the same amount of candy.

The iterations of the transformation defined above eventually eliminate the differences in the numbers of candies the children have; so we see that the transformation is smoothing.

In the next problem we have a transformation which can make the differences larger and larger, instead of smoothing the data.

PROBLEM 6.2 [BMO 1963]. Suppose 2^n pennies, where n is an arbitrarily given positive integer, are randomly distributed into several boxes. Take any two boxes A and B with p and q pennies, respectively. If $p \geq q$ you are allowed to remove q pennies from box A and put them into box B, and this action is called an *operation*. Show that regardless of the original distribution of pennies, a finite number of these operations can move all the pennies into one box.

SOLUTION. Suppose there are m boxes numbered $1, 2, \ldots, m$, and that the ith box contains p_i pennies. Then the specified operation is a transformation defined on m-tuples $(p_1, p_2, \ldots, p_m)$ with nonnegative integer components satisfying $\sum p_i = 2^n$. To achieve the desired result we may need different operations at different steps. Hence this may not be an iterative process according to the definition in Chapter 1. However, the operations we repeatedly apply are of the same type, so the problem still falls within the scope of this book. We shall meet similar problems in the sequel without making further comment.

We prove the desired result by mathematical induction on the power n of 2.

For $n = 1$, the number of pennies is 2. If these two pennies are already in one box, then no operation is needed. If they are in two different boxes, it is obvious that one operation can bring them into a single box.

Assume that all 2^n pennies can be put into one box by a finite number of operations, and now consider the case of 2^{n+1} pennies. We note first that the number of boxes containing an odd number of pennies is even, since otherwise the total number of pennies would be odd and hence not a power of 2. Let $2s$ be the number of boxes containing an odd number of pennies. If $s > 0$, we can find two boxes A and B each containing an odd number of pennies, say p and q, respectively. If $p \geq q$, then we move q pennies from A to B and are left with $p - q$ and $2q$ pennies in boxes A and B, respectively. Both these numbers are even. Hence after the transformation, the number of boxes containing an odd number of pennies is reduced to $2(s-1)$. Therefore, after s operations have been applied, each box contains an even number of pennies.

If in each box all pennies are paired, and if the two members of each pair are glued together to form a 'double coin', then we have 2^n such double coins. By the induction hypotheses, we can move them into a single box by applying a finite number of the permitted operations. This completes the proof.

The restriction of the number of pennies to 2^n is necessary for the conclusion. If the number is not of this special form, the conclusion may fail. For example, if we put three pennies into two boxes, one into one of them and two into the other, we see clearly that the operation does not change the pattern of the penny distribution.

Exercises

6.1 Suppose there are 4 children, and the initial distribution of candy is $0, 2, 4, 6$. Apply the operation defined in Problem 6.1 to this distribution to verify the conclusion. From this numerical example, you can see that the maximum number of candies may remain unchanged in the whole process and that the minimum number of candies need not increase in each iteration.

6.2 Suppose the number of pennies in Problem 6.2 is not a power of 2, while the rule of manipulation remains unchanged. Show that we can always get all the pennies into *one or two* boxes.

6.3 [Putnam Competition 1973] Let $x_1, x_2, \ldots, x_{2n+1}$ be integers such that, if any one of them is removed, the remaining ones can be divided into two sets of n integers with the same sum. Prove that $x_1 = x_2 = \cdots = x_{2n+1}$.

CHAPTER SEVEN

Sugar Rather Than Candy

Modifying the problem in Chapter 6, suppose that the children are given sugar rather than candy. The transformation now consists of each child simultaneously giving half of his or her sugar to the neighbor on the right. It is not surprising that, in the limit, each child will have the same amount of sugar after an infinite number of transformations. The proof is instructive.

Suppose there are n children sitting in a circle, facing the center; number them $1, 2, \ldots, n$ consecutively clockwise. Denoting by

$$x_i^{(j)}$$

the amount of sugar that the ith child has after j transformations, we see that

$$x_i^{(j)} = \tfrac{1}{2}\left(x_i^{(j-1)} + x_{i+1}^{(j-1)}\right), \quad i = 1, 2, \ldots, n, \tag{7.1}$$

where the subscripts are taken modulo n, so that $x_{n+1}^{(j)} = x_1^{(j)}$. The conclusion comes from the following theorem.

THEOREM 7.1. *Let $x_1,\ x_2, \ldots, x_n$ be arbitrary real numbers, whose arithmetic mean is $\overline{x}$. Define $x_i^{(0)} = x_i$ and*

$$x_i^{(j)} = \tfrac{1}{2}\left(x_i^{(j-1)} + x_{i+1}^{(j-1)}\right), \quad i = 1, 2, \ldots, n.$$

Then, $x_i^{(j)}$ tends to $\overline{x}$ as j tends to infinity.

PROOF. First, we make two remarks.

1. Each transformation leaves the arithmetic mean unchanged:

$$\sum_{i=1}^{n} x_i^{(j)} = \sum_{i=1}^{n} \frac{1}{2}\left(x_i^{(j-1)} + x_{i+1}^{(j-1)}\right) = \sum_{i=1}^{n} x_i^{(j-1)}.$$

   This is intuitively clear if all $x_i \geq 0$, since the total amount of sugar and the total number of children do not change in this transformation.

2. If the initial amount of sugar each child has is bounded by M, i.e., if

$$|x_i| \leq M, \quad i = 1, 2, \ldots, n, \tag{7.2}$$

then

$$\left|x_i^{(j)}\right| \le M, \quad i = 1, 2, \ldots, n,$$

for $j = 1, 2, 3, \ldots$. This is also clear since each child receives at most $M/2$ from one neighbor and has at most $M/2$ of her own sugar left after giving half to her other neighbor. We show this formally by induction:

$$\left|x_i^{(j+1)}\right| = \tfrac{1}{2}\left|x_i^{(j)} + x_{i+1}^{(j)}\right| \le \tfrac{1}{2}\left(\left|x_i^{(j)}\right| + \left|x_{i+1}^{(j)}\right|\right) \le \tfrac{1}{2}(M + M) = M.$$

The main idea of the proof of Theorem 7.1 is to show how the maximum deviations from the mean, $\max_i \left|x_i^{(j)} - \overline{x}\right|$, tend to 0 as j tends to infinity.

First consider the special case in which $\overline{x} = 0$, in other words, $x_1 + x_2 + \cdots + x_n = 0$. Denote the largest absolute value of the x_i by M, so inequality (7.2) holds. From equation (7.1) we obtain

$$\begin{aligned} x_1^{(1)} &= \tfrac{1}{2}(x_1 + x_2) \\ x_1^{(2)} &= \tfrac{1}{2^2}(x_1 + 2x_2 + x_3) \\ x_1^{(3)} &= \tfrac{1}{2^3}(x_1 + 3x_2 + 3x_3 + x_4), \end{aligned}$$

and finally

$$x_1^{(n-1)} = \frac{1}{2^{n-1}} \sum_{i=1}^{n} \binom{n-1}{i-1} x_i. \tag{7.3}$$

By assumption $x_1 + x_2 + \cdots + x_n = 0$. Divide this sum by 2^{n-1} and subtract it from (7.3) to get

$$x_1^{(n-1)} = \frac{1}{2^{n-1}} \sum_{i=1}^{n} \left[\binom{n-1}{i-1} - 1\right] x_i.$$

By (7.2) we have

$$\left|x_1^{(n-1)}\right| \le \frac{M}{2^{n-1}} \sum_{i=1}^{n} \left[\binom{n-1}{i-1} - 1\right] = \frac{M}{2^{n-1}}(2^{n-1} - n);$$

thus

$$\left|x_1^{(n-1)}\right| \le M\left(1 - \frac{n}{2^{n-1}}\right).$$

By symmetry, this bound applies to all x_i, $i = 1, 2, \ldots, n$:

$$\left|x_i^{(n-1)}\right| \le M\left(1 - \frac{n}{2^{n-1}}\right), \quad i = 1, 2, \ldots, n. \tag{7.4}$$

After another set of $n - 1$ transformations, the bound $M(1 - n/2^{n-1})$ contracts by another factor $(1 - n/2^{n-1})$, so

$$\left|x_i^{(2n-2)}\right| \le M\left(1 - \frac{n}{2^{n-1}}\right)^2, \quad i = 1, 2, \ldots, n. \tag{7.5}$$

In general we have

$$(7.6)\quad \left|x_i^{(kn-k)}\right| \le M\left(1-\frac{n}{2^{n-1}}\right)^k, \quad i=1,2,\ldots,n, \quad k=1,2,\ldots.$$

According to the second remark, for all $j \ge kn-k$ we have

$$(7.7)\qquad \left|x_i^{(j)}\right| \le M\left(1-\frac{n}{2^{n-1}}\right)^k, \quad i=1,2,\ldots,n.$$

Since $0 \le 1-n/2^{n-1} < 1$, we conclude from (7.7) that $x_i^{(j)}$ approaches zero as j tends to infinity.

We now remove the restriction $\overline{x}=0$ and consider the numbers $y_1 = x_1-\overline{x}$, $y_2 = x_2-\overline{x}, \ldots, y_n = x_n-\overline{x}$. Note that the arithmetic mean of the y_i is zero. Our linear transformation applied to these numbers yields

$$y_i^{(1)} = x_i^{(1)} - \overline{x}, \quad i=1,2,\ldots,n,$$

and after the jth iteration, we have

$$y_i^{(j)} = x_i^{(j)} - \overline{x}, \quad i=1,2,\ldots,n.$$

By what we have proved above, each of these numbers tends to zero as j tends to infinity: $\lim_{j\to\infty} y_k^{(j)} = \lim_{j\to\infty}(x_k^{(j)}-\overline{x}) = 0$. This implies that

$$\lim_{j\to\infty} x_k^{(j)} = \overline{x}, \quad k=1,2,\ldots,n$$

which completes the proof.

Exercise

7.1 One way to prove Theorem 7.1 is to find a measure of the deviation of the n-tuple $(x_1,\ldots,x_n)$ from $(\overline{x},\overline{x},\ldots,\overline{x})$ which decreases at each step and can not tend to a finite limit other than 0.

Show that $A^{(j)} = \sum_i |x_i^{(j)} - \overline{x}|$ does not increase, but need not decrease at each step but

$$S^{(j)} = \sum_i \left(x_i^{(j)} - \overline{x}\right)^2$$

does decrease at each step and tends to 0.

CHAPTER EIGHT

Checkers on a Circle

Let n be any positive integer. Place n red and black checkers, in any order, along the circumference of a circle. We define a transformation which maps the n-tuple of red and black checkers into a new n-tuple of red and black checkers: If two adjacent checkers are of the same color, place a red checker between them; if they are of different colors, place a black checker between them. Then remove all the original checkers.

We are interested in what happens when this process is repeated. The answer depends on n in a surprising way.

PROBLEM 8.1. Show that if $n = 2^m$ then after a finite number of transformations all checkers on the circle become red.

This is a reformulation, in nontechnical language, of a problem which has occurred in many competitions and has been discussed in several books and papers.

Vectors and Operators

Before we tackle the problem and its extensions, we shall spend some time presenting algebraic concepts which have many applications. The above transformation can be described algebraically as follows: Assign to each red checker the number 0, and to each black checker the number 1. Denote the initial distribution of checkers by $\mathbf{x} = (x_1, x_2, \ldots, x_n)$, where $x_i = 0$ or 1 for a red or a black checker, respectively. We regard a 1-index array of n 0's and 1's as a *vector* and denote it by a bold face lower case letter. Its *coordinates* are the n 0's and 1's denoted by corresponding ordinary lower case letters.

The reader has worked with vectors already. What is probably new in this chapter is the use of mod 2 arithmetic with vectors.

DEFINITION. Two integers h and k are said to be *congruent* modulo an integer p if their difference is divisible by p. In symbols: $h \equiv k (\bmod\ p) \leftrightarrow p \mid (h-k)$.

If we use this arithmetic then our transformation is given by the simple formula

$$(8.1) \qquad T\mathbf{x} = (x_1 + x_2,\ x_2 + x_3, \ldots,\ x_{n-1} + x_n,\ x_n + x_1),$$

where each sum $x_i + x_{i+1}$ is to be taken mod 2. Our T could be called a *transformation, mapping* or *function*. In this chapter we shall use yet another name, *operator*, for this type of object. A mapping may be called an operator only if its range is a subset of its domain, not necessarily a proper subset.

The operator T satisfies, for any scalars c, d,

$$T(c\mathbf{x} + d\mathbf{y}) = c\,T\mathbf{x} + d\,T\mathbf{y}. \tag{8.2}$$

An operator with this property is called *linear*. We might note that in the context of mod 2 arithmetic which we are using in the present chapter, we could have omitted the c and the d in equation (8.2): the only scalars in our arithmetic are 0 and 1, and hence (8.2) is equivalent to just the additive property $T(\mathbf{x} + \mathbf{y}) = T\mathbf{x} + T\mathbf{y}$.

We digress to note an inconsistency in mathematical terminology. A linear *function* $y = ax + b$ can be regarded as an operator mapping the set of real numbers into itself, but it is not a linear *operator* because (8.2) is not satisfied unless $b = 0$.

In (8.1) which defines our transformation T, the last entry has a form different from the others. We shall avoid this by the seemingly wasteful procedure of defining our x_i for *all* integer values of i, so that they form a sequence with period n, i.e., $x_{i+n} = x_i$, for all i. Now the transformation $\mathbf{y} = T\mathbf{x}$ can be described so that the same formula covers all terms:

$$y_i = x_i + x_{i+1}, \quad i = \ldots, -2, -1, 0, 1, 2, \ldots. \tag{8.3}$$

It is useful to decompose our operator T into the sum of two simpler operators I and E. I is the *identity operator*, i.e., $I\mathbf{x} = \mathbf{x}$, and E is the *shift operator* defined by

$$E\mathbf{x} = \mathbf{z}, \text{ where } z_i = x_{i+1}, \quad i = \ldots, -2, -1, 0, 1, 2, \ldots.$$

We suggest the reader spend a moment to think about the fact that E causes sequence $\mathbf{x}$ to be shifted *backwards* by one place.

We make here some rather obvious but very useful definitions which will allow us to apply many of the rules of algebra to operators.

The *sum* of two operators P, Q is the operator

$$(P + Q)\mathbf{x} \stackrel{\text{def}}{=} P\mathbf{x} + Q\mathbf{x}.$$

To be precise, we should say that the sum is defined only if the sum of two elements of the domain is defined. The *product* or *composite* PQ of two operators P and Q is the result applying first Q and then P:

$$(PQ)\mathbf{x} \stackrel{\text{def}}{=} P(Q\mathbf{x}).$$

The term "product" is used because this construct has some of the algebraic properties of multiplication, but it does not have all of them. For instance, in general, $QP \neq PQ$; that is, composition is *not* commutative.

When $P = Q$, we write $PP\mathbf{x} = P^2\mathbf{x}$; applying (or iterating) P k times is abbreviated as P^k.

We list the identities which are used in algebraic calculations. They are obvious consequences of the definitons.

1. Addition is commutative and associative:

$$P + Q = Q + P, \qquad (P + Q) + R = P + (Q + R).$$

2. Let $\mathbf{o}$ be the additive identity, i.e., the element such that $\mathbf{o} + \mathbf{x} = \mathbf{x} + \mathbf{o} = \mathbf{x}$ for all $\mathbf{x}$. (If the domain of the operators consists of sequences, as in applications discussed in the present chapter, $\mathbf{o}$ is the sequence whose terms are all 0.)
3. Operator multiplication is associative:

$$P(QR) = (PQ)R.$$

4. The identity operator I, defined by $I\mathbf{x} = \mathbf{x}$, acts as a multiplicative identity for operators, i.e., for every operator P,

$$IP = PI = P.$$

5. The distributive law is not fully valid for operators. If we multiply a sum from the right, then the distributive law

$$(Q + R)P \equiv QP + RP$$

is equivalent to the definition of the sum $(Q + R)$, so this always holds. On the other hand, for a given P the identity

$$P(Q + R) \equiv PQ + PR \tag{8.4}$$

is not necessarily true for all Q and R; the distributive law (8.4) holds for a given P and all Q and R if and only if P is linear.
6. Sums and products of linear operators are linear.

Let us return now to our operators I, E, and $T = I + E$. Note that I and E both obviously commute with I and with E. Also, I and E are linear operators and so is any polynomial in these operators. Hence, as long as all our operators are polynomials in I and E, we can use the distributive and commutative laws.

The arithmetic we are using in this chapter is integer arithmetic mod 2. Hence in this context operator addition also obeys the mod 2 addition table. For example, $E + E = 0 = I + I$, and making use of the distributive law, we get

$$\begin{aligned} T^2 &= (I + E)^2 = (I + E)(I + E) = II + EI + IE + EE \\ &= I^2 + E + E + E^2 = I + E^2 \end{aligned} \tag{8.5}$$

and

$$T^3 = (I+E)^3 = (I+E)(I+E^2) = I^2 + IE^2 + EI + E^3 = I + E + E^2 + E^3.$$

For n-tuples with n even, (8.5) has a very simple interpretation to which we need to refer later on, so we state it as a theorem.

THEOREM 8.1. *When n is even, T^2 has the same effect as T acting separately on the sequence of the even numbered elements, and the sequence of the odd numbered elements.*

Theorem 8.1 enables us to solve Problem 8.1 very quickly. For $n = 2$ we get all 0's in at most 2 steps, as one can verify by trying out all 3 possibilities. Hence for $n = 4$ we get all 0's in at most 4 steps, for $n = 8$, in at most 8 steps, etc.

Suppose now that n has an odd factor: $n = 2^m q$, where $q > 1$ is odd. By Theorem 8.1, the action of T^{2^m} is the same as that of T acting separately on each of the 2^m vectors of q elements, where each vector consists of entries of the original vector whose indices differ by integer multiples of 2^m. In this way, questions involving repeated applications of T will be reduced to the case when n is odd.

If n is odd and the initial distribution contains checkers with different colors, then after one transformation there are still checkers with different colors. For, if all had the same color after the transformation, then the red and black checkers must have been alternately distributed on the circle before the transformation, and this is possible only if there are equal numbers of red and black checkers, which can not happen when the number of checkers is odd. In other words: If n is odd, the only distributions changed into all zeros by repeated application of the transformation T are the one consisting of all ones and the one consisting of all zeros.

Even if we do not get all 0's by repeated applications of T to $\mathbf{x}$, after a while we must get a result which occurred earlier, and from then on the sequence $T^k\mathbf{x}$ is periodic in k. It is clear from (8.1) that $\mathbf{o}$ is the only vector which is unchanged by T. Another way of saying this is that if the periodic part of the sequence contains nonzero vectors, the period is >1.

At this stage, we need to discuss an ambiguity contained in our discussion so far.

In the original formulation of the transformation, a circular array was transformed into another circular array. The new counters were not in the same places as the original counters have been. Therefore it would be reasonable to consider a circular arrangement to be the same as any other which can be obtained from it by a rotation. On the other hand, when we formulated the process in terms of sequences, we assigned in $T\mathbf{x}$ the index i to the "checker" put between the original checkers with indexes i and $i+1$. Also, we did not regard two patterns as equal if one can be obtained from the other by a rotation or, in the periodic sequence representation, a shift. This did not matter in the proof that if n has no odd factor, we ultimately get all 0's. However, it does make a difference when we discuss periods. As a very simple example, when $n = 3$, there is only one pattern consisting of two 1's and one 0. If we apply T to this pattern, we again get two 1's and one 0, but the indices where these values stand are different. If we do not distinguish between patterns which differ only by a rotation, the period of our transformation is 1. On the other hand, the transformation defined by (8.3) has period 3.

First we prove the following simple fact:

THEOREM 8.2. *Suppose n is odd, and let $\mathbf{x}$ be a vector such that $\sum_{j=1}^{n} x_j \equiv 0 \pmod 2$; i.e., an even number of its components are 1's.* (Since we extended the definition of the x_i to all integers i as a sequence of period n, we should really say the number of 1's per period is even.) *Then the sequence of iterates $T^k\mathbf{x}$ is purely periodic as a function of k; that is, there exists a positive integer p such that*

$$T^p\mathbf{x} = \mathbf{x}.$$

We shall call such a p a T-period of $\mathbf{x}$. The proof is based on the following

LEMMA. *Denote the set of all n-tuples with an even number of 1's by X_0. If n is odd, T maps X_0 one-to-one onto itself.*

PROOF OF THE LEMMA. First we note that $\mathbf{y} = T\mathbf{x}$ is always in X_0 because if we go round the circle and come back to the original element, the number of color changes is even. Conversely, for any $\mathbf{y}$ with an even number of 1's, we can find an $\mathbf{x}$ such that $T\mathbf{x} = \mathbf{y}$ by setting $x_1 = 0,\ x_{i+1} = y_i - x_i$ for $i = 1, 2, \ldots, n-1$. The definition ensures that for $i < n$ the ith component of $T\mathbf{x}$ is y_i. Since both $T\mathbf{x}$ and $\mathbf{y}$ contain an even number of 1's, the nth components of these vectors must also be the same.

If we start with $x_1 = 1$, we get the other preimage of $\mathbf{y}$; it is related to the previous one by changing every 0 to 1 and every 1 to 0. Since n is odd, exactly one of these two preimages is in X_0. Thus T has an inverse on X_0, and the lemma is proved.

To prove Theorem 8.2, let $T^h\mathbf{x}$ be a vector which is repeated:

$$T^{h+p}\mathbf{x} = T^h\mathbf{x}. \tag{8.6}$$

We want to show that this equation holds also with $h = 0$. This follows from the fact that all our vectors are in the set X_0 where T has an inverse. So we can apply the inverse h times to (8.6) and reduce h to 0. This completes the proof of Theorem 8.2.

To readers who know basic linear algebra, we point out that X_0 is a vector space, and in the proof of the lemma we could have used the following theorem:

A linear mapping of a finite dimensional vector space into itself which annihilates only the zero vector is invertible.

T maps X_0 into itself and the only nonzero vector annihilated by T, the vector with $x_i = 1$ for all i, is not in X_0 because it has an odd number of 1's. So we can conclude that the restriction of T to X_0 has an inverse without calculating it. The proof of the lemma was easy enough even without this shortcut, but in more complicated problems a similar shortcut can be very useful.

Let us call a p such that $T^p\mathbf{x}$ represents the same circular pattern as $\mathbf{x}$, which in terms of sequences means $T^p\mathbf{x} = E^j\mathbf{x}$ for some j, a *pattern period* of $\mathbf{x}$.

The next theorem is quite general and useful.

THEOREM 8.3. *Every period p of a periodic sequence is an integral multiple of the smallest positive period s.*

To see this, note that as the difference of two periods p, s is also a period. Thus, if dividing p by s leaves a positive remainder r, then r is also a period because r is obtained by successive subtractions of s from p. Hence, if s is the smallest positive period, the remainder r must be 0.

THEOREM 8.4. *Let* $\mathbf{u}$ *denote the n-component vector* $(0, 0, \ldots, 0, 1)$. *(In the periodic infinite sequence form, $u_i = 1$ if i is a multiple of n and 0 otherwise.) Denote the shortest T-period of the periodic part of the sequence*

$$\mathbf{u},\ T\mathbf{u},\ T^2\mathbf{u}, \ldots, \tag{8.7}$$

by $c(n)$. Then the shortest T-period of every vector $\mathbf{x}$ *is a divisor of $c(n)$.*

PROOF. The translates $E^i\mathbf{u}$ also ultimately have the T-period $c(n)$ and since any $\mathbf{x}$ is a linear combination of $\mathbf{u}$ and its translates, $c(n)$ is also a T-period for $\mathbf{x}$, and hence $c(n)$ is a multiple of the smallest T-period of $\mathbf{x}$.

Next we note the analogous facts about the first reappearance of the original pattern, possibly in a shifted position.

Let $\check{c}(n)$ denote the smallest positive pattern period of the periodic part of the sequence (8.7), and let $\check{s}(n)$ be the shift associated with it, in the sense that

$$T^{\check{c}(n)}\mathbf{v} = E^{\check{s}(n)}\mathbf{v}.$$

$\check{s}(n)$ is defined only mod n. Please keep in mind that $E\mathbf{x}$ is the sequence $\mathbf{x}$ shifted *backwards* by 1 place, or forward by $n-1$ places. Thus if the shift produced by $T^{\check{c}(n)}$ is a forward shift by 1 place, which is the case for many values of n, we have $\check{s}(n) = -1$. If the first and hence every reappearance of the original pattern is in its original position then $\check{c}(n) = c(n)$ and $\check{s}(n) = 0$.

Since every vector is a linear combination of vectors obtained by shifting $\mathbf{u}$, $\check{c}(n)$ is a pattern period for every T-periodic sequence. The number of times we have to shift an unsymmetric pattern by i places until it returns to its original position is $n/\gcd(i, n)$. Thus

$$c(n) = \check{c}(n)\, n/\gcd(\check{s}(n), n) \tag{8.8}$$

This formula remains true when the first pattern recurrence is in an unshifted position, since $\gcd(0, n) = n$.

If n is odd then by Theorem 8.2 the sequence (8.7) is periodic from the second term $\mathbf{w} = (0, \ldots, 0, 1, 1)$ on, and $c(n)$, $\check{c}(n)$ are just the smallest positive T-period and pattern-period of $\mathbf{w}$.

We reproduce here the table of $c(n)$ in [Engel '93], p. 224, to which we have added the values of $\check{c}(n)$ and $\check{s}(n)$. The values for even n can be deduced from Problem 8.1 and Theorems 8.1 and 8.6, so they are not listed.

n	3	5	7	9	11	13	15	17	19	21	23	25	27	29	31
$c(n)$	3	15	7	63	341	819	15	255	9709	63	2047	25575	13797	475107	31
$\check{c}(n)$	1	3	7	7	31	63	15	15	511	63	2047	1023	511	16383	31
$\check{s}(n)$	−1	−1	0	−1	−1	−1	0	−1	−1	0	0	−1	−1	−1	0

n	33	35	37	39	41	43	45	47	49	51
$c(n)$	1023	4095	3233097	4095	41943	5461	4095	8388607	2097151	255
$\check{c}(n)$	31	4095	87381	4095	1023	127	4095	8388607	2097151	255
$\check{s}(n)$	−1	0	12	0	−1	−1	0	0	0	0

We see intriguing regularities. For instance, all the values of the pattern recurrence period $\check{c}(n)$ are of the form $2^m - 1$—except for $n = 37$. The nonzero $\check{s}(n)$ are all -1, except for $n = 37$ again.

We make here a digression about drawing general conclusions from patterns observed in tables. Usually, if a pattern is observed in a number of instances then it persists throughout. However, one can not be sure without a rigorous proof. The earliest instance of being misled by tables was the inference by ancient Chinese mathematicians, that $2^n - 2$ is divisible by n if *and only if* n is a prime. Pierre de Fermat (1601–1665) proved and generalized the *if* part. The great Gottfried Wilhelm Leibniz (1646–1716) once thought he had a proof of the *only if* part. However, he had made a mistake. In fact, there are composite numbers n which divide $2^n - 2$. The smallest is $11 \times 31 = 341$. Although $2^{341} - 2$ is enormous, one can verify in a few minutes with paper and pencil and using modular arithmetic that it is divisible by 11 and by 31.

We devote the rest of this chapter to proving some of the properties suggested by the table.

THEOREM 8.5. *Let n be odd.*

a) *If $d \mid n$, then $c(d) \mid c(n)$ and $\check{c}(d) \mid \check{c}(n)$.*
b) *If* $\gcd(\check{s}(n), n) = 1$, *then* $\gcd(\check{s}(d), d) = 1$. *The most useful special case of this is: If $\check{s}(n) = 1$ then $\check{s}(d) \neq 0$.*

PROOF.

a) Since d is a divisor of n, a sequence $\mathbf{x}$ with index-period d is also a sequence with index period n. Hence $c(n)$ is a T-period of $\mathbf{x}$ and by Theorem 8.3 the shortest T-period of $\mathbf{x}$ is a divisor of $c(n)$. A similar argument holds for $\check{c}(d)$.

b) Since a sequence of period d is also a sequence of period n, $T^{\check{c}(n)}$ will shift a sequence of period d with an even number of 1's per period by an amount $\check{s}(n)$, and this shift is relatively prime to d.

One might wonder whether $\check{s}(d)$ can be different from $\check{s}(n)$ if $d \mid n$. The lone different item in our table shows that this is possible. We have $\check{s}(37) = 25$. 37 is a factor of $2^{18} + 1$ and we shall prove that $\check{s}(2^m + 1)$ is always 1.

THEOREM 8.6. *If n has an odd factor, then*

$$a) \quad c(2n) = 2c(n) \qquad \text{and} \qquad b) \quad \check{c}(2n) = 2\check{c}(n). \tag{8.9}$$

Note that this formula gives us the value of $c(k)$ for all arguments in terms of the values for odd arguments.

PROOF.

a) We have seen (see proof of Theorem 8.2) that if n is odd, or if it just has an odd factor, there are vectors other than zero whose transforms form a purely periodic sequence. The T-period of such a sequence is >1. Hence $c(n) > 1$. By Theorem 8.1, $2c(n)$ is a T-period for any $2n$-dimensional initial vector. We have to show that there is an $\mathbf{x}$ which has no shorter T-period.

Let $\mathbf{x}'$ be a nonzero n-component vector such that the sequence $\mathbf{x}'$, $T\mathbf{x}'$, $T^2\mathbf{x}', \ldots$ is purely periodic, with least period p. Let $\mathbf{x}$ be the $2n$-dimensional vector whose odd numbered coordinates are those of $\mathbf{x}'$ while the even numbered coordinates are 0. We want to show that the least T-period of $\mathbf{x}$ is $2p$.

For any integer k, the odd-numbered coordinates of $T^{2k}\mathbf{x}$ are the same as the coordinates of $T^k\mathbf{x}'$ and the even numbered coordinates of $T^{2k}\mathbf{x}$ are 0, by Theorem 8.1. Applying T to such a vector leaves the odd numbered coordinates unchanged and every even numbered coordinate becomes equal to the preceding odd numbered one. This makes it clear that the shortest T-period of $\mathbf{x}$ is $2p$. It follows that $c(2n) \geq 2c(n)$. The same construction also gives b).

We shall need the following property of binomial coefficients:

THEOREM 8.7. *All entries in the* 2^m*th row* ($m = 0, 1, 2, 3, \ldots$) *of the Pascal triangle* mod 2 *are* 1.

PROOF. We remind the reader that in Pascal's triangle of binomial coefficients each entry, except the 1 in the top row, is the sum of the two closest entries in the previous row:

```
                1
              1   1
            1   0   1
          1   1   1   1
        1   0   0   0   1
      1   1   0   0   1   1
    1   0   1   0   1   0   1
  1   1   1   1   1   1   1   1
. . . . . . . . . . . . . . . . .
```

The proof is based on the construction of the Pascal triangle mod 2 in chunks of $1, 2, 4, 8, \ldots, 2^m, \ldots$ rows, as follows:

Suppose we already have the subtriangle consisting of the first 2^m rows, and the bottom of this triangle consists of 2^m 1's. Then the next row will consist of two 1's at the ends, separated by $(2^m - 1)$ 0's. Place two copies of the first 2^m rows so that these two 1's are their tips, and fill the inverted triangle between the two copies with 0's. In the enlarged triangle each entry is again the sum (mod 2) of the two entries diagonally above it, so that we now have the first 2^{m+1} rows of the Pascal triangle mod 2. The bottom row of each subtriangle we copy consists of 2^m 1's, so that the 2^{m+1}th row consists of all 1's, as claimed.

The construction of the Pascal triangle mod 2 we have just given enables us to read off the following fact which we shall need shortly:

THEOREM 8.8. *We have*

$$\binom{k}{i} = 0 \pmod 2 \quad \text{for} \quad 0 < i < k \tag{8.10}$$

if and only if k is a power of 2.

The following result shows that $c(n)$ can be as small as n infinitely often:

THEOREM 8.9. *We have*

$$c(n) \geq n, \quad \textit{with equality if and only if} \quad n = 2^m - 1. \tag{8.11}$$

PROOF. We have for any vector $\mathbf{v}$

$$T^k \mathbf{v}|_i = \sum_{j=0}^{k} \binom{k}{j} v_{i+j}. \tag{8.12}$$

(Here $|_i$ denotes the coordinate with index i.) If $k = 2^m$, this simplifies to

$$T^{2^m} \mathbf{v}|_i = v_i + v_{i+2^m} \quad \text{or} \quad T^{2^m} \mathbf{v} = \mathbf{v} + E^{2^m} \mathbf{v}. \tag{8.13}$$

We apply this to the vector $\mathbf{u}$ defined in Theorem 8.4 and with $n = 2^m - 1$. Using the fact that E^n leaves a sequence with period n unchanged, we get

$$T^{2^m} \mathbf{u} = (I + E^{n+1})\mathbf{u} = (I + E)\mathbf{u} = T\mathbf{u}. \tag{8.14}$$

Hence n is a T-period of $\mathbf{w} = T\mathbf{u}$. We have to show that a proper divisor k of n is not a T-period of $\mathbf{w}$. One can easily see from the definition of T that for $1 < k < n - 1$, the first 1 in $T^k\mathbf{w}$ has index $n - k - 1$, whereas $w_{n-k} = 0$ for these values. This proves that a proper divisor k of n is not a T-period of $\mathbf{w}$ and hence equality holds in (8.11) if $n = 2^m - 1$.

It remains to be shown that if $n + 1$ is not a power of 2 then (8.11) is false. It is easy to derive from (8.12) that $c(n) \geq n$, but since this follows from Theorem 8.11 below, which is proved from first principles, we omit this. To show that $c(n) \neq n$ if $n + 1$ is even but $\neq 2^m$, we use Theorem 8.7. According to this theorem, for such a value of n there is at least one value of i such that $0 < i < n + 1$ and $\binom{n+1}{i} = 1 \bmod 2$. We have $\binom{n+1}{1} = \binom{n+1}{n} = 0 \bmod 2$. So $\binom{n}{i} = 1$ for some i in $\{2, 3, \ldots, n - 1\}$. Hence by (8.12), $T^{n+1}\mathbf{u}|_{n-i} = 1$ for this i, and hence $T^n\mathbf{w} \neq \mathbf{w}$.

One can analyze the case $n = 2^m + 1$ in a similar way. One gets

$$T^{n-2}\mathbf{w} = T^{n-1}\mathbf{u} = \mathbf{u} + E^{n-1}\mathbf{u} = \mathbf{u} + E^{-1}\mathbf{u} = E^{-1}\mathbf{w}.$$

Thus in this case any pattern with an even number of 1's is reproduced after $n - 2$ steps, but the pattern is shifted forward by 1 place. We can argue as before that

$n-2$ is the smallest exponent k such that $T^k\mathbf{w}$ is $\mathbf{w}$ shifted by some amount. The amount of shift is -1, as measured by exponents of E, or one place forward. Thus we get

THEOREM 8.10. *We have*

$$\check{c}(n) \geq n-2 \quad \textit{with equality if and only if} \quad n = 2^m + 1, \tag{8.15}$$

and for $n = 2^m + 1$, $\check{s}(n) = 1$, *and* $c(n) = (n-2)n$.

THEOREM 8.11. $c(n)$ *is a multiple of* n.

This property of $c(n)$ was proposed as a problem in [Engel '93], p. 227, on the basis of numerical observations, but neither the author nor the editors of that book could prove it. The authors of the present book were asked to try to prove it and they did, by means of binomial identities. The problem was assigned in the Chinese Mathematical Olympiad in Hefei in January 1995. Contestants had 4 hours to solve 3 problems. Two out of 80 contestants, Liu Cong and a Russian guest Sergei Norin, discovered the following very simple solution.

Go back to the original formulation of the problem: if two adjacent checkers have the same color, place a red checker between them; if they have different colors, place a black checker between them. Then remove all the original checkers. Suppose the starting pattern has an axis of symmetry, say the vertical. After we put in the additional n checkers and even after we remove the original checkers, we still have the same axis of symmetry. Now rotate the pattern to get the new checkers into the places where the original ones had been. The formula $(T\mathbf{x})_i = x_i + x_{i+1}$ assigns the index i in the new pattern to the checker which was placed between the checkers with index i and index $i+1$ in the previous pattern. This corresponds to a rotation through $-360°/2n = -180°/n$. Thus the axis of symmetry is rotated by this amount. If the original pattern has only one axis of symmetry, it can return to its original position only when the axis of symmetry is back in its original position and hence the number of steps must be a multiple of n.

Some of the values of $c(n)$ are very large. The observation of Liu Cong and Sergei Norin gives as an upper bound for $\check{c}(n)$, n odd, the number of symmetric patterns which have a positive even number of black checkers. It is easy to count these patterns, as follows.

For an odd value of n the axis of symmetry must go through exactly one of the checkers. That checker must be red because the number of black checkers is even. The number of ways black checkers can be placed in one of the remaining half-circles is $2^{(n-1)/2} - 1$. Hence we have

THEOREM 8.12. *The following upper bounds hold for pattern periods and T-periods for odd n:*

(8.16) $$\breve{c}(n) \leq 2^{(n-1)/2} - 1 \quad \textit{and} \quad c(n) \leq n(2^{(n-1)/2} - 1) \quad \textit{for odd } n.$$

We see from the table that up to $n = 51$, the bound for $\breve{c}(n)$ is attained for $n = 5,\ 7,\ 11,\ 13,\ 19,\ 23,\ 29,\ 47$ and of these, the bound for $c(n)$ is also attained for $5,\ 11,\ 13,\ 19,\ 29$.

Let us gather in one place the exact values, divisibility properties, and inequalities we have derived for the functions $\breve{c}(n)$ and $c(n)$:

a) If q is odd, then $c(2^m q) = 2^m c(q)$ and $\breve{c}(2^m q) = 2^m \breve{c}(q)$.
b) $\breve{c}(2^m - 1) = c(2^m - 1) = 2^m - 1$.
c) $\breve{c}(2^m + 1) = 2^m - 1, \quad c(2^m + 1) = (2^m - 1)(2^m + 1) = 2^{2m} - 1$.
d) If $d \mid n$ then $\breve{c}(d) \mid \breve{c}(n)$ and $c(d) \mid c(n)$.
e) If $d \mid n$ and $\gcd(\breve{s}(n), n) = 1$ then $\gcd(\breve{s}(d),\ d) = 1$.
f) If n has an odd factor then $n \mid c(n)$.
g) $n - 2 \leq \breve{c}(n) \leq 2^{(n-1)/2} - 1$ and $c(n) \leq n(2^{(n-1)/2} - 1)$.

We have exact values when n has the form $2^m \pm 1$. We get some information from d) if we find a multiple of n which is of the form $2^m \pm 1$, especially if it is a small multiple. It is therefore relevant to know that

THEOREM 8.13. *For any odd number n, there is an exponent $m \leq n - 1$ such that $n \mid 2^m - 1$.*

PROOF. None of the n numbers $1, 2, 2^2, \ldots, 2^{n-1}$ is divisible by n. There are only $n - 1$ different residue classes other than $0 \bmod n$, so two numbers in the above list are equivalent $\bmod n$. Thus $n \mid 2^k - 2^h = 2^h(2^{k-h} - 1)$ for some $0 \leq h < k \leq n - 1$. Since n is odd, it must divide the second factor.

For information about the question of whether a given n has a multiple of the form $2^m + 1$, i.e., whether $2^m \equiv -1(\text{mod})2$ has a solution, we refer to books on number theory. Further information about cycle lengths can be found in [Furno '81].

Our relations enable us to evaluate our functions exactly for some n which are not of the form $2^m \pm 1$. For example, to evaluate $c(1365)$, we note that $3 \times 1365 = 4095 = 2^{12} - 1$. By d) and f) above, $c(1365)$ can only be one of 1365 and 4095. Since 1365 is not of the form $2^m - 1$, $c(n) > n$ by theorem 8.8 and we conclude that $c(1365) = 4095$.

Exercises

8.1 Let $d(n)$ be the number of 1's in the base 2 representation of n. Show that the number of odd binomial coefficients $\binom{n}{k}$ equals $2^{d(n)}$.

8.2 Show that in any row of the Pascal triangle the number of even entries does not equal the number of odd entries.

8.3 Let p be a prime number and k any positive integer. Show that $\binom{p^k}{i}$, $i = 1, 2, \ldots, p^k - 1$, is divisible by p, and that $\binom{p^k-1}{i}$, $i = 0, 1, 2, \ldots, p^k - 1$, is not divisible by p.

8.4 Evaluate $c(341)$.

8.5 Give the missing details of the proof of Theorem 8.9.

8.6 The set of integers $\{1, 2, \ldots, 2n\}$ is arbitrarily divided into two subsets each of which consists of n integers $\{a_i : a_1 < a_2 < \cdots < a_n\}$ and $\{b_i : b_1 > b_2 > \cdots > b_n\}$. Calculate the sum $\sum_{i=1}^{n} |b_i - a_i|$.

8.7 Let n red and black checkers be placed equidistantly along the circumference of a circle. The transformation is the same as defined in Problem 8.1. After the transformation is carried out k-times ($k \geq 1$), we find that the numbers of red and black checkers are equal. Show that n is a multiple of 4.

8.8 The first author was the leader of the Chinese Olympiad team at the 29th IMO in July 1988 in Canberra, Australia. Problem pamphlets written in Russian were distributed as presents from the Soviet team. One of the problems was the following.

Let $a_1, a_2, \ldots, a_n$ be arbitrarily chosen real numbers. For another arbitrary real number α, form

$$|a_1 - \alpha|, |a_2 - \alpha|, \ldots, |a_n - \alpha|.$$

Thus, we have defined a transformation which carries one n-tuple into another n-tuple. The transformation may be applied again and again with different values of α. Two questions are posed regarding this sequence of transformations:

1. Show that a finite sequence of transformations exists which results in an n-tuple whose elements are all zero (or, which *annihilates* the n-tuple).
2. What is the shortest sequence of transformations which can accomplish this in all cases?

CHAPTER NINE

Decreasing Sets of Positive Integers

We begin with *Euclid's algorithm* for finding the greatest common divisor of two integers.

Suppose that a, b are integers. One says that b *divides* a (written $b \mid a$) if there exists an integer c such that $a = bc$. In this case b is referred to as a *divisor* of a, and a is called a *multiple* of b. It is clear that if $c \mid b$ and $b \mid a$ then $c \mid a$.

We shall frequently appeal to *division with remainder*: For any given integer a (the dividend) and any given positive integer b (the divisor) there exist integers q (the quotient) and r (the remainder) such that

$$a = bq + r, \quad 0 \le r < b.$$

Let $\lfloor x \rfloor$ denote the *floor* function, i.e., the greatest integer which is $\le x$. The function is also called the greatest integer function and in many books it is denoted by $[x]$. Then $q = \lfloor a/b \rfloor$ and $r = a - bq$. Since $q \le a/b < q + 1$, we have $bq \le a < bq + b$, i.e., $0 \le r < b$. Keep in mind that for noninteger negative values of x, the greatest integer function of x has a greater absolute value than x. For example, if $a = -9$ and $b = 4$, the integer part of the quotient is $q = -3$ and the remainder is $r = 3$.

Let a and b be two integers, not both 0. An integer d is called a *common divisor* of a and b if $d \mid a$ and $d \mid b$. By the *greatest common divisor* (or GCD) of a and b we mean the greatest among all common divisors of a and b. Clearly, the GCD of two integers is unique. It is customary to denote the GCD of a and b by (a, b). We make the convention that $(0, 0) = 0$.

We are now in a position to find the greatest common divisor in a constructive way.

The following method for finding d, the GCD of a and b with $b > 0$, was described by Euclid, who worked in Alexandria approximately 303–275 B.C.

EUCLID'S ALGORITHM. By division with remainder there exist integers q_1, r_1 such that

$$a = bq_1 + r_1, \quad 0 \le r_1 < b.$$

If $r_1 \neq 0$ then there exist integers q_2, r_2 such that

$$b = r_1 q_2 + r_2, \quad 0 \leq r_2 < r_1.$$

If $r_2 \neq 0$ then there exist integers r_3, q_3 such that

$$r_1 = r_2 q_3 + r_3, \quad 0 \leq r_3 < r_2.$$

Continuing thus, one obtains a decreasing sequence of integers $r_1, r_2, \ldots$ satisfying

$$r_{j-2} = r_{j-1} q_j + r_j.$$

Since a decreasing sequence of positive integers cannot contain infinitely many terms, we must have $r_{k+1} = 0$ for some k; i.e., $r_{k-1} = r_k q_{k+1}$. It is then readily verified that the GCD of a and b is $d = r_k$. Indeed it is evident from the equations above that every common divisor of a and b divides $r_1, r_2, \ldots, r_k$; and moreover, viewing the equations in the reverse order, it is clear that r_k divides each r_j and hence also divides b and a.

Euclid's GCD algorithm can be viewed as an iteration of division with remainder. Many readers will be familiar with it, and we include it here to illustrate an important principle employed in its derivation: *any decreasing sequence of positive integers must terminate.*

Pythagoras allegedly proved that the square root of 2 is an irrational number. Plato's dialogue *Theaetetos* refers to proofs of the irrationality of square roots of 2, 3, 5, 6, 7, 8, 10, 11, 12, 13, 14, 15 and possibly 17 by the mathematician Theodorus of Cyrene. Why he stopped at 17 is a subject of lively speculation, but the general tone of modern writers is to sympathize with the early giants of the classical age for not having had a proof of the uniqueness of factorization of integers into prime factors. If they had been able to use the fact (a proof of which can be found in a Euclid's *Elements* written about 100 years later), that if a prime p divides a^2, then p divides a, they would have seen through the problem at a glance.

In preparing a talk to high school students, Sagher (see [Sagher '88]) came up with the following proof for

PROBLEM 9.1. If a positive integer k is not a square, then $\sqrt{k}$ is irrational.

SOLUTION. The proof does not depend on properties of prime numbers, and so was fully accessible to the classical Greek mathematicians as well as to modern high school students after one year of algebra. (See also [Zippin '62])

Suppose $\sqrt{k} = m/n$, where m and n are positive integers with no common divisor. Then $k = m^2/n^2$. If k is not a square, then m/n is not an integer. There exists an integer q such that

$$q < \frac{m}{n} < q + 1.$$

Since $\frac{kn}{m} = \frac{m}{n}$, we have $\sqrt{k} = \frac{m}{n} = \frac{kn-qm}{m-qn}$. From $q < \frac{m}{n} < q+1$ we get $0 < m - qn < n$. Thus we have obtained a new fraction for $\sqrt{k}$ with a denominator smaller than the one in the original fraction. Continuing, we can create an infinite decreasing sequence of positive integers, an impossibility.

The following is a problem from the 27th IMO held in Warsaw, Poland.

PROBLEM 9.2 [IMO 1986/6]. To each vertex of a pentagon an integer is assigned such that the sum of all the five numbers is positive. If three consecutive vertices are assigned the numbers x, y, z respectively and $y < 0$ then the following operation is allowed: the numbers x, y, z are replaced by $x + y, -y, z + y$ respectively. Such an operation is performed repeatedly as long as at least one of the five numbers is negative. Determine whether this procedure necessarily comes to an end after a finite number of steps.

The answer to this question is: the five integers *will* be nonnegative after a finite number of operations.

SOLUTION. Let $x_1, x_2, \ldots, x_5$ be the five numbers. To preserve symmetry we define x_n for all integers n by means of the equation $x_m = x_n$ if $m \equiv n \pmod 5$. Define also $s = x_1 + \cdots + x_5$.

It is useful to look at the "sum sequences", i.e., the sequences such that $X_n - X_{n-1} = x_n$ for $n = \ldots - 2, -1, 0, 1, 2, \ldots$. Any such sequence can be obtained from any other by adding a constant to each term. Any one of the sum sequences will do for our purposes and we do not specify which one $\ldots X_{-1},\ X_0,\ X_1, \ldots$ is. We have $X_{n+5} = X_n + s$ for all n. So, in the long run, the terms of a sum sequence are at a bounded distance from the linear increasing sequence $sn/5$. Suppose now that $x_2 < 0$ and we perform the sign-changing operation on x_2. Then the modified sequence is $x_1' = x_1 + x_2, x_2' = -x_2, x_3' = x_3 + x_2, \ldots$. Hence a sum sequence of the modified sequence is $\ldots X_1' = X_2, X_2' = X_1, X_3' = X_3, \ldots$, i.e., a sum sequence of the modified sequence is obtained by interchanging X_{5k+1} and X_{5k+2}, $(k = \ldots, -1, 0, 1, \ldots)$. We can do this only when the terms which are moved backward are smaller than the terms with which they are interchanged. Since the long term trend of each of the sum sequences is up with slope $s/5$, this means our transformation, in some sense, smoothes the sum sequence. We need to find a function of $x_1, \ldots, x_5$ which measures the roughness of the sequence and will decrease with each application of the transformation.

The following function f was devised by U.S. team member Joseph Keane. He was awarded the only special prize of the 1986 Olympiad for it. We define Keane's f in terms of a sum sequence rather than the original sequence because the expression is somewhat simpler. Put

$$f = \sum_{i=1}^{5} \sum_{j=1}^{4} |X_{i+j} - X_i|.$$

Note that the summand is unchanged if we replace i by $i \pm 5$. So i can be summed over any 5 values which represent all congruence classes modulo 5. Thus all values of i enter Keane's function on an equal footing. Thus it suffices to prove that the sign changing operation applied, say, to X_2 decreases the value of f.

We saw that the new values X'_n are a certain permutation of the previous values. Consequently, most of the terms in the double summation are just permuted. The only term in f whose two constituents are moved apart too much to enter the sum f' is the former $|X_6 - X_2|$ which is $|X'_7 - X'_1|$. The new term in f' is $|X'_6 - X'_2| = |X_7 - X_1|$. Thus

$$f' - f = |X_7 - X_1| - |X_6 - X_2| = |s + x_2| - |s - x_2| < 0,$$

because s is positive and x_2 is negative. Since the value of f is a positive integer and decreases with each application of the transformation, the process must terminate.

Keane did not use the sum sequence. Having introduced it, Bernard Chazelle of Princeton (see *Mathematics Magazine* **60** (1987), p. 58) was able to spot an even better function f:

$$f = \text{the number of pairs } i, j \text{ such that } 1 \le i \le 5, i < j, X_i > X_j.$$

He noted that this function decreases by one at each step and shows not only that the number of steps is finite but also that it does not depend on the choices we make when there is more than one negative x. Interestingly enough, his method proves that the process terminates even if the x's are not integers.

There are many choices for the functions with the desired properties. For example, define

$$f(x_1, \ldots, x_5) = (x_1 - x_2)^2 + (x_2 - x_3)^2 + (x_3 - x_4)^2 + (x_4 - x_5)^2 + (x_5 - x_1)^2,$$

and suppose $x_4 < 0$. The function value decreases by $-2sx_4$, a positive integer, at each step. The rest of the argument remains as before.

Exercise

9.1 Show that $(x, y, z) = (0, 0, 0)$ is the unique integer solution to the equation

$$x^2 + y^2 + z^2 = 2xyz.$$

CHAPTER TEN

Matrix Manipulations

A set of mn numbers arranged in a rectangular array

$$\begin{bmatrix} a_{11} & a_{12} & \dots & a_{1n} \\ a_{21} & a_{22} & \dots & a_{2n} \\ \dots & \dots & \dots & \dots \\ a_{m1} & a_{m2} & \dots & a_{mn} \end{bmatrix} \tag{10.1}$$

of m rows and n columns is called an $m \times n$ (read "m by n") *matrix*. The matrix (10.1) can be denoted by the capital letter A or by the condensed symbol (a_{ij}); a_{ij} represents the entry in the ith row and jth column of the matrix (10.1). An $n \times n$ matrix is called a *square matrix of order* n. If all $a_{ij} = 0$, then the matrix A is called a *zero matrix*.

Often problems and their solutions become more understandable if expressed in terms of matrices, as we shall see from the following two problems.

PROBLEM 10.1 [BMO 57]. Along a road there are six stops, numbered consecutively $1, 2, \dots, 6$. A van runs from the first stop to the sixth. Passengers can get on and off at each stop. At any time, at most five passengers are allowed to occupy the van. Show that there exist four different stops i, j, k, l such that no passenger who gets on at i gets off at j, and no passenger who gets on at k gets off at l.

This was the hardest problem of the 1957 Beijing Mathematical Olympiad. Only a few students presented correct solutions. After the competition, a lengthy solution to this problem appeared in a Chinese intermediate mathematics journal. One year later, a new solution given by two university students was published in the same journal. The following is the matrix version of their solution.

SOLUTION. The number of passengers who get on the van at the ith stop and get off at the jth stop is denoted by d_{ij}, where $i < j$. Form the triangular array with these entries d_{ij}:

$$\begin{array}{ccccc} d_{12} & d_{13} & d_{14} & d_{15} & d_{16} \\ & d_{23} & d_{24} & d_{25} & d_{26} \\ & & d_{34} & d_{35} & d_{36} \\ & & & d_{45} & d_{46} \\ & & & & d_{56} \end{array}$$

Using the 9 numbers in the upper right corner of the triangular array we form the following 3×3 matrix:

$$D = \begin{bmatrix} d_{14} & d_{15} & d_{16} \\ d_{24} & d_{25} & d_{26} \\ d_{34} & d_{35} & d_{36} \end{bmatrix}$$

The sum of all the entries in D is the number of passengers on the van when it runs from the third stop to the fourth. By assumption, this sum cannot exceed 5. Hence at least 4 entries of matrix D must be zero. Among these four zeros, we can always find two zeros that are neither in the same row nor in the same column. For instance, if $d_{14} = 0$ and $d_{26} = 0$ then we see that no passenger gets on at the first and off at the fourth stop, and that no passenger gets on at the second and off at the last stop.

PROBLEM 10.2 (PUTNAM COMPETITION 1965). At a party, no boy dances with every girl, but each girl dances with at least one boy. Prove there are two couples bg and $b'g'$ who dance, but b does not dance with g' nor g with b'.

SOLUTION. Let m and n be the numbers of boys and girls respectively. Form an $m \times n$ matrix $A = (a_{ij})$ in which $a_{ij} = 1$ if the ith boy dances with the jth girl; otherwise $a_{ij} = 0$. In terms of this matrix the given conditions are

(a) every row contains at least one 0.

(b) every column contains at least one 1.

We wish to prove that there are two rows and two columns in A such that the entries at their intersection points have one of the patterns

$$\begin{bmatrix} \cdots & 1 & \cdots & 0 & \cdots \\ \cdots & \cdots & \cdots & \cdots & \cdots \\ \cdots & 0 & \cdots & 1 & \cdots \end{bmatrix} \quad \text{or} \quad \begin{bmatrix} \cdots & 0 & \cdots & 1 & \cdots \\ \cdots & \cdots & \cdots & \cdots & \cdots \\ \cdots & 1 & \cdots & 0 & \cdots \end{bmatrix}$$

Suppose row h contains a maximal number of 1's. According to condition (a), there must be a 0 entry in this row, say in column k. From condition (b), we know that column k has at least one entry of 1, say in row s. Now we conclude that there is a column which contains a 1 in row h and a 0 in row s, since otherwise row s would have more 1's than row h, a contradiction.

The next problem is a rephrasing of one in [BMO, 1988].

PROBLEM 10.3. Arrange 64 symbols 1 and -1 in a 8×8 square array. Such an array is called *desirable* if its first row and first column contain 1's only, and if in any two different rows (columns) there are exactly 4 pairs of identical symbols in the same column (row). Find a desirable array.

SOLUTION. We consider the following matrices (where $-H_i$ denotes the matrix obtained from H_i by replacing all its entries by their negatives):

$$H_1 = (1)$$

$$H_2 = \begin{bmatrix} H_1 & H_1 \\ H_1 & -H_1 \end{bmatrix} = \begin{bmatrix} 1 & 1 \\ 1 & -1 \end{bmatrix}$$

$$H_4 = \begin{bmatrix} H_2 & H_2 \\ H_2 & -H_2 \end{bmatrix} = \begin{bmatrix} 1 & 1 & 1 & 1 \\ 1 & -1 & 1 & -1 \\ 1 & 1 & -1 & -1 \\ 1 & -1 & -1 & 1 \end{bmatrix}$$

$$H_8 = \begin{bmatrix} H_4 & H_4 \\ H_4 & -H_4 \end{bmatrix}$$

It is easy to see that a square array corresponding to H_8 is a desirable array.

This problem is related to *Hadamard matrices*. A *Hadamard matrix* $A = (a_{ij})$ of order n has all its elements $a_{ij} = \pm 1$ and satisfies

$$\sum_{k=1}^{n} a_{ik}a_{jk} = 0 \quad \text{for all} \quad i \neq j.$$

It can be shown that this implies

$$\sum_{k=1}^{n} a_{ki}a_{kj} = 0 \quad \text{for all} \quad i \neq j.$$

In mathematical language, these two equations express the fact that different rows (or different columns) of A are *orthogonal*. Since the left side of each equation is a sum of n terms, either 1 or -1, it follows that n must be even. In fact, one can show that if $n > 2$ then it has to be a multiple of 4. The existence of Hadamard matrices of order $4k$ (k an arbitrary positive integer) remains an open problem. However, we can show that Hadamard matrices of order 2^k exist, by means of the induction used in the above solution to define H_1, H_2, H_4, and H_8.

PROBLEM 10.4. Let A be an $m \times n$ matrix. A row sum (column sum) is defined as the sum of all the numbers in a row (column). If A has a negative row sum or a negative column sum, then each number in that row or column may be multiplied by -1. Such an operation may be performed repeatedly as long as one of the row or column sums is negative. Show that this procedure necessarily comes to an end after a finite number of steps.

SOLUTION. The sum of all the entries of the matrix increases with each operation, since the sum of the row or column which was negative becomes positive and the sum of the other entries does not change. Thus we never get the same matrix again. But there are only finitely many different matrices one can get by changing signs in A.

Exercises

10.1 Let $A = \{a_{ij}\}$ be an $n \times n$ matrix. The entries $a_{11}, a_{22}, \ldots, a_{nn}$ form the *main diagonal* of A. If $a_{ij} = a_{ji}$ for all i and j, then A is called a *symmetric matrix*. Let A be a symmetric $n \times n$ matrix, where n is odd. Show that if each row and each column of A is a permutation of $1, 2, \ldots, n$, then the main diagonal of A is also a permutation of $1, 2, \ldots, n$.

10.2 [IMO 71 Problem 6]. Let $A = (a_{ij})$, $(i, j = 1, 2, \ldots, n)$ be a matrix whose entries are nonnegative integers. Suppose that whenever an entry $a_{ij} = 0$, the sum of the entries in the ith row plus the sum of those in the jth column is $\geq n$. Prove that the sum of all the entries of the matrix is $\geq n^2/2$.

CHAPTER ELEVEN

Nested Triangles

In 1956, the following problem was proposed in The American Mathematical Monthly:

PROBLEM 11.1. Let ABC be an arbitrary triangle. Let the points of contact with its incircle be A_1, B_1, C_1, see (Fig. 11.1). Let the points of contact of $\triangle A_1B_1C_1$ with its incircle be A_2, B_2, C_2, and so on. Let a_n, b_n, c_n be the sides of triangle $A_nB_nC_n$, and let r_n be the radius of its incircle. Show that

$$\lim_{n\to\infty} \frac{r_n}{a_n} = \lim_{n\to\infty} \frac{r_n}{b_n} = \lim_{n\to\infty} \frac{r_n}{c_n} = \frac{\sqrt{3}}{6}. \tag{11.1}$$

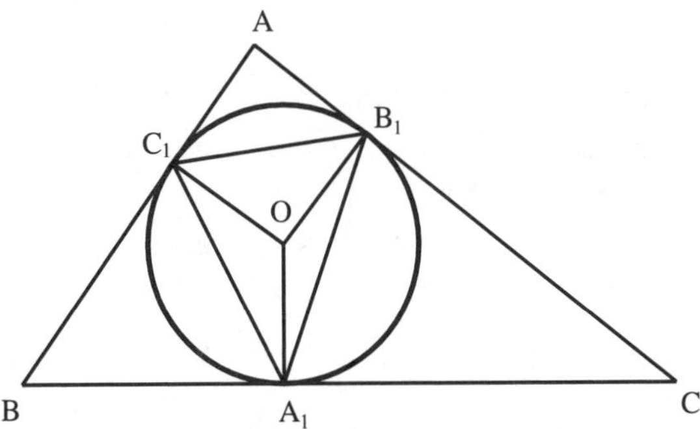

Figure 11.1. Incircle.

SOLUTION. Equation (11.1) is clearly equivalent to the statement that the angles A_n, B_n, and C_n approach $\pi/3$ as n approaches infinity. This indicates that the transformation from $\triangle ABC$ to $\triangle A_1B_1C_1$ is smoothing.

Let O be the incenter of triangle ABC. In the quadrilateral OB_1AC_1, the angles at B_1 and C_1 are right angles; thus $\angle B_1OC_1 = \pi - A$. Since $\angle B_1A_1C_1 = \angle B_1OC_1/2$, we have $A + 2A_1 = \pi$. In general,

$$A_{n-1} + 2A_n = \pi, \quad n = 1, 2, 3, \ldots \tag{11.2}$$

and hence

$$A_n - \pi/3 = -\tfrac{1}{2}(A_{n-1} - \pi/3), \tag{11.3}$$

We see that A_n rapidly approaches $\pi/3$ as n tends to infinity. The same holds for B_n and C_n, and the proof is complete.

Let us denote our transformation by T. If we apply T to an equilateral triangle, the result is an equilateral triangle with half the linear dimensions. This suggests that we consider a modified transformation T^* which consists of T, followed by a doubling of the linear dimension of the resulting triangle. (The position of $T^*\triangle ABC$ in the plane is unimportant; any two congruent triangles are equivalent for our purpose.) The transformation T^* leaves an equilateral triangle unchanged. Since the angles of our triangles rapidly approach 60° under repeated applications of T or T^*, it is plausible and can easily be proved that under repeated applications of T^* the sides all tend to the same limit, yielding a limit equilateral triangle of a fixed size. What is surprising is that there is a simple formula for the limit side length, which we will now derive.

We go back to the original transformation T. Let K, K_1 be the areas of $\triangle ABC$, $\triangle A_1B_1C_1$. Let R denote the circumradius (the radius of the circumscribed circle) of triangle ABC. Its area K, side lengths a, b, c, and circumradius R are related by the formula

$$K = \tfrac{1}{2}ab \sin C = 2R^2 \sin A \sin B \sin C. \tag{11.4}$$

We can also express the area in terms of the radius r of the inscribed circle, as a sum of areas of triangles with bases a, b, and c:

$$K = r\frac{a+b+c}{2} = rR(\sin A + \sin B + \sin C). \tag{11.5}$$

To proceed further, we note first that by (11.2), $\frac{1}{2}A + A_1 = 90°$ and hence

$$\sin A_1 = \cos \tfrac{1}{2}A, \qquad \sin B_1 = \cos \tfrac{1}{2}B, \qquad \sin C_1 = \cos \tfrac{1}{2}C.$$

We can make further progress thanks to the fortunate circumstance that when $A + B + C = 180°$, the following identity holds:

$$\begin{aligned} \sin A + \sin B + \sin C &= 4\cos\left(\frac{A}{2}\right)\cos\left(\frac{B}{2}\right)\cos\left(\frac{C}{2}\right) \\ &= 4 \sin A_1 \sin B_1 \sin C_1. \end{aligned} \tag{11.6}$$

To verify this, use $\angle C = 180° - (A + B)$, so $\sin C = \sin(A + B)$, $\cos C = -\cos(A + B)$, and apply repeatedly the addition theorems of trigonometry.

Substituting (11.6) in (11.5) we get

$$K = 4Rr \sin A_1 \sin B_1 \sin C_1. \tag{11.7}$$

Note now that $r = R_1 =$ radius of circumcircle of $\triangle A_1B_1C_1$. We apply (11.4) to the small triangle to get $r \sin A_1 \sin B_1 \sin C_1 = K_1/2r = K_1/2R_1$. Substituting this in (11.7) we get

$$K = 2R\frac{K_1}{R_1} \quad \text{or} \quad \frac{K}{R} = 2\frac{K_1}{R_1}. \tag{11.8}$$

Now we are ready to discuss T^*. Let K_1^*, R_1^* be the area and the circumradius of the image of $\triangle ABC$ under T^*. We have $K_1^* = 4K_1$ and $R_1^* = 2R_1$. Thus (11.8) becomes

$$\frac{K}{R} = \frac{K_1^*}{R_1^*}, \tag{11.9}$$

i.e., area/circumradius is invariant under the transformation T^*! Thus the limit our triangles approach under iterations of T^* is the equilateral triangle with area/circumradius $= K/R$. An equilateral triangle with side s, has area/circumradius $= \frac{3}{4}s$. Hence the side of the equilateral limit triangle must be

$$s = \frac{4K}{3R} = \frac{8}{3}R \sin A \sin B \sin C.$$

The next problem, though not involving nested triangles, is closely related to the previous example in the sense that the relations between the angles of successive triangles are the same.

PROBLEM 11.2. Lines l and m are parallel edges of a strip of paper, and P_1, Q_1 are points on l and m, respectively (see Fig.11.2). Fold the paper so that P_1Q_1 lies along l, making a crease P_1Q_2. Next, fold the paper so that P_1Q_2 lies along m, obtaining a new crease P_2Q_2. Next fold so that P_2Q_2 lies along l and get the crease P_2Q_3. Show that, if the process is continued indefinitely, the triangles $P_nP_{n+1}Q_{n+1}$ become more and more nearly equilateral.

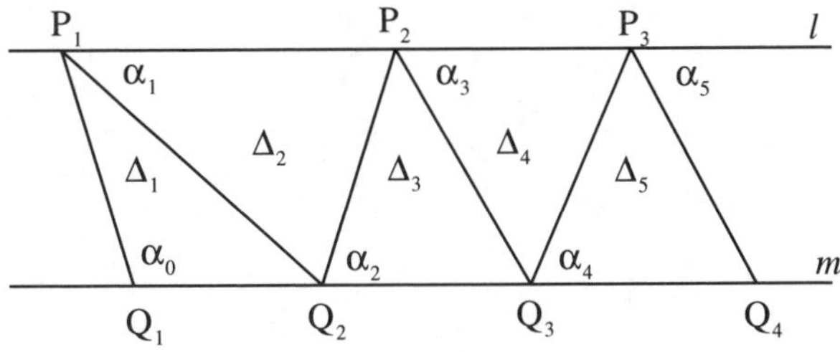

Figure 11.2. Folding a paper strip.

(The problem was proposed by Daniel Pedoe in a Canadian mathematical journal [Pedoe '79] with the following remark: "I don't know the origin of this problem. A student asked me to prove it many years ago").

SOLUTION. (given by Jeremy D. Primer, a student of Columbia High School, Maplewood, New Jersey. He also mentioned that an equivalent problem appeared in the October, 1978 issue of *The Mathematics Student*.)

Let α_0, $0 < \alpha_0 < \pi$, be the angle that P_1Q_1 makes with line m on the side where the folds are made; and let $\alpha_1, \alpha_2, \ldots$ be the angles, and $\Delta_1, \Delta_2, \ldots$ the triangles, formed by the successive creases, as shown in Fig. 11.2. Since lines l and m are parallel and the angles adjacent to a crease are equal, the angles of Δ_1 are $\alpha_0, \alpha_1, \alpha_1$, and it follows by induction that the angles of triangle Δ_n are $\alpha_{n-1}, \alpha_n, \alpha_n$. Hence we have

$$\alpha_{n-1} + 2\alpha_n = \pi, \quad n = 1, 2, 3, \ldots$$

which has the same pattern as (11.3). Thus we conclude that α tends to $\pi/3$ as n goes to infinity.

Exercise

11.1 Draw a circle with radius 1, and inscribe an equilateral triangle in it. In the triangle inscribe another circle; in the second circle, inscribe a regular hexagon. Continue with a circle in this hexagon, and follow with a regular 12-sided figure in the circle. Repeat this procedure, each time doubling the number of sides of the previous regular polygon. Let r_n be the circumradius of the nth polygon. Find the limit of r_n as n tends to infinity.

CHAPTER TWELVE

Morley's Theorem and Napoleon's Theorem

This chapter discusses three transformations which carry an arbitrary triangle to an equilateral triangle in a single iteration.

Frank Morley (1860–1937) made the remarkable discovery about 1899 that the intersections of adjacent pairs of angle trisectors in a triangle are vertices of an equilateral triangle (Fig. 12.1).

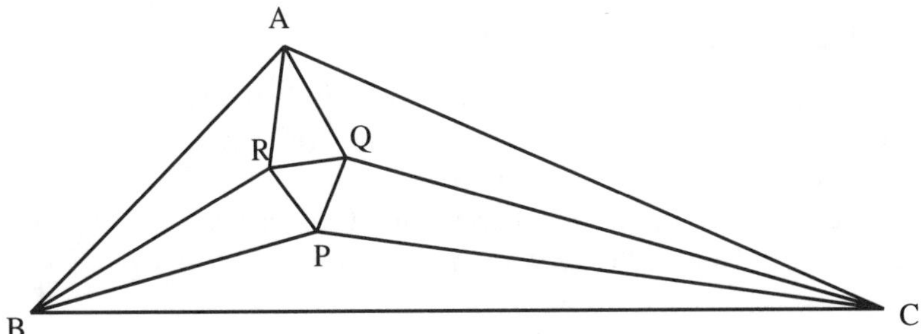

Figure 12.1. Morley's Theorem.

This is one of the most surprising theorems in elementary geometry, illustrating that many theorems of geometry are entertaining indeed. Impressed with its beauty, Morley showed this theorem to his friends, and it quickly spread to the rest of the mathematical world as an interesting item for gossip. However, the proof of the theorem was not published until 1914 [Coxeter '69; Kay '69]. Various proofs of Morley's Theorem can now be found in the literature.

THEOREM 12.1 (MORLEY). *In any triangle, the three points of intersection of adjacent angle trisectors form an equilateral triangle.*

PROOF. The treatment given here appeared first in 1922, see [Coxeter '69]. We first state a simple fact to be used in the proof.

LEMMA. *In $\triangle ABC$, let the exterior angle at A (the supplement of $\angle BAC$) be ϵ. If I is a point on the angle bisector of $\angle BAC$, and if the supplement of $\angle BIC$ is $\epsilon/2$, then I is the incenter of $\triangle ABC$.*

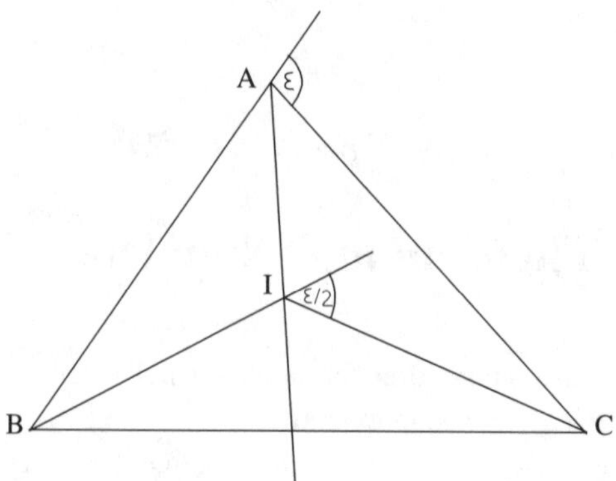

Figure 12.2. Lemma.

The proof is an easy exercise (see Fig. 12.2).

We prove Morley's Theorem by working backwards, beginning with an equilateral triangle and building up a general triangle to be identified afterwards with the given triangle ABC.

On the sides QR, RP, PQ of a given equilateral triangle PQR (Fig. 12.3), erect isosceles triangles $P'QR$, $Q'RP$, $R'PQ$ whose base angles α, β, γ satisfy

$$\alpha + \beta + \gamma = 120^\circ,$$

$$\alpha < 60^\circ, \quad \beta < 60^\circ, \quad \gamma < 60^\circ.$$

Extend the sides of the isosceles triangles beyond their bases until they meet again in points A, B, C. Since $\alpha + \beta + \gamma + 60^\circ = 180^\circ$, we can immediately infer the measures of some other angles, as marked in Fig. 12.3.

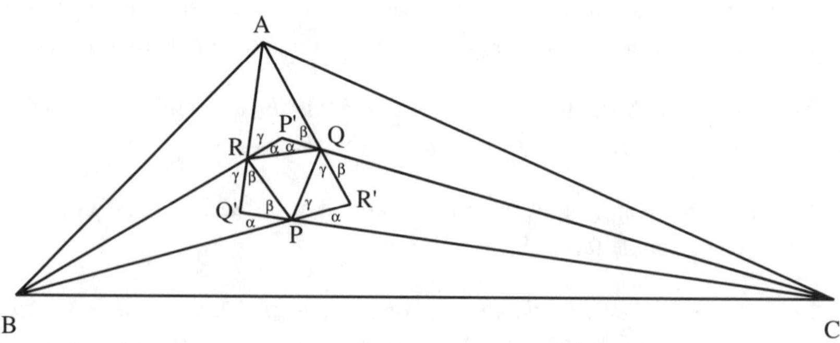

Figure 12.3. Proof of Morley's Theorem.

Consider $\triangle P'BC$. It is clear that PP' bisects the angle at P'. The exterior angle at P' is 2α and the supplement of $\angle BPC$ is α. By the Lemma P is the incenter of $\triangle P'BC$. Likewise Q is the incenter of $\triangle Q'CA$, and R of $\triangle R'AB$. In other words, the angles of $\triangle ABC$ are trisected. In $\triangle ARQ$, we see that $A/3 = \pi - (\gamma + \alpha + \alpha + \beta) = 60^\circ - \alpha$, Similarly $B/3 = 60^\circ - \beta$ and $C/3 = 60^\circ - \gamma$. Now suppose $\triangle ABC$ is given. By choosing

$$\alpha = 60^\circ - \frac{A}{3}, \qquad \beta = 60^\circ - \frac{B}{3}, \qquad \gamma = 60^\circ - \frac{C}{3},$$

we can ensure that the above procedure yields a triangle similar to it.

Next, we turn the discussion to Napoleon's Theorem.

Let ABC be an arbitrary triangle whose area is K. Sides BC, CA and AB have lengths a, b, and c, respectively. With these sides as bases, construct three isosceles triangles BCA', CAB', ABC' each having base angles $\pi/6$, and each directed outward from ABC. Triangle $A'B'C'$ is called the *outer Napoleon triangle* of triangle ABC (Fig. 12.4).

If we do the same thing inwardly, the *inner Napoleon triangle* is obtained. Hence we have defined two geometric transformations which carry any triangle into its outer Napoleon triangle and inner Napoleon triangle. We will show that these two transformations are smoothing by verifying that the Napoleon triangles are always equilateral. In this case smoothing is immediate; equilateral triangles are arrived at after a single iteration.

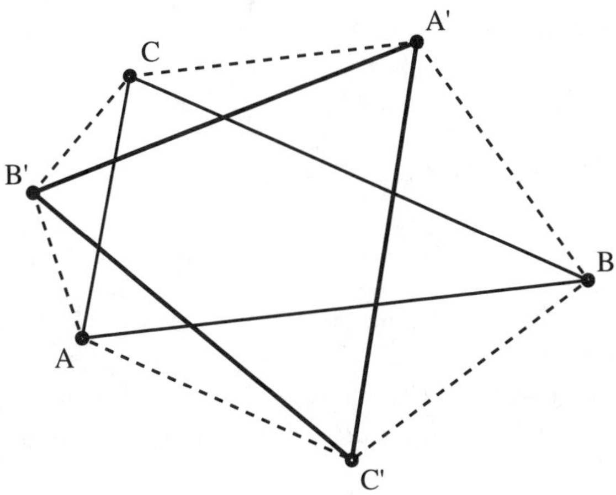

Figure 12.4. Outer Napoleon Triangle.

In Fig. 12.4 we have $A'B = a/\sqrt{3}$ and $BC' = c/\sqrt{3}$ by the law of sines, applied to triangles $A'BC$ and ABC'. Hence, by the law of cosines in $\Delta A'BC'$,

$$(C'A')^2 = (A'B)^2 + (BC')^2 - 2(A'B)(BC')\cos\left(B + \frac{\pi}{3}\right)$$
$$= \frac{a^2+c^2}{3} - \frac{2}{3}ac\left(\cos B\cos\left(\frac{\pi}{3}\right) - \sin B\sin\left(\frac{\pi}{3}\right)\right)$$
$$= \frac{a^2+c^2}{3} - \frac{ac}{3}\cos B + \frac{\sqrt{3}}{3}ac\sin B.$$

Since $ac\cos B = (a^2+c^2-b^2)/2$ and $ac\sin B = 2K$, we deduce

$$(C'A')^2 = \frac{a^2+b^2+c^2}{6} + \frac{2\sqrt{3}}{3}K. \tag{12.1}$$

Note that the right side of (12.1) is invariant under permutations of a, b, c so the left side must be invariant under permutations of A, B, C. Thus $C'A' = A'B' = B'C'$, and triangle $A'B'C'$ is equilateral.

If $A''B''C''$ is the inner Napoleon triangle of ABC, we obtain similarly

$$(C''A'')^2 = \frac{a^2+b^2+c^2}{6} - \frac{2\sqrt{3}}{3}K, \tag{12.2}$$

which implies that the inner Napoleon triangle is also equilateral.

Many interesting results can easily be derived from (12.1) and (12.2). First of all, by adding (12.1) and (12.2), we find that the sum of the squared sides of ΔABC is equal to the sum of the squared sides of the two Napoleon triangles. Next, the area of the outer Napoleon triangle is

$$\frac{\sqrt{3}}{24}(a^2+b^2+c^2+4\sqrt{3}K), \tag{12.3}$$

and the area of the inner Napoleon triangle is

$$\frac{\sqrt{3}}{24}(a^2+b^2+c^2-4\sqrt{3}K); \tag{12.4}$$

their difference is K, the area of the original triangle. The above results lead to

THEOREM 12.2 (NAPOLEON THEOREM). *The outer and inner Napoleon triangles of any triangle are equilateral; the difference of their areas is the area of the original triangle.*

There are additional interesting consequences: The right side of (12.4), being the area of the inner Napoleon triangle, must be nonnegative. If it is zero, then the inner Napoleon triangle reduces to a single point; this can occur only if the original triangle is equilateral. Thus we have a new proof for

PROBLEM 12.1 [IMO 1961/2]. Let a, b, c be the sides of a triangle and K its area. Prove:

$$a^2 + b^2 + c^2 \geq 4\sqrt{3}K,$$

In what case does equality hold?

PROBLEM 12.2. A triangle $A_1A_2A_3$ with positive (counterclockwise) orientation and a point P_0 are given. We construct three points P_1, P_2, P_3 such that P_{k+1} is the image of P_k under a clockwise rotation with center A_{k+1} through an angle of 120° (for $k = 0, 1, 2$). Prove that if $P_0 = P_3$, then the triangle $A_1A_2A_3$ is equilateral.

SOLUTION. Consider the triangle $P_{k-1}A_kP_k$, $k = 1, 2, 3$. By construction, these are isosceles triangles with base angles $\pi/6$. Hence, $A_1A_2A_3$, the outer Napoleon triangle of $P_1P_2P_3$, is of course equilateral.

The above problem is closely related to [IMO 1986/3] which will be discussed in Chapter 16. For the discussions and solutions to this and other problems, it will be convenient to use complex numbers as vectors in the plane. This topic is discussed in Chapter 13.

Exercises

12.1 Prove the following by working backwards: Inside a square $ABDE$, take a point C such that CDE is an isosceles triangle with angles of 15° at D and E. Show that the triangle ABC is equilateral.

12.2 Consider the proof of Morley's Theorem. What values of α, β, γ will make triangle ABC:

1. equilateral?
2. right-angled isosceles?

Sketch the figure in each case.

12.3 Let a, b, c be the sides of a triangle and let K be its area. Show that

$$a^2 + b^2 + c^2 \geq 4\sqrt{3}K + (b - c)^2 + (c - a)^2 + (a - b)^2$$

with equality if and only if the triangle is equilateral.

CHAPTER THIRTEEN

Complex Numbers in Geometry

We start with an explanation of complex numbers, for readers unfamiliar with them.

Around the year 1520 the Italian mathematician Scipione del Ferro discovered how to solve a cubic equation in terms of square roots and cube roots. He refused to publish the formula, as did Niccolo Tartaglia who managed to figure out what it was. Girolamo Cardano, betraying Tartaglia's trust, published it and the formula came to be called Cardano's formula. Every cubic equation can be simplified to $x^3 + px + q = 0$ with a linear substitution, and for the last equation Cardano's formula is

$$x = \sqrt[3]{-\frac{1}{2}q + \sqrt{\frac{1}{4}q^2 + \frac{1}{27}p^3}} + \sqrt[3]{-\frac{1}{2}q - \sqrt{\frac{1}{4}q^2 + \frac{1}{27}p^3}}.$$

The formula has a puzzling deficiency. Whenever p and q are such that the equation has 3 distinct (real) solutions, e.g., for $p = -1$ and $q = 0$, the expression under the square root is negative.

Over the next two centuries, mathematicians came to believe that it is permissible to include the square roots of negative numbers, which they called imaginary numbers, in computations. Nevertheless, simply assuming the existence of a thing with a property we would like it to have could lead to false conclusions. Wishful thinking often does. So it is reassuring that Jean Robert Argand (1768–1822) managed to extend the real number system to one in which negative numbers have square roots and which still forms a *commutative field,* by using only simple and familiar mathematical concepts.

Commutative field means that the properties we apply when we perform algebraic computations are valid in our system. These properties are:

Addition and multiplication are associative and commutative and they obey the distributive law.

Addition of 0 and multiplication by 1 leaves every number unchanged.

Every number has an additive inverse (negative) and every number other than 0 has a multiplicative inverse (reciprocal).

Real numbers can be thought of as points on a number line. Multiplication by -1 is a 180° rotation about the origin. A 180° rotation can be achieved by two 90° rotations. So a 90° rotation should represent a multiplication by $\sqrt{-1}$. This suggests that in order to create a square root for -1, we should extend the number line to a number plane. We are now going to describe Argand's algebra of vectors in a plane.

Take a plane with a rectangular coordinate system. To simplify the notation, we will not always distinguish between the point (x, y) and the vector (x, y). The latter is represented by a directed line segment from the origin to the point (x, y).

The sum $z_1 + z_2$ of two vectors z_1, z_2 is defined as the diagonal of the parallelogram with sides z_1 and z_2; see Fig. 13.1. It can be constructed by displacing z_2 in the direction of, and by the length of, z_1; the line segment from the origin to the endpoint of the displaced vector is the vector $z_1 + z_2$. Equivalently, the diagonal from the origin of the parallelogram with sides z_1 and z_2 represents $z_1 + z_2$. If we interchange the summands we get the same result, so addition is commutative.

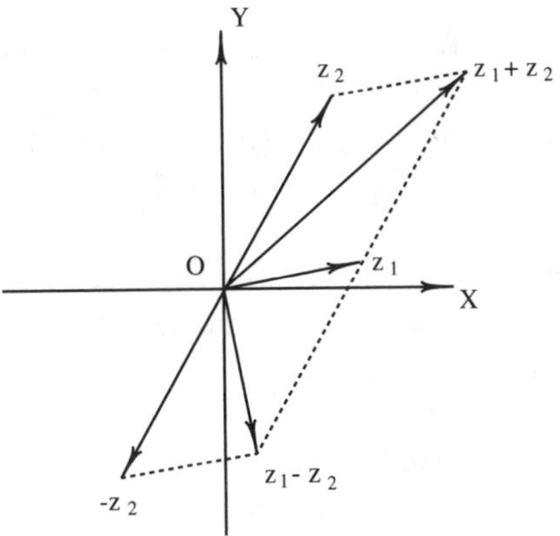

Figure 13.1. Addition of complex numbers.

The vector $(0, 0)$ will be denoted by 0. For every $z = (x, y)$, we define $-z$ to be $(-x, -y)$. We have $z + (-z) = 0$.

The *absolute value* $|z|$ of z is just its length $\sqrt{x^2 + y^2}$. The direction of z may be specified by the angle θ from the positive x-axis to z. (Fig. 13.2). The angle θ is called the *argument* of z and is denoted by $\arg z$. For $z = 0$ we can use any angle as argument. When z is specified by giving $r = |z|$ and $\theta = \arg z$, we say it is given in *polar coordinates.*

We come to the definition of a product of our vectors. Let z_1, z_2 have polar coordinates (r_1, θ_1) and (r_2, θ_2). We define the *product* $z = z_1 z_2$ as the vector

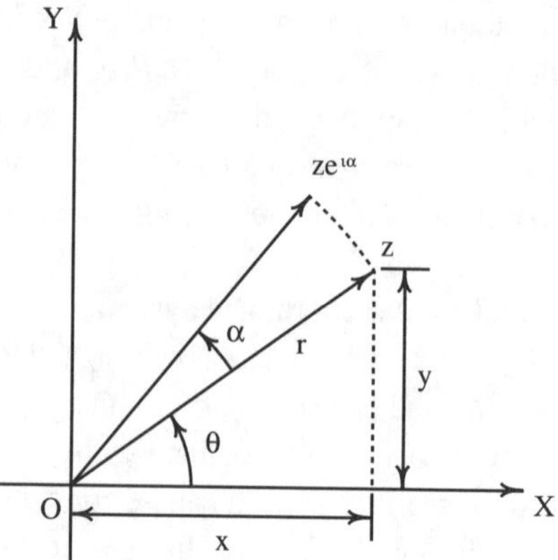

Figure 13.2. Polar form of complex numbers.

with polar coordinates $(r_1 r_2,\ \theta_1 + \theta_2)$. In words, add the arguments and multiply the absolute values.

The reader may know of two other constructs which are also called products of vectors, the scalar product (dot product), and the vector product (cross product, exterior product). The product we just defined is different from both. We shall call the set of vectors, with the sum and product defined above, the set of *complex numbers.*

A few words about units of angles. The customary and convenient unit in elementary geometry, geography and astronomy is the degree. It came down to us from Babylonian scholars who used it by 1700 B.C. In contexts involving calculus a more convenient unit of angle is the *radian*. One radian is an angle which subtends an arc of length 1 in a circle of radius 1. It is $\left(\frac{180}{\pi}\right)^{\circ} \approx 57.3^{\circ}$. In formulas which involve arclengths, using radians saves us a factor of proportionality but we incur the disadvantage that the measure of a right angle in radians is the irrational number $\frac{\pi}{2}$ and the measure of a complete angle, i.e., a full turn, is 2π.

Arguments of complex numbers are customarily expressed in radians, and that is the unit we shall use. There is an additional complication involving arguments which causes more difficulty than the matter of units. An angle is defined only up to a multiple of a complete angle, since a complete turn brings a directed line back to its original position. This means that, in radians, $\arg(z)$ is defined only up to a multiple of 2π. We say that the expression $\arg(z)$ is defined only *modulo* 2π. $\operatorname{Arg}(z)$ used to be called a multi-valued function, but now the term "function" is single-valued by definition. Many authors select the range $-\pi < \arg(z) \leq \pi$ for the $\arg(z)$, to make it single-valued. (Some use $0 \leq \arg(z) < 2\pi$.) This does not eliminate the complication inherent in the situation. Instead of a multi-valued function, we have a function which has an arbitrarily created discontinuity along the negative or the positive real axis.

With these definitions we can regard the x-axis to be the (real) number line. Indeed, for two complex numbers $z_1 = (x_1, 0)$, $z_2 = (x_2, 0)$ which are on the x-axis, the above definitions give $z_1 + z_2 = (x_1 + x_2, 0)$ and $z_1 z_2 = (x_1 x_2, 0)$, which agree with the usual definitions of addition and multiplication for real numbers. We call the x-axis the *real axis*, and the complex numbers which lie on it will be referred to by the usual notation. In particular, the vector $(1, 0)$ will be denoted by 1.

Multiplying a complex number z by a positive real number p stretches it by the factor p but does not change its direction. Multiplying by a negative real number n reverses the direction of z and stretches it by $|n|$.

It is easy to check that addition and multiplication of complex numbers have the associative, commutative, and distributive properties. Here is how we see the validity of the distributive property

$$\text{(D)} \qquad (z_1 + z_2)z_3 = z_1 z_3 + z_2 z_3,$$

which is somewhat less obvious than the others. Take the upper parallellogram in Fig. 13.1. Enlarge it by the factor $|z_3|$ and turn it by $\arg(z_3)$. The diagonal of the resulting parallellogram represents the left side of (D) and the two sides represent the right side.

We verify the remaining commutative field properties for the system of complex numbers. Adding 0 to any complex number leaves it unchanged, as does multiplying it by 1. We already introduced $-z$, the additive inverse of z. The multiplicative inverse (reciprocal) of a nonzero complex number with polar representation (r, θ) is $(\frac{1}{r}, -\theta)$. This completes our check that the complex numbers form a commutative field. (We point out in passing that in three or more dimensions it is not possible to define a product of vectors so that they form a commutative field.)

The vector $(0, 1)$ has modulus 1 and argument $\pi/2$. The square of this number is -1; it is called i. The number $-i$ is the other square root of -1.

We have

$$z = (x, y) = (x, 0) + (0, y) = x + (0, 1)y = x + iy.$$

The expression $x + iy$, in which x and y are real numbers, is called the *standard form* or *rectangular form* of a complex number. The real numbers x and y are called the *real* and *imaginary* parts of z respectively, and are denoted by

$$x = \text{Re}(z), \qquad y = \text{Im}(z).$$

The complex numbers with real part zero are called *pure imaginary* numbers.

Let $z_1 = x_1 + iy_1$ and $z_2 = x_2 + iy_2$ be complex numbers in rectangular form. We see directly from the definition of vector addition that the rectangular form of the sum is

$$z_1 + z_2 = (x_1 + x_2) + i(y_1 + y_2).$$

We compute the rectangular form of the product $z_1 z_2$ by using the distributive, associative and commutative laws:

$$\begin{aligned}(x_1 + iy_1)(x_2 + iy_2) &= x_1(x_2 + iy_2) + iy_1(x_2 + iy_2) \\ &= x_1x_2 - y_1y_2 + i(x_1y_2 + y_1x_2)\end{aligned} \tag{13.1}$$

The rectangular coordinates (x, y) of a complex number are related to its polar coordinates (r, θ) by

$$x = r\cos\theta, \qquad y = r\sin\theta.$$

Substituting the above two equations into $z = x + iy$ and recalling the notation $|z|$ for the length of the vector, we get

$$z = |z|(\cos\theta + i\sin\theta).$$

The above expression is the *polar form* of the complex number z.

A vector u which satisfies $|u| = 1$ is called a *unit vector*. The polar form represents z as the product of a nonnegative real number and a unit vector. If $z \neq 0$ then the unit vector in polar form is $z/|z| = \cos\theta + i\sin\theta$.

Let $z_1 = \cos\theta + i\sin\theta$ and $z_2 = \cos\phi + i\sin\phi$. By definition of the product, we add the arguments of the factors when we multiply two complex numbers, hence

$$z_1z_2 = \cos(\theta + \phi) + i\sin(\theta + \phi).$$

On the other hand, by (13.1),

$$z_1z_2 = \cos\theta\cos\phi - \sin\theta\sin\phi + i(\cos\theta\sin\phi + \sin\theta\sin\phi).$$

Comparing the last two expressions for z_1z_2, we find that the algebra of complex numbers has provided us with a neat derivation of the addition theorems for the sin and cos functions.

We now have square roots for negative numbers; but do we have square roots for all the new numbers we have introduced? We do.

Given any complex number z with polar coordinates (r, θ), the complex number with polar coordinates $(\sqrt{r}, \frac{1}{2}\theta)$ is a square root of z, as one can verify by squaring. In the above, θ could be any of the values of $\arg(z)$, which is a set of numbers consisting of a certain number and all the numbers which differ from it by an integral multiple of 2π. Hence the angle occurring in the square root is defined only modulo π. Adding an even multiple of π to the argument gives us the same complex number, but adding an odd multiple changes the number into its negative. This tells us what we already know from the laws of algebra: the negative of a square root is another square root.

The same argument can be used to obtain the nth roots of a nonzero complex number z. They are the numbers with polar form $(\sqrt[n]{|z|}, \frac{1}{n}\arg(z))$; here $\frac{1}{n}\arg(z)$ represents a set consisting of an angle and all other angles which differ from it by integral multiples of $\frac{2\pi}{n}$. This gives n different nth roots which form the vertices of a regular n-gon with center at 0. For example, the cube roots of 1 are $\cos(0) + i\sin(0) = 1$, $\cos(2\pi/3) + i\sin(2\pi/3) = -\frac{1}{2}(1 + i\sqrt{3})$ and $\cos(4\pi/3) + i\sin(4\pi/3) = -\frac{1}{2}(1 - i\sqrt{3})$. These three numbers are called the *third roots of unity.*

We observe that the property

$$\arg(zw) = \arg(z) + \arg(w) \tag{13.2}$$

is very similar to the property

$$\log zw = \log z + \log w \tag{13.3}$$

of the logarithm. For positive real numbers (13.3) is true but useless since for all these numbers $\arg(z) = 0 \pmod{2\pi}$. However, for complex numbers of modulus 1, multiplication of two numbers is by definition equivalent to addition their arguments. With the analogy between arguments and logarithms in mind, the reader will not be dumbfounded by the following formula of Euler, which can be derived from Taylor's series:

$$e^{i\theta} = \cos\theta + i\sin\theta; \tag{13.4}$$

i.e., $e^{i\theta}$ is the unit vector with argument θ.

A consequence of Euler's formula is $e^{2\pi i} = 1$, an unexpected relation between the three most notable constants encountered in higher mathematics.

We can define e^z for any complex value $z = x + iy$ by setting

$$e^{x+iy} = e^x(\cos y + i\sin y). \tag{13.5}$$

It is easy to verify that for real values of the exponent this agrees with the usual definition. Also, the basic law of exponents $e^{z+w} = e^z e^w$ holds with this definition for any complex exponents z, w. For the purposes of this book we can regard (13.5), of which Euler's formula is a special case, as the definition of powers of e with complex exponents. The question of why e is the appropriate base, and the radian the appropriate measure of angle for defining powers with complex exponents becomes clear only if one discusses derivatives. We have to refer to books on functions of complex variables for further information.

Multiplying z by $e^{i\phi}$ has the effect of rotating z counterclockwise by the angle ϕ (Fig. 13.2). For instance, since $i = e^{i\pi/2}$ and $-i = e^{-i\pi/2}$, iz and $-iz$ are obtained by rotating z by 90° counterclockwise and clockwise respectively.

It is customary to write the complex number z with polar coordinates (r, θ) in the *exponential form*

$$z = |z|\frac{z}{|z|} = re^{i\theta}.$$

With this notation the nth roots of unity can be written in the form $e^{2\pi ik/n}$, $k = 0, 1, \ldots, n-1$. If we set $\omega = e^{i2\pi/n}$, the nth roots of unity can also be written as

$$\omega^0,\ \omega^1,\ \omega^2,\ \ldots,\ \omega^{n-1}.$$

They are equally distributed on the *unit circle*—the circle with unit radius centered at the origin. The case $n = 6$ is shown in Fig. 13.3.

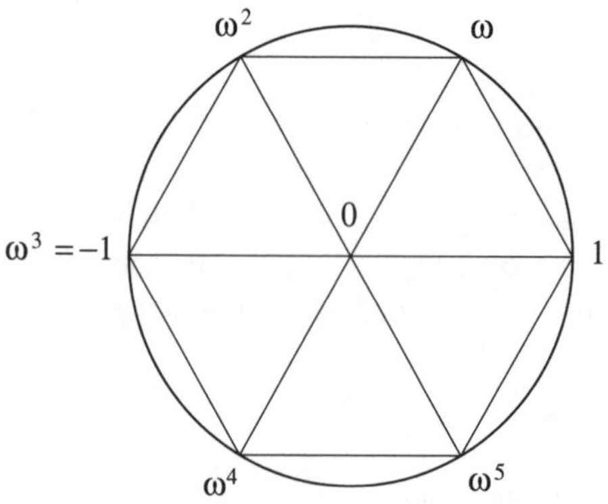

Figure 13.3. 6th roots of unity.

The complex number $\overline{z} = x - iy$ is called the *complex conjugate* of $z = x + iy$. The vectors z and $\overline{z}$ have the same length; they are symmetric with respect to the x-axis. If $z = |z|e^{i\theta}$ then

(13.6) $$\overline{z} = |z|e^{-i\theta}.$$

We shall frequently use that

(13.7) $$\operatorname{Re} z = \frac{1}{2}(z+\overline{z}), \qquad \operatorname{Im} z = \frac{1}{2i}(z-\overline{z}), \qquad |z|^2 = z\,\overline{z}, \qquad \frac{1}{i} = -i$$

and

$$\overline{z} + \overline{w} = \overline{z + w}, \qquad \overline{z}\,\overline{w} = \overline{zw}, \qquad \overline{z}/\overline{w} = \overline{z/w}, \qquad \overline{\overline{z}} = z.$$

If z_1 and z_2 are not zero, then we have

$$|z_1/z_2| = |z_1|/|z_2|, \qquad \arg(z_1/z_2) = \arg(z_1) - \arg(z_2).$$

The last equation says that $\arg(z_1/z_2)$ is the angle from z_2 to z_1. Generally speaking, if z_1, z_2, z_3 are points in the complex plane, the angle from the directed segment $\overrightarrow{z_1 z_2}$ to $\overrightarrow{z_1 z_3}$, i.e., the angle from the vector $z_2 - z_1$ to $z_3 - z_1$, is $\arg[(z_3 - z_1)/(z_2 - z_1)]$, with counterclockwise being the positive direction of rotation.

Three more statements will help us solve some geometric problems using complex number representation.

1. Triangle inequality. One side of a triangle can not be longer than the sum of the other two sides. This implies that for any two complex numbers,

$$|z_1 + z_2| \le |z_1| + |z_2|.$$

Equality holds if and only if one of z_1 and z_2 is a nonnegative multiple of the other. We can show by induction that

$$|z_1 + z_2 + \cdots + z_n| \le |z_1| + |z_2| + \cdots + |z_n|,$$

with the equality holding if and only if the points $z_1, z_2, \ldots, z_n$ all lie on the same ray starting at the origin.

2. Similar triangles. We shall use the notation $\Delta z_1 z_2 z_3$ to denote a triangle whose vertices are ordered z_1, z_2, z_3. A triangle whose vertices are ordered in a counterclockwise direction will be referred to as a *positive* triangle, whereas a negative triangle is one with clockwise orientation (Fig. 13.4). Thus $\Delta z_1 z_2 z_3$ and $\Delta z_2 z_1 z_3$ are two triangles with opposite orientations.

Let $\Delta z_1 z_2 z_3$ and $\Delta w_1 w_2 w_3$ be two triangles. The complex equation

$$\frac{z_3 - z_1}{z_2 - z_1} = \frac{w_3 - w_1}{w_2 - w_1}, \tag{13.8}$$

is equivalent to the following two real equalities:

$$\frac{|z_3 - z_1|}{|z_2 - z_1|} = \frac{|w_3 - w_1|}{|w_2 - w_1|}$$

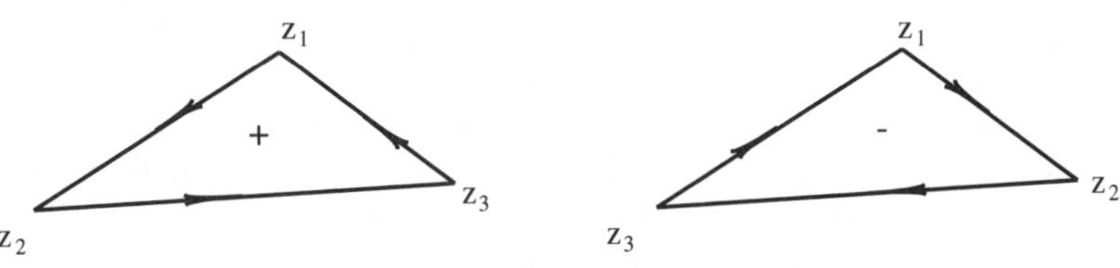

Figure 13.4. Triangle orientations.

and

$$\arg\left(\frac{z_3 - z_1}{z_2 - z_1}\right) = \arg\left(\frac{w_3 - w_1}{w_2 - w_1}\right).$$

Thus equation (13.8) expresses three conditions:

1. corresponding sides of the triangles are proportional,
2. corresponding angles are equal, and
3. the two triangles have the same orientation.

Two such triangles are called *directly similar*, and direct similarity is denoted by

$$\Delta z_1 z_2 z_3 \sim \Delta w_1 w_2 w_3.$$

3. Signed areas of triangles. Let $\Delta 0ab$ be a triangle with one vertex at the origin, and let $\phi = \arg(b) - \arg(a)$ be the signed angle through which a must be rotated about the origin to reach the direction of b. We have $\phi = \arg(b\overline{a})$. The *signed area* of the triangle $0ab$ is given by

$$\frac{1}{2}|a||b|\sin\phi = \frac{1}{2}\mathrm{Im}(\overline{a}b) = \frac{1}{4i}(\overline{a}b - a\overline{b}) = \frac{i}{4}\begin{vmatrix} a & b \\ \overline{a} & \overline{b} \end{vmatrix};$$

in the last step we used that $-1/i = i$. The signed area is positive if and only if the triangle $0ab$ has positive orientation.

We will denote the signed area of $\Delta 0ab$ by $[0ab]$. Clearly $[0ab] = -[0ba]$. The signed area of a triangle with vertices z_1, z_2, z_3, denoted by $[z_1 z_2 z_3]$, is obtained by substituting $z_2 - z_1$ and $z_3 - z_1$ for a and b, respectively, into the above equation. It is customary to write the area in a more symmetric form as a 3×3 determinant:

$$(13.9) \qquad [z_1 z_2 z_3] = \frac{i}{4}\begin{vmatrix} z_2 - z_1 & z_3 - z_1 \\ \overline{z}_2 - \overline{z}_1 & \overline{z}_3 - \overline{z}_1 \end{vmatrix} = \frac{i}{4}\begin{vmatrix} 1 & 1 & 1 \\ z_1 & z_2 & z_3 \\ \overline{z}_1 & \overline{z}_2 & \overline{z}_3 \end{vmatrix}$$

The 3×3 determinant can be reduced to the 2×2 determinant by subtracting the first column from the other two columns and then expanding in terms of the first row.

One of the advantages of signed areas is that for any four points a, b, c and d in the complex plane, we have

$$(13.10) \qquad [abc] = [dbc] + [dca] + [dab].$$

If d is inside Δabc, this triangle can be dissected into the three triangles occurring on the right-hand side of (13.10); with our sign convention, the identity continues to hold when d is outside Δabc.

We are ready to present examples of the usefulness of complex numbers in geometry.

PROBLEM 13.1 (A GENERALIZATION OF PTOLEMY'S THEOREM). Let z_1, z_2, z_3, z_4 be any four points in the complex plane. Then

$$|z_3 - z_2||z_4 - z_1| + |z_4 - z_3||z_2 - z_1| \geq |z_3 - z_1||z_4 - z_2|$$

with equality if z_1, z_2, z_3, and z_4 lie, in this order, on a circle or a straight line.

SOLUTION. It is easy to verify the identity:

$$(z_3 - z_2)(z_4 - z_1) + (z_4 - z_3)(z_2 - z_1) = (z_3 - z_1)(z_4 - z_2).$$

Take absolute values on both sides and use the triangle inequality to get the inequality stated in the problem.

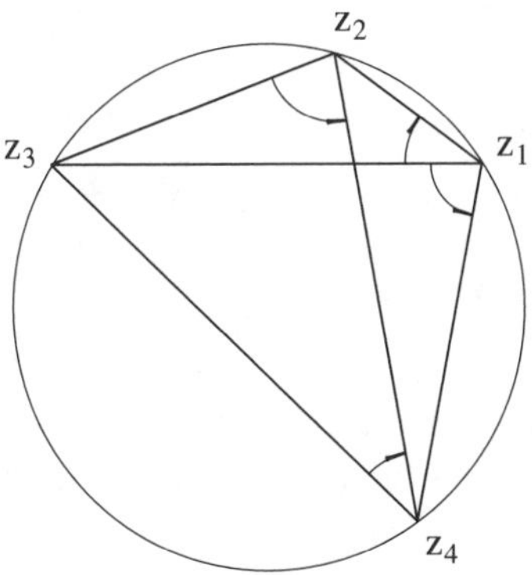

Figure 13.5. Ptolemy's theorem.

If z_1, z_2, z_3, z_4 are vertices of a quadrilateral, then the above inequality says: the product of the two diagonals of any quadrilateral is less than or equal to the sum of the two products of its opposite sides.

If z_1, z_2, z_3, z_4 are concyclic in the designated order, then the first term on the left side of our identity is a positive multiple of the right side. Indeed, dividing the first term by the right side we get

$$\frac{z_3 - z_2}{z_4 - z_2} \frac{z_4 - z_1}{z_3 - z_1},$$

and the theorem about the angles subtended by points on a circle tells us that the arguments of the two fractions have the same magnitude and opposite signs. Similarly, the second term on the left-hand side is also a positive multiple of the right-hand side. Hence equality holds in the inequality in this case, and we get one of the famous theorems of geometry known as *Ptolemy's theorem,* see [Pedoe '70]. Since the equality of the angles subtended holds only for cyclic quadrilaterals, it follows that equality holds only for cyclic quadrilaterals. (The great Alexandrian astronomer Claudius Ptolemy, about 85–165 A.D., used the theorem to obtain trigonometric identities.)

PROBLEM 13.2. A triangle and a point inside it are arbitrarily given. Joining the point and the three vertices of the triangle by line segments, we obtain three subtriangles. Show that the sum of the products of the three sides of each subtriangle is greater than or equal to the product of the three sides of the original triangle.

SOLUTION. Let the given interior point be the origin of the complex plane, and let the three vertices of the triangle be the complex numbers z_1, z_2, and z_3. The identity

$$z_1z_2(z_1 - z_2) + z_2z_3(z_2 - z_3) + z_3z_1(z_3 - z_1) = (z_1 - z_2)(z_2 - z_3)(z_1 - z_3)$$

can be easily verified. By the triangle inequality we get

$$|z_1||z_2||z_1 - z_2| + |z_2||z_3||z_2 - z_3| + |z_3||z_1||z_3 - z_1| \geq |z_1 - z_2||z_2 - z_3||z_3 - z_1|,$$

which is what we want.

The following problem of solid geometry can be reduced to a planar problem and then solved by means of complex numbers. In [Stolarsky '71] it was conjectured:

PROBLEM 13.3. If $a, a'; b, b'; c, c'$ are the lengths of the pairs of skew edges of a tetrahedron T with base abc, then

$$\frac{b'c'}{bc} + \frac{c'a'}{ca} + \frac{a'b'}{ab} \geq 1.$$

SOLUTION. It is clear that if the vertex V of T opposite the base abc and the edges that meet at V are projected vertically down to the plane of the base, then the lengths a', b', c' all decrease, so the left side of the desired inequality also decreases. Hence we can assume that the vertex V is in the plane of abc. Then the problem becomes planar and the desired inequality can be obtained by dividing both sides of the inequality in the previous problem by $|z_1 - z_2||z_2 - z_3||z_3 - z_1|$.

PROBLEM 13.4. If $z_1, z_2, \ldots, z_7, z_8$ are eight points such that the triangles $z_1z_2z_4$, $z_2z_3z_5$, and $z_6z_7z_8$ are directly similar, and $z_1z_2z_6$ is directly similar to $z_2z_3z_7$, show that the last two triangles are also similar to $z_4z_5z_8$ (see Fig. 13.6, where only the subscripts of the points have been marked).

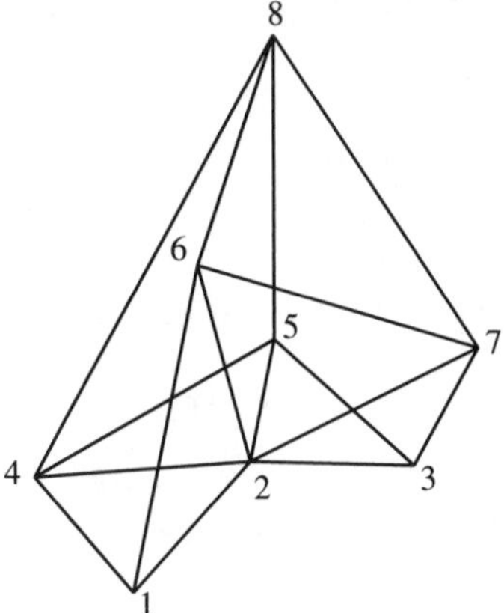

Figure 13.6. Six triangles.

SOLUTION. $z_1, z_2, \ldots, z_8$, the symbols for the eight points, are regarded as complex numbers. The given conditions can then be written in the form:

$$\frac{z_4 - z_1}{z_2 - z_1} = \frac{z_5 - z_2}{z_3 - z_2} = \frac{z_8 - z_6}{z_7 - z_6} = u, \tag{13.11}$$

$$\frac{z_6 - z_1}{z_2 - z_1} = \frac{z_7 - z_2}{z_3 - z_2} = v. \tag{13.12}$$

From (13.11) we derive

$$z_4 = (1-u)z_1 + uz_2,$$

$$z_5 = (1-u)z_2 + uz_3,$$

$$z_8 = (1-u)z_6 + uz_7.$$

Thus

$$\frac{z_8 - z_4}{z_5 - z_4} = \frac{(1-u)(z_6 - z_1) \; + \; u(z_7 - z_2)}{(1-u)(z_2 - z_1) \; + \; u(z_3 - z_2)}.$$

The fraction on the right-hand side can be seen to be equal to the quantity v of (13.11), which is what we had to prove.

While Fig. 13.6 helps us visualize the problem, it is of no help in our purely algebraic solution.

PROBLEM 13.5 [USAMO 1977/2]. ABC and $A'B'C'$ are two triangles in the same plane such that the lines AA', BB', CC' are parallel. Prove that the signed area of the triangle, denoted by $[ABC]$, satisfies the equation

$$3([ABC]+[A'B'C']) \\ = [AB'C']+[A'BC']+[A'B'C]+[A'BC]+[AB'C]+[ABC'].$$

SOLUTION. By (13.10) we have

$$[ABC]=[A'BC]+[A'CA]+[A'AB], \tag{13.13}$$

$$[A'B'C']=[AB'C']+[AC'A']+[AA'B']. \tag{13.14}$$

By assumption $CC'||AA'$, thus $[AC'A']=ACA'=-[A'CA]$; similarly $AA'||BB'$ implies $[AA'B'] = -[A'AB]$. By adding (13.13) and (13.14) we obtain

$$[ABC]+[A'B'C']=[A'BC]+[AB'C'].$$

We get two similar equations by cyclically permuting A, B, C. Adding all three equations we obtain the desired result.

For other solutions of this problem, the reader is referred to [Klamkin '88].

The next problem, involving two triangles, concerns a beautiful, symmetric inequality. If one of the two triangles in it, say $A'B'C'$, is equilateral, then the inequality reduces to the inequality [IMO 1961/2] in Chapter 12. This inequality is discussed in [Pedoe '43] and [Pedoe '63], though it was partially proved in [Neuberg 1891].

Professor Pedoe prized this result. In [Pedoe '70] he called his inequality "the first interesting inequality for two triangles" and told a long story about his discovery, involving the illustrious names Hadwiger and Finsler.

There are now more than twenty different proofs of the Neuberg-Pedoe inequality. It has been widely discussed, generalized, and even extended to higher dimensional space [Yang & Zhang '83]. We present a proof using complex numbers [Chang '82].

PROBLEM 13.6 (NEUBERG-PEDOE INEQUALITY). Let a, b, c, a', b', c' denote the sides of triangles ABC, $A'B'C'$, and let F and F' denote their areas. Then

$$a'^2(-a^2+b^2+c^2)+b'^2(a^2-b^2+c^2)+c'^2(a^2+b^2-c^2) \geq 16FF' \tag{13.15}$$

with equality if and only if the triangles are similar.

SOLUTION. Let u, v be the complex numbers representing the vectors $\overrightarrow{CA}$, $\overrightarrow{CB}$. Then

$$a=|v|, \qquad b=|u|, \qquad c=|v-u|.$$

Dealing with the triangle $A'B'C'$ similarly, we have

$$a' = |v'|, \qquad b' = |u'|, \qquad c' = |v' - u'|.$$

Straightforward computation yields

$$\begin{aligned} a'^2(-a^2 + b^2 + c^2) &= v'\,\overline{v'}\,(2u\,\overline{u} - (u\,\overline{v} + \overline{u}\,v)), \\ b'^2(a^2 - b^2 + c^2) &= u'\,\overline{u'}\,(2v\,\overline{v} - (u\,\overline{v} + \overline{u}\,v)), \\ c'^2(a^2 + b^2 - c^2) &= (u'\,\overline{u'} + v'\,\overline{v'} - (u'\,\overline{v'} + \overline{u'}\,v'))(u\,\overline{v} + \overline{u}\,v). \end{aligned}$$

Adding the three equations, we obtain

$$H = 2(|u'|^2|v|^2 + |u|^2|v'|^2) - (u\,\overline{v} + \overline{u}\,v)(u'\,\overline{v'} + \overline{u'}\,v'),$$

where H denotes the quantity on the left side of (13.15).

The (unsigned) area F of triangle ABC is

$$F = \tfrac{1}{2}\,|\mathrm{Im}(\overline{u}\,v)| = \tfrac{1}{2}\,|\mathrm{Im}(u\,\overline{v})| = \tfrac{1}{4}\,|u\,\overline{v} - \overline{u}\,v|$$

and similarly for F'. Hence $16F\,F' = \pm(u\,\overline{v} - \overline{u}\,v)(u'\,\overline{v'} - \overline{u'}\,v')$, the sign being chosen to make the expression positive. Then

$$H - 16FF' = 2|u\,v' - u'\,v|^2 \quad \text{or} \quad 2|u\,\overline{v'} - \overline{u'}\,v|^2,$$

depending on whether triangles ABC and $A'B'C'$ have the same or opposite orientations. Thus $H \geq 16F\,F'$, with equality holding if and only if

$$\frac{v}{u} = \frac{v'}{u'} \quad \text{or} \quad \frac{v}{u} = \overline{\left(\frac{v'}{u'}\right)}.$$

These two equations are say respectively that triangles ABC and $A'B'C'$ are directly or oppositely similar.

Exercise 10 below is an interesting consequence of the Neuberg-Pedoe inequality.

Exercises

13.1 Show that in a parallelogram the sum of the squares of the lengths of the diagonals equals the sum of the squares of the lengths of the sides.

13.2 Show that triangle $z_1z_2z_3$ is equilateral if and only if

$$z_1^2 + z_2^2 + z_3^2 = z_1z_2 + z_2z_3 + z_3z_1.$$

13.3 Show that if $|z_1| = |z_2| = |z_3|$ and $z_1 + z_2 + z_3 = 0$ then $z_1z_2z_3$ is an equilateral triangle.

13.4 Suppose that $|z_1| = |z_2| = |z_3|$ and that $z_1z_2 + z_2z_3 + z_3z_1 = 0$. Show that the triangle $z_1z_2z_3$ is equilateral.

13.5 Let a, b, c be the sides of triangle ABC, and P any point in the plane of the triangle. Show that

$$a|PA|^2 + b|PB|^2 + c|PC|^2 \geq abc.$$

13.6 Let P and Q be two points in the plane of triangle ABC. Show that

$$a|PA|\,|QA| + b|PB|\,|QB| + c|PC|\,|QC| \geq abc.$$

13.7 Let $ABCD$ be a convex quadrilateral. If the sum of the line segments joining the midpoints of opposite sides is half the perimeter of the quadrilateral, show that $ABCD$ is a parallelogram.

13.8 Five points A, B, C, U and V lie in the same plane. Suppose triangles AUV, UBV and UVC are directly similar to one another. Show that triangle ABC is directly similar to each of them.

13.9 Let triangles $z_1z_2z_3$ and $w_1w_2w_3$ be directly similar. For any $t \in [0, 1]$ define

$$u_j = (1-t)z_j + tw_j, \quad j = 1, 2, 3.$$

Show that the triangle $u_1u_2u_3$ is directly similar to the triangle $z_1z_2z_3$.

13.10 Derive the following consequence of the Neuberg-Pedoe inequality:

$$a^2(a')^2 + b^2(b')^2 + c^2(c')^2 \geq 16FF', \tag{13.16}$$

with equality if and only if the triangles ABC and $A'B'C'$ are both equilateral.

CHAPTER FOURTEEN

Birth of an IMO Problem

In May, 1985 a comprehensive academic competition was held for young people all over China. One of the mathematical problems on the competition was:

PROBLEM 14.1. A, B, C are arbitrary points in a plane. A frog starts from a point P_0 and travels directly toward A. Upon reaching A, the frog continues in the same direction to the point P_1 such that $P_0A = AP_1$. Next, the frog travels from P_1 directly through B to the point P_2 such that $P_1B = BP_2$. The frog then starts from P_2 and travels through C to the point P_3 such that $P_2C = CP_3$. Next from P_3, the frog repeats the same action with respect to A, B, C cyclically, generating a sequence of points $P_1, P_2, P_3, P_4, \ldots$. Assuming that the distance from P_0 to C is 27cm, what is the distance between P_0 and P_{1985}?

SOLUTION. To solve the problem, (see Fig. 14.1), let A, B, and C have Cartesian coordinates (a_1, a_2), (b_1, b_2), and (c_1, c_2) respectively. Let $P_i = (x_i, y_i)$, $i = 0, 1, 2, 3, \ldots$. Since A is the midpoint of P_0P_1, we have $a_1 = (x_0 + x_1)/2$ or

$$x_1 = 2a_1 - x_0,$$

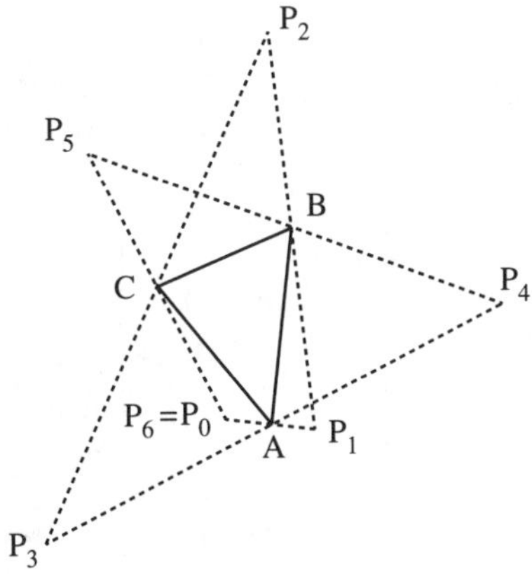

Figure 14.1. Sequence of points with period 6.

and since B, C are midpoints of P_1P_2 and P_2P_3, respectively, we have

$$x_2 = 2b_1 - x_1 = 2(b_1 - a_1) + x_0$$

and

$$x_3 = 2c_1 - x_2 = 2(c_1 - b_1 + a_1) - x_0.$$

From the last equation, we see that

$$x_6 = 2(c_1 - b_1 + a_1) - x_3 = 2(c_1 - b_1 + a_1) - 2(c_1 - b_1 + a_1) + x_0,$$

so that

$$x_6 = x_0.$$

Similarly, $y_6 = y_0$. Hence $P_0 = P_6 = P_{12} = P_{18} = \dots$. This means that the sequence of points $\{P_i\}$ is *periodic with period 6*, see Fig. 14.1. Since $1985 = 6 \times 33 + 5$, we have $P_{1985} = P_5$. Finally, we get the distance

$$P_0P_{1985} = P_0P_5 = P_5P_6 = 2P_0C = 54 \text{ cm}.$$

Professor Qi Dongxu (Chairman of the Coordinating Committee at the 31st IMO, held in Beijing, 1990), along with the first author of this book, considered possible generalizations of this problem such as: if the frog turns by an angle θ at each of the points A, B and C, then what happens?

Suppose that there are three arbitrarily given points A, B, and C in the plane. An ant starts from a point, P_0, and travels directly toward A. Upon reaching A, the ant turns clockwise through an angle θ, then travels to a point P_1 whose distance from A is equal to the distance from A to P_0. This is the first action of the ant. If it repeats the same action with respect to $B, C, A, B, \dots$ cyclically, a sequence of points $P_1, P_2, P_3, \dots$ is generated. (See Fig. 14.2, where $\theta = 120°$.)

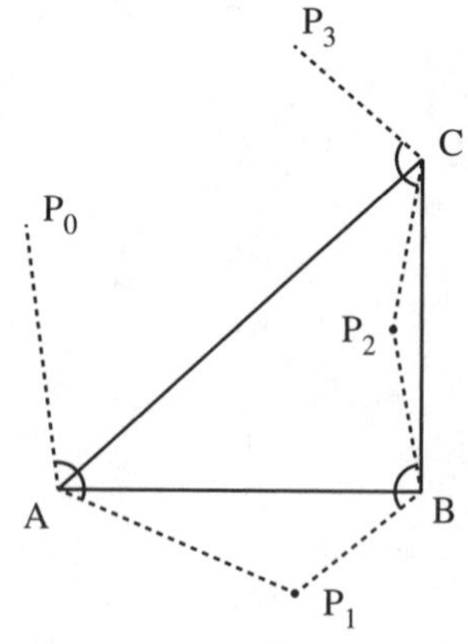

Figure 14.2. Movement of the ant.

What is the path of the ant? Under what circumstances will it return to P_0 after six steps? In other words, when is $P_6 = P_0$?

If the angle $\theta = 0$, then the problem reduces to the 'frog problem', and the answer is 'always'. The simplest case is $\theta = \pi$, in which the ant returns to the point P_0 at each step. The point P_0 is a fixed point of the motion, so $P_0 = P_1 = P_2 = \dots$. Now suppose $u = e^{i\theta}$ is any one of the 6th roots of unity. Interpreting points $A, B, C, P_0, P_1, \dots$ as complex numbers, we have $P_1 - A = (A - P_0)u$, that is,

$$P_1 = (1+u)A - uP_0.$$

Similarly

$$P_2 = (1+u)B - uP_1 = (1+u)(B - uA) + u^2 P_0,$$
$$P_3 = (1+u)C - uP_2 = (1+u)(u^2 A - uB + C) - u^3 P_0.$$

To get P_6 from P_3, we add 3 to the subscripts in the last equation

$$P_6 = (1+u)(u^2 A - uB + C) - u^3 P_3.$$

Hence

$$P_6 = (1-u^3)(1+u)(u^2 A - uB + C) + u^6 P_0.$$

Since $u^6 = 1$, we have

$$P_6 = (1-u^3)(1+u)(u^2 A - uB + C) + P_0.$$

Hence $P_6 = P_0$ if and only if

$$(1-u^3)(1+u)(u^2 A - uB + C) = 0.$$

If $u^3 - 1 = 0$ or $u - 1 = 0$, the above equation is automatically satisfied, no matter where the points A, B and C are located. Since the cube roots of one are $u = 1,\ e^{2\pi i/3},\ e^{-2\pi i/3}$, and since $e^{\pi i} = -1$, this happens for $\theta = 0, 2\pi/3, -2\pi/3$ and π. The other sixth roots of one are $u = e^{\pi i/3}$ and $e^{-\pi i/3}$. In these cases the above equation can hold only if

$$u^2 A - uB + C = 0.$$

Now $u = e^{\pi i/3}$ and $e^{-\pi i/3}$ satisfy $u^3 + 1 = 0$, i.e., $(u+1)(u^2 - u + 1) = 0$. Since $u + 1 \neq 0$, we have $u^2 = u - 1$. Thus, $u^2 A - uB + C = 0$ becomes

$$C - A = (B - A)u,$$

which says that AC is obtained from AB by rotating it through the angle θ.

When $\theta = \pi/3$, ABC is a positively oriented equilateral triangle; and if $\theta = -\pi/3$, ABC is a negatively oriented equilateral triangle. Since 1986 is a multiple of 6, the first of these two conclusions is what IMO contestants were asked to prove in

PROBLEM 14.2 [IMO 1986/3]. A triangle $A_1A_2A_3$ and a point P_0 are given in the plane. We define $A_s = A_{s-3}$ for all $s \geq 4$. We construct a sequence of points $P_1, P_2, P_3, \ldots$ such that P_{k+1} is the image of P_k under the clockwise rotation with center A_{k+1} through $120°$ (for $k = 0, 1, 2, \ldots$). Prove that if $P_{1986} = P_0$, then the triangle $A_1A_2A_3$ is equilateral.

We can make further extensions. Let t points $A_1, A_2, \ldots, A_t$ and another point P_0 be given in the plane. The ant now moves with respect to these t points in the same manner in which it moved in the case $t = 3$. Again a sequence of points $P_1, P_2, P_3, \ldots$ is constructed. Set $u = e^{i\theta}$ again. We have the recursion formula

$$P_k = (1+u)A_k - uP_{k-1}, \quad k = 1, 2, 3, \ldots$$

in which the subscripts of A_k should be counted in arithmetic modulo t. By repeatedly using the recursion relation, we get

$$\begin{aligned} P_n = {} & (1+u)\big(A_n - uA_{n-1} + u^2A_{n-2} - u^3A_{n-3} \\ & + \cdots + (-u)^{n-1}A_1\big) + (-u)^n P_0. \end{aligned}$$

It is obvious that if $\theta = \pi$, that is, $u = -1$, then

$$P_0 = P_1 = P_2 = \cdots.$$

Now assume $u \neq -1$. The sequence $\{P_n\}$ has period t if and only if $P_t = P_0$, that is,

$$(1+u)\big(A_t - uA_{t-1} + u^2A_{t-2} - u^3A_{t-3} + \cdots + (-u)^{t-1}A_1\big) + (-u)^t P_0 = P_0.$$

The above equation is satisfied for the following simple choice:

$$(-u)^t = 1, \tag{14.1}$$

and

$$A_t - uA_{t-1} + u^2A_{t-2} - u^3A_{t-3} + \cdots + (-u)^{t-1}A_1 = 0. \tag{14.2}$$

Equation (14.1) says that $(-u)$ is a tth root of unity. If in addition equation (14.2) is satisfied by $A_1, A_2, \ldots, A_t$, then the sequence $\{P_n\}$ is periodic.

For example, put $t = 4$. The 4th roots of unity are $1, i, -1, -i$. We skip the trivial case $u = -1$. For $u = 1$ we have $\theta = 0$, and (14.2) becomes

$$A_4 - A_3 = A_1 - A_2.$$

This equation can be interpreted geometrically as follows: points A_1, A_2, A_3, A_4 are successive vertices of a parallelogram. In this case, an ant starting from any point of the plane and travelling according to the rule given at the beginning of this chapter (i.e., $\theta = 0$) will come back to its starting point after 4 motions.

If $u = i$, then equation (14.2) gives

$$i(A_1 - A_3) = A_2 - A_4.$$

This says that rotating the segment A_1A_3 by a positive angle of 90° about A_1 yields a segment parallel and of the same length as A_2A_4. In particular, the condition will be satisfied when A_1, A_2, A_3, and A_4 are successive vertices of a positively oriented square. In this case, starting from any point P_0 in the plane an ant will return to P_0 after turning through 90° at each of the four points A_1, A_2, A_3, and A_4 (Fig. 14.3).

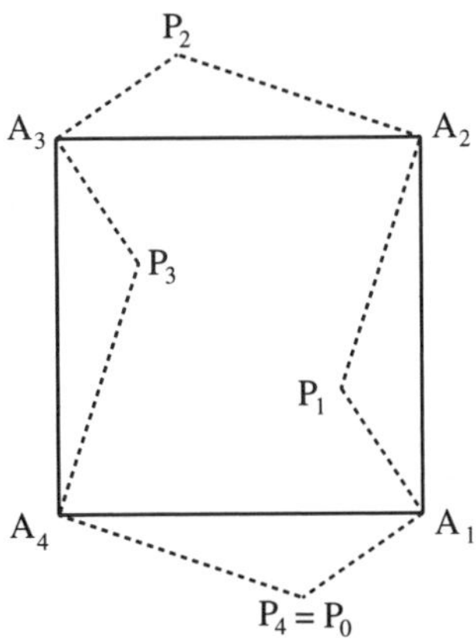

Figure 14.3. A positively oriented square.

In this manner, numerous geometric propositions can be formulated by specifying a positive integer t and then a tth root of unity for $(-u)$.

Exercises

14.1 Prove Napoleon's Theorem of Chapter 12 by means of complex numbers.

14.2 [CMC 1988] A given point $z_0 \neq 0$ lies in the complex plane. A moving point z_1 satisfies $|z_1 - z_0| = |z_1|$. Another moving point z is defined by the equation $zz_1 = -1$. Determine the geometric figure described by z.

14.3 [CMO 1987] Let n be a positive integer. Show that the equation

$$z^{n+1} - z^n - 1 = 0$$

has a solution on the unit circle in the complex plane if and only if $n+2$ is divisible by 6.

14.4 [Putnam Competition 1989] Prove that if

$$11z^{10} + 10iz^9 + 10iz - 11 = 0, \quad \text{then } |z| = 1.$$

14.5 Let ABC be a fixed horizontal triangle with angles α, β, γ. The vertices are listed counterclockwise. Let Π be a movable flat sheet lying on top of the plane of ABC. If we turn Π clockwise by 2α around the point A and then turn it clockwise by 2β around the point B, the resulting position of Π will be the same as if we had turned Π counterclockwise by 2γ around C. (This reduction of two successive rotations to a single rotation is valid for spherical as well as planar motions, and is due to Olinde Rodrigues (1794–1851) and Sir William Rowan Hamilton (1805–1865)).

CHAPTER FIFTEEN

Barycentric Coordinates

Let ABC be an arbitrarily given triangle in the complex plane, and let P be a point inside the triangle. Join points A and P by a line segment and then extend the segment to meet side BC at P', as shown in Fig. 15.1. Since P' is on the line BC, we have

(15.1) $$P' = (1-\lambda)B + \lambda C, \quad 0 \le \lambda \le 1,$$

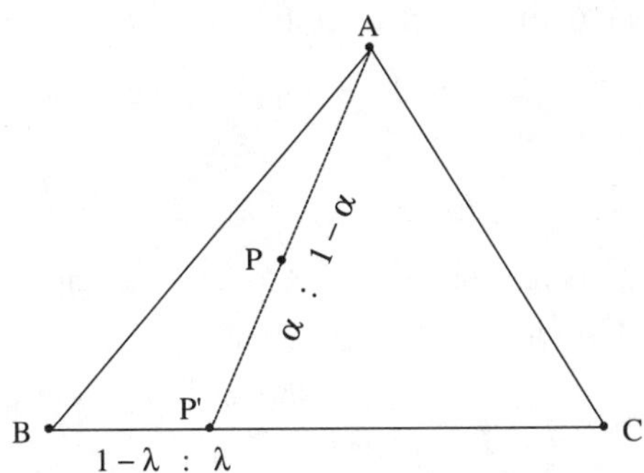

Figure 15.1. A point in a triangle.

where λ is the ratio of lengths BP' to BC. Similarly

(15.2) $$P = (1-\mu)A + \mu P', \quad 0 \le \mu \le 1.$$

Substituting (15.1) for P' into (15.2), we have

(15.3) $$P = (1-\mu)A + (1-\lambda)\mu B + \lambda\mu C.$$

Setting

(15.4) $$\alpha = 1-\mu, \qquad \beta = (1-\lambda)\mu, \qquad \gamma = \lambda\mu,$$

equation (15.3) becomes

(15.5) $$P = \alpha A + \beta B + \gamma C,$$

where α, β, and γ are nonnegative numbers satisfying

(15.6) $$\alpha + \beta + \gamma = 1.$$

The numbers α, β, and γ are called the *barycentric coordinates* of P with respect to the *coordinate triangle* ABC. We will at times drop the distinction between the point P and its barycentric coordinates and write $P = (\alpha, \beta, \gamma)$.

It will be useful to write α as a ratio of signed areas of triangles. By 15.2 and 15.4, α is the signed ratio of the collinear line segments $P'P$ and $P'A$. Since P' is on BC, α, and similarly β and γ, can be expressed as

$$(15.7) \qquad \alpha = \frac{[PBC]}{[ABC]}, \qquad \beta = \frac{[APC]}{[ABC]}, \qquad \gamma = \frac{[ABP]}{[ABC]}.$$

All of the above considerations still hold when P lies outside ABC, except that at least one of the barycentric coordinates α, β, γ is negative. Equations (15.6) and (15.7) continue to hold. Note that if P is on the line determined by B and C (C and A or A and B), then $\alpha = 0$ ($\beta = 0$ or $\gamma = 0$). The barycentric coordinates of the vertices of the coordinate triangle are

$$A = (1, 0, 0), \qquad B = (0, 1, 0), \qquad C = (0, 0, 1).$$

Suppose that P_k, $k = 1, 2, 3$, are three points in the complex plane. Let P_k have the barycentric coordinates

$$P_k = (\alpha_k, \beta_k, \gamma_k), \quad k = 1, 2, 3,$$

with respect to the coordinate triangle ABC. By using the same technique employed to prove (15.7), we can show that

$$(15.8) \qquad [P_1P_2P_3] = \begin{vmatrix} \alpha_1 & \beta_1 & \gamma_1 \\ \alpha_2 & \beta_2 & \gamma_2 \\ \alpha_3 & \beta_3 & \gamma_3 \end{vmatrix} [ABC].$$

Adding the second and the third columns to the first column of the determinant, we obtain

$$(15.9) \qquad [P_1P_2P_3] = \begin{vmatrix} 1 & \beta_1 & \gamma_1 \\ 1 & \beta_2 & \gamma_2 \\ 1 & \beta_3 & \gamma_3 \end{vmatrix} [ABC].$$

In particular, three points P_1, P_2, and P_3 are collinear if and only

$$(15.10) \qquad \begin{vmatrix} 1 & \beta_1 & \gamma_1 \\ 1 & \beta_2 & \gamma_2 \\ 1 & \beta_3 & \gamma_3 \end{vmatrix} = 0.$$

Barycentric coordinates are widely used not only in geometry but also in many branches of mathematics, such as topology, game theory, partial differential equations, finite element analysis and computer aided geometric design. Here, we are interested only in their application to elementary geometry. We use them again in the chapters on the representation of surfaces by polynomials.

PROBLEM 15.1. For each point P inside a triangle ABC, let D, E and F be the points of intersection of the lines AP, BP, and CP with the sides opposite to A, B, and C, respectively. Determine P in such a way that the area of triangle DEF is as large as possible. (see [Klamkin '86, p. 86]).

SOLUTION. Take ABC as coordinate triangle of the barycentric coordinate system and let $P = (\alpha, \beta, \gamma)$. Since D plays the role of P' in equation (15.1), we have

$$D = (1-\lambda)B + \lambda C.$$

From (15.4) we get

$$\lambda = \gamma/\mu = \gamma/(1-\alpha)\,, \qquad (1-\lambda) = \beta/\mu = \beta/(1-\alpha)\,.$$

Thus

$$D = \frac{B\beta}{1-\alpha} + \frac{C\gamma}{1-\alpha}.$$

We can similarly obtain the barycentric coordinates of D, E, and F:

$$D = \left(0, \frac{\beta}{1-\alpha}, \frac{\gamma}{1-\alpha}\right),$$

$$E = \left(\frac{\alpha}{1-\beta}, 0, \frac{\gamma}{1-\beta}\right),$$

$$F = \left(\frac{\alpha}{1-\gamma}, \frac{\beta}{1-\gamma}, 0\right).$$

By (15.8), we have

$$\begin{aligned}
[DEF] &= \frac{1}{(1-\alpha)(1-\beta)(1-\gamma)} \begin{vmatrix} 0 & \beta & \gamma \\ \alpha & 0 & \gamma \\ \alpha & \beta & 0 \end{vmatrix} [ABC] \\
&= \frac{2\alpha\beta\gamma[ABC]}{(\beta+\gamma)(\gamma+\alpha)(\alpha+\beta)}\,.
\end{aligned}$$

It follows from the arithmetic-geometric mean inequality that

$$\beta + \gamma \geq 2\sqrt{\beta\gamma}, \qquad \gamma + \alpha \geq 2\sqrt{\gamma\alpha}, \qquad \alpha + \beta \geq 2\sqrt{\alpha\beta},$$

so that $[DEF] \leq [ABC]/4$ with equality only if $\alpha = \beta = \gamma = 1/3$. Thus $\max[DEF] = [ABC]/4$ and is reached if and only if $P = (A+B+C)/3$, i.e., if and only if P is the centroid of triangle ABC.

PROBLEM 15.2 [CMC 1983]. In Fig. 15.2, let $ABCD$ be a quadrilateral in which

$$[ABD] : [BCD] : [ABC] = 3 : 4 : 1.$$

Points M and N lie on AC and CD respectively such that $AM : AC = CN : CD$. In addition, B, M, N are collinear. Show that M and N bisect AC and CD respectively.

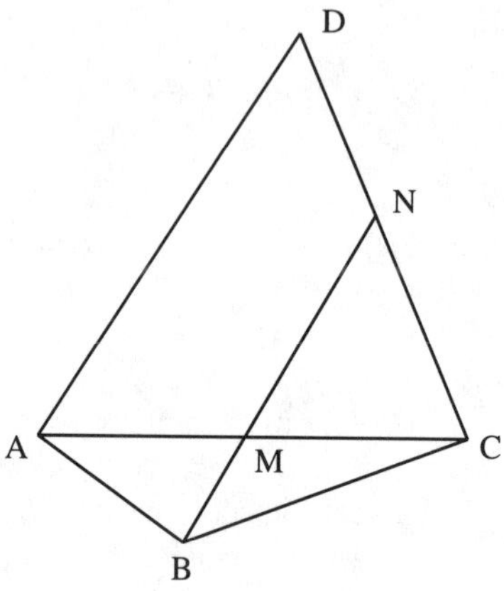

Figure 15.2. Area ratios 3 : 4 : 1.

SOLUTION. Let u, v, w be the barycentric coordinates of $D = (u, v, w)$ with respect to triangle ABC. From the given relations of triangle areas, we see immediately that $u = 4$ and $w = 3$. Hence $v = 1 - u - w = -6$. Let $AM : AC = CN : CD = r : 1$, then we have

$$M = (1-r)A + rC = (1-r)(1, 0, 0) + r(0, 0, 1) = (1-r, 0, r),$$

$$N = (1-r)C + rD = (1-r)(0, 0, 1) + r(4, -6, 3) = (4r, -6r, 1+2r).$$

Since M, N and B are collinear, then by (15.10)

$$\begin{vmatrix} 1 & 0 & r \\ 1 & -6r & 1+2r \\ 1 & 1 & 0 \end{vmatrix} = 0.$$

Expanding the determinant we get $6r^2 - r - 1 = 0$. In the interval $[0, 1]$, the equation has the unique solution $r = 1/2$. [The negative root $-1/3$ of the equation would put points M, N outside the segments AC, DC, respectively.] This completes the proof.

One of the main purposes of this section is to determine the rate of convergence to zero of the areas of the following set of nested triangles (see [Davis '79, p. 3]):

PROBLEM 15.3. Let P be an arbitrary point lying in triangle ABC, and let $A_1B_1C_1$, $A_2B_2C_2$, ... be determined as in Fig. 15.3. Show that $4^n[A_nB_nC_n]$ tends to a finite nonzero limit as $n \to \infty$ and determine that limit.

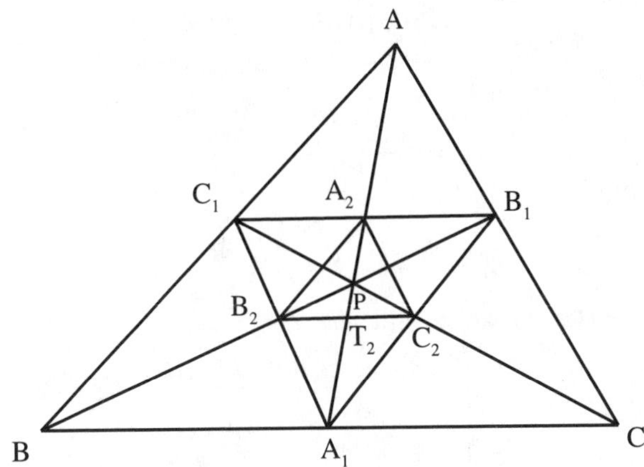

Figure 15.3. Nested triangles.

SOLUTION. Take triangle ABC as the coordinate triangle of a barycentric system. Let $P = (\alpha, \beta, \gamma)$. From the solution of Problem 15.1 we know that

$$A_1 = \left(0, \frac{\beta}{1-\alpha}, \frac{\gamma}{1-\alpha}\right),$$

$$B_1 = \left(\frac{\alpha}{1-\beta}, 0, \frac{\gamma}{1-\beta}\right),$$

$$C_1 = \left(\frac{\alpha}{1-\gamma}, \frac{\beta}{1-\gamma}, 0\right),$$

and that

$$[A_1B_1C_1] = \frac{2\alpha\beta\gamma[ABC]}{(1-\alpha)(1-\beta)(1-\gamma)}. \tag{15.11}$$

Now we compute the barycentric coordinates of P with respect to triangle $A_1B_1C_1$. These barycentric coordinates are denoted by $(\alpha_1, \beta_1, \gamma_1)$. From equation (15.8), it follows that

$$[PB_1C_1] = \frac{1}{(1-\beta)(1-\gamma)} \begin{vmatrix} \alpha & \beta & \gamma \\ \alpha & 0 & \gamma \\ \alpha & \beta & 0 \end{vmatrix} [ABC] = \frac{\alpha\beta\gamma[ABC]}{(1-\beta)(1-\gamma)}$$

and that

$$[A_1B_1C_1] = \frac{1}{(1-\alpha)(1-\beta)(1-\gamma)} \begin{vmatrix} 0 & \beta & \gamma \\ \alpha & 0 & \gamma \\ \alpha & \beta & 0 \end{vmatrix} [ABC]$$

$$= \frac{2\alpha\beta\gamma[ABC]}{(1-\alpha)(1-\beta)(1-\gamma)}.$$

Therefore

$$\alpha_1 = \frac{[PB_1C_1]}{[A_1B_1C_1]} = \frac{1-\alpha}{2}.$$

By symmetry,

$$\beta_1 = \frac{1-\beta}{2}, \qquad \gamma_1 = \frac{1-\gamma}{2}.$$

Hence equation (15.11) can be written as

(15.12) $$[ABC]\alpha\beta\gamma = 4\alpha_1\beta_1\gamma_1[A_1B_1C_1].$$

Setting

(15.13) $$\alpha_n = \frac{1-\alpha_{n-1}}{2}, \qquad \beta_n = \frac{1-\beta_{n-1}}{2}, \qquad \gamma_n = \frac{1-\gamma_{n-1}}{2}, \qquad n = 1, 2, \ldots$$

and using (15.12) repeatedly, we get

(15.14) $$\alpha\beta\gamma[ABC] = 4^n\alpha_n\beta_n\gamma_n[A_nB_nC_n].$$

To evaluate the limit of α_n as n approaches infinity, note that $\alpha_{i+1} - 1/3 = -\frac{1}{2}(\alpha - 1/3)$. Therefore $\lim_{n\to\infty} \alpha_n = 1/3$ and similarly, $\beta_i \to 1/3$ and $\gamma_i \to 1/3$. From (15.14) we conclude that

$$\lim_{n\to\infty} 4^n[A_nB_nC_n] = 27\alpha\beta\gamma[ABC].$$

This problem and its solution have been generalized to higher dimensional simplices by [Chang and Davis '83].

Exercises

15.1 Let M be a point inside a triangle ABC. Draw AM, BM, CM and extend them to meet the opposite sides at A', B', and C' respectively. Suppose that the triangles MAC', MBA' and MCB' have the same area. Show that then M is the centroid of triangle ABC.

15.2 [IMO 1961/4] Consider triangle $P_1P_2P_3$ and a point P inside it. Lines P_1P, P_2P, P_3P intersect the opposite sides in points Q_1, Q_2, Q_3 respectively. Prove that, of the numbers

$$\frac{P_1P}{PQ_1}, \qquad \frac{P_2P}{PQ_2}, \qquad \frac{P_3P}{PQ_3},$$

at least one is ≤ 2 and at least one is ≥ 2.

15.3 [IMO 1966/6] In the interior of sides BC, CA, AB of triangle ABC, select any points K, L, M, respectively. Prove that the area of at least one of the triangles AML, BKM, CLK is less than or equal to one quarter of the area of triangle ABC.

15.4 Any line passing through the centroid of a triangle ABC divides it into two parts. Show that the areas of these two parts differ by at most one ninth of the area of ABC.

15.5 [BMO, 1962] Let $ABCD$ be a parallelogram. Show that there is a unique point M inside $ABCD$ such that $[MAB], [MBC], [MCD]$ and $[MDA]$ form a geometric progression. (Hint: If triangle ABC is the coordinate triangle, then $D = (1, -1, 1)$)

15.6 Let $ABCD$ be a parallelogram. Show that there is a unique point M inside $ABCD$ such that $[MAB]$, $[MBC]$, $[MCD]$, and $[MDA]$ form an arithmetic progression.

15.7 Let A, B, and C be vertices of a triangle, and let D, E, and F be points on the sides BC, CA, and AB respectively. Let U, X, V, Y, W, Z be the respective midpoints of BD, DC, CE, EA, AF, FB. Prove that

$$[UVW] + [XYZ] - \tfrac{1}{2}[DEF] = \tfrac{3}{4}[ABC]$$

([Demir '91]).

CHAPTER SIXTEEN

Douglas-Neumann Theorem

Napoleon's Theorem discussed in Chapter 12 can be summarized as follows: on the three sides of any given triangle ABC, isosceles triangles with base angles $\pi/6$ are erected, either outside, when they are denoted by BCA', CAB', ABC', or inside, when they are denoted by BCA'', CAB'', ABC''. Triangles $A'B'C'$ and $A''B''C''$ are always equilateral, and their areas differ by the area of ABC.

A number of other interesting relations between the three triangles can be deduced.

A beautiful generalization of Napoleon's Theorem was discovered independently by Jesse Douglas and B. H. Neumann about 1940 on opposite sides of the Atlantic, (see [Douglas '40] and [Neumann '41]). On the sides of an arbitrary n-sided polygon, we draw similar isosceles triangles whose shape will be specified later. Their far vertices form a new polygon. We repeat the construction on the resulting new polygon. If we repeat it $n-2$ times and choose the shapes of the isosceles triangles suitably at each stage, we obtain a regular polygon.

The construction of Douglas and Neumann can be viewed as a geometric transformation which is smoothing. These constructions deserve to be better known, first for their intrinsic geometric value, and second because the proofs are an excellent illustration of the power of complex numbers in geometry at a level suitable for undergraduate students.

In Douglas' paper, the theory of circulant matrices is used; and in Neumann's paper, "symmetrical components" are introduced. In a published note, H. F. Baker gave a direct proof of the construction of a regular polygon from an arbitrary polygon, avoiding the symmetrical components [Baker '42]. The most elementary proof was given by Neumann himself, see [Neumann '42]. We present the proof using "symmetrical components", i.e., an algebraic representation of an arbitrary n-gon in terms of regular star (n,k)-gons, because it will enable us to read off another interesting result and it is a good illustration of important ideas in linear algebra.

Let $z_0, z_1, \dots, z_{n-1}$ be n complex numbers, representing n points in the complex plane, not necessarily distinct. The indices will be understood $\bmod n$, i.e., we extend the definition of the z_j to be a periodic sequence of period n. Joining points z_j and z_{j+1}, $j = 0, 1, 2, \dots, n$, by line segments we obtain a closed

n-gon and denote it by $(z_0, z_1, \ldots, z_{n-1})$. Points $z_0, z_1, \ldots, z_{n-1}$ are the *vertices* of the n-gon; some of them may coincide. The n-gon may not be convex and may have self-intersections. The segments (z_j, z_{j+1}) are called *sides* of the n-gon.

DEFINITION. A *regular star* (n, k)*-gon* is a closed n-gon $\mathbf{z} = (z_1, z_2, \ldots, z_n)$ such that the lengths of all sides are equal and the angles between all pairs (z_{j-1}, z_j) and (z_j, z_{j+1}) of adjacent sides are equal. The integer k is the number of complete turns one makes going round the n-gon. Expressed algebraically, $\mathbf{z}$ is a regular star (n, k)-gon if

$$\frac{z_3 - z_2}{z_2 - z_1} = \frac{z_4 - z_3}{z_3 - z_2} = \cdots = \frac{z_1 - z_n}{z_n - z_{n-1}} = \frac{z_2 - z_1}{z_1 - z_n} = e^{k2\pi i/n}. \tag{16.1}$$

In the exceptional case $z_1 = z_2 = \cdots = z_n$, the (n, k)-gon degenerates to a single point. It is reasonable to think that the smoothest of all n-gons are the regular ones.

This definition of a regular star (n, k)-gon is convenient for the purposes of the present chapter. The usual definition of a regular star (n, k)-gon refers just to the geometric figures without orientation and excludes degenerate cases. Since changing the sign of k corresponds merely to reversing the direction of traversal, k is always positive in the usual definition and always $< n/2$.

As an example, a regular star (5, 2)-gon is obtained by taking every second vertex of a regular pentagon in the anticlockwise direction until one gets back to the original vertex. Our regular star 6-gons include one obtained by taking every second vertex of a regular hexagon and going around twice. The geometrical figure here is a triangle. Our regular star (6, 3)-gon is even more degenerate: it is obtained by going back and forth along a line segment 3 times.

Before we come to the Douglas-Neumann Theorem, we derive a smoothing property of the simple transformation T which maps the n-gon $\mathbf{z} = (z_1, z_2, \ldots, z_n)$ to the n-gon formed from the midpoints of its sides. Thus $T\,\mathbf{z}$ has vertices $(z_1 + z_2)/2,\ (z_2 + z_3)/2, \ldots$. It follows from Theorem 7.1 that the sequence of polygons obtained by iterating this map tends to the point $(z_1 + z_2 + \cdots + z_n)/n$. If we scale the iterated images so that they retain a finite size, we find that the n-gons come closer and closer to being inscribed in an ellipse, see Fig. 16.1. (From [Davis '79]).

Our n-gons are n-tuples of complex numbers. It will be useful to consider them to be n-component vectors. In case the reader has not encountered vectors with complex components before, we note that addition of vectors and multiplication of vectors by scalars, which can now be complex, are defined as in the real case. Dot products are defined somewhat differently, but we will not need them.

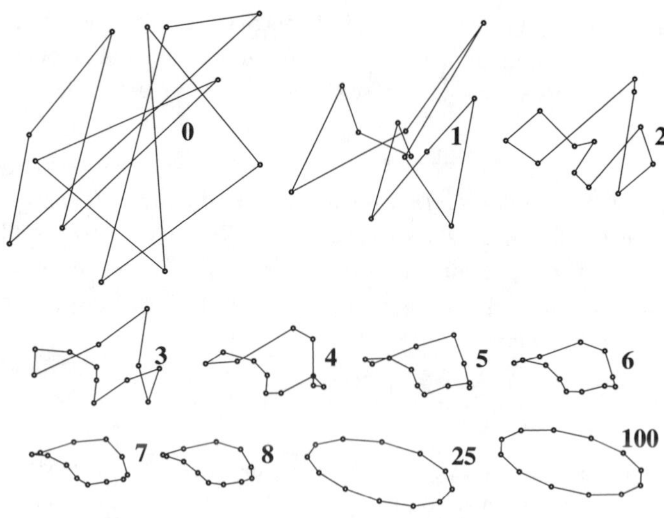

Figure 16.1. Order out of chaos.

In Chapter 8 we used the shift operator E. We use it here too. If $\mathbf{w} = E\mathbf{z}$ then $w_j = z_{j+1}$ $j = 1, 2, \ldots, n$; because the indices are to be taken mod n, this means $w_n = z_1$.

We have $T = \frac{1}{2}(I + E)$, where I is the identity operator. As in Chapter 8, it will be very useful that our operators are *linear*, i.e., $T(a\mathbf{z}+b\mathbf{w}) = aT\mathbf{z}+bT\mathbf{w}$, with similar equations for E and I. In Chapter 8 we studied the iteration of $I+E$ and that is what we are going to do here too, but here our scalars are complex numbers instead of just 0 and 1 as in Chapter 8. Surprisingly, this will make our work easier.

Basic to the study of iterations of a linear operator A are eigenvectors. A *nonzero* vector $\mathbf{e}$ is an *eigenvector* of A with *eigenvalue* λ if $A\mathbf{e} = \lambda\mathbf{e}$. Here λ is a scalar which may have any complex value, including 0. Keep in mind that any nonzero scalar multiple of an eigenvector is an eigenvector with the same eigenvalue. The result of applying A repeatedly to an eigenvector $\mathbf{e}$ can be written down immediately:

(16.2) $$A^k\mathbf{e} = \lambda^k\mathbf{e}.$$

More generally, if $\mathbf{e}^{(1)}, \mathbf{e}^{(2)}, \ldots$ are eigenvectors with eigenvalues $\lambda_1, \lambda_2, \ldots$, and

(16.3) $$\mathbf{v} = c_1\mathbf{e}^{(1)} + c_2\mathbf{e}^{(2)} + \cdots, \quad \text{then } A^k\mathbf{v} = c_1\lambda_1^k\mathbf{e}^{(1)} + c_2\lambda_2^k\mathbf{e}^{(2)} + \cdots.$$

This is indeed a simple formula for $A^k\mathbf{v}$ but it applies only to vectors $\mathbf{v}$ which are linear combinations of eigenvectors. Amazingly, for most linear operators A, *every* vector can be so represented when the scalars are the complex numbers.

If $\mathbf{e}$ is an eigenvector of the operator A with eigenvalue λ, then for any scalars c_0, c_1, $\mathbf{e}$ is clearly also an eigenvector of $c_0 I + c_1 A$, with eigenvalue $c_0 + c_1\lambda$. Hence, to find the eigenvectors of T, it suffices to find those of the shift operator E.

The scalar form of the vector equation $E\mathbf{e} = \lambda\mathbf{e}$ is

$$e_2 = \lambda e_1,\ e_3 = \lambda e_2,\ \ldots,\ e_1 = \lambda e_n. \tag{16.4}$$

If we set $e_1 = 1$, the first $n-1$ equations give $e_k = \lambda^{k-1}$, $k = 2, \ldots, n$ and the last equation gives $\lambda^n = 1$. We see that the eigenvalues are the nth roots of unity. If we set $\omega = e^{2\pi i/n}$, we can write the eigenvectors as

$$\mathbf{e}^{(k)} = (1,\ \omega^k,\ \omega^{2k}, \ldots, \omega^{(n-1)k}), \quad k = 1, 2, \ldots, n. \tag{16.5}$$

To follow the usual convention in matrix algebra, we should have written the vectors as columns but since we are not going to use the matrix representations of our operators in this chapter, we do not need to be strict about distinguishing row and column vectors.

We have $T\mathbf{e}^{(k)} = \frac{1}{2}(I + E)\mathbf{e}^{(k)} = \frac{1}{2}(1 + \omega^k)\mathbf{e}^{(k)}$. In other words, $\mathbf{e}^{(k)}$ is an eigenvector of T with eigenvalue $\mu_k = \frac{1}{2}(1 + \omega^k)$. We want to write these eigenvalues in polar form.

The point representing the complex number $(e^{i\theta} + 1)/2$ is the midpoint of the base (the segment connecting 1 and $e^{i\theta}$) of the isosceles triangle with vertices $1,\ 0,\ e^{i\theta}$. This means

$$\tfrac{1}{2}\left(e^{i\theta} + 1\right) = \cos\left(\tfrac{1}{2}\theta\right)e^{i\theta/2}.$$

Hence

$$\mu_k = \cos\left(\frac{k\pi}{n}\right)e^{ik\pi/n}. \tag{16.6}$$

We have listed the eigenvectors and eigenvalues of T and are now ready to prove that an arbitrary vector $\mathbf{z}$ is a linear combination of eigenvectors. The most natural way to come to this conclusion is to use the theorem of linear algebra that eigenvectors with distinct eigenvalues are linearly independent. However, we can also rely on the Lagrange interpolation formula of Chapter 17, as follows.

We write the vector equation

$$\mathbf{z} = \sum_{k=1}^{n} a_k \mathbf{e}^{(k)} \tag{16.7}$$

in scalar form:

$$z_j = a_0 + a_1\omega^j + a_2\omega^{2j} + \cdots + a_{n-1}\omega^{(n-1)j} \quad j = 1, 2\ldots, n. \tag{16.8}$$

Satisfying this set of equations is the same as finding a polynomial $p(z) = a_0 + a_1 z + \cdots + a_{n-1}z^{n-1}$ such that $p(\omega^j) = z_j$ for $j = 1, 2, \ldots, n$. We have shown that this is possible when we derived the Lagrange interpolation formula.

We note now that the right-hand side of (16.5) is defined for all integers k, and is periodic in k with period n. Thus, in (16.7) the range of summation can be any complete set of residues mod n. The following discussion will be somewhat simpler if instead of the range $1 \le k \le n$ we use $\lfloor -n/2 \rfloor + 1 \le k \le \lfloor n/2 \rfloor$. Thus we have

$$\mathbf{z} = \sum_{k=\lfloor -n/2 \rfloor+1}^{\lfloor n/2 \rfloor} a_k \mathbf{e}^{(k)}. \tag{16.9}$$

For a value of k other than 0 in this range, the n-gon $\mathbf{e}^{(k)}$ is a regular star (n, k)-gon with center at the origin and vertices on the unit circle. For $k = 0$ we get the degenerate polygon, all of whose vertices are at $z = 1$. The centers of mass of all the other polygons $\mathbf{e}^{(k)}$ are at the origin. For the purposes of our geometrical problems we may assume without loss of generality that the center of mass of the polygon $\mathbf{z}$ is the origin. Then the degenerate polygon $a_0\mathbf{e}^{(0)}$ will not occur in (16.9).

Even with the modified indexing of the eigenvectors, the eigenvalues are given by Formula (16.6). Thus $|\mu_k| = \cos(k\pi/n)$, and $|k| \le n/2$. Assume for the time being that at least one of a_1, a_{-1} is not 0. Each application of T shrinks the $\mathbf{e}^{(k)}$ with $|k| > 1$ more than it shrinks $\mathbf{e}^{(1)}$ and $\mathbf{e}^{(-1)}$. Introduce the scaled transformation

$$\widetilde{T} = \frac{1}{\cos(\pi/n)} T. \tag{16.10}$$

This operator transforms the n-gons $\mathbf{e}^{(1)}$, $\mathbf{e}^{(-1)}$ into n-gons of equal size but for $|k| > 1$ it multiplies the sizes of a star n-gon in (16.9) by a factor < 1. Thus, for large h,

$$\widetilde{T}^h \mathbf{z} = e^{hi\pi/n} a_1 \mathbf{e}^{(1)} + e^{-hi\pi/n} a_{-1} \mathbf{e}^{(-1)} + \text{small terms}. \tag{16.11}$$

We want to show that the first two terms in (16.11) represent an n-gon inscribed in an ellipse.

DEFINITION. An *affine transformation* of the plane is a mapping $(x, y) \to (x', y')$ of the plane into itself or another plane, given by

$$x' = ax + by + e, \qquad y' = cx + dy + f, \qquad ad - bc \ne 0 \tag{16.12}$$

where all the coefficients are real.

Translations, rotations, and similarity transformations are examples of affine transformations. Another example is given by the formulas $x' = x$, $y' = \frac{1}{2}y$ which squeezes the plane in the y-direction. An affine transformation maps parallel lines into parallel lines. The condition $ad - bc \ne 0$ ensures that the transformation is invertible. One can prove that every invertible transformation of a plane onto a plane which maps straight lines into straight lines is affine, but we shall not need this.

We shall use that the ellipse $x^2/a^2 + y^2/b^2 = 1$ is obtained from the circle $x^2 + y^2 = a^2$ by the affine transformation $x' = x,\ y' = \frac{b}{a}y$. Geometrically we can express this by saying that if a circle is squeezed or stretched perpendicularly to a diameter it becomes an ellipse.

LEMMA. *Let a, b be complex constants. The locus*

$$z = ae^{it} + be^{-it}, \qquad |a| \neq |b|, \qquad 0 \leq t \leq 2\pi \tag{16.13}$$

is an ellipse with axes $|a| + |b|,\ \ ||a| - |b||$, and there is an affine mapping in which the point $ae^{it} + be^{-it}$ of the ellipse corresponds to the point $(|a| + |b|)e^{it}$ of the circle whose diameter is the major axis of the ellipse.

PROOF. The two terms in (16.13) are vectors rotating in opposite directions through 2π as t goes from 0 to 2π. For some value t_0 of t they are therefore parallel. Take this direction as the direction of the positive x-axis of a coordinate system. In this coordinate system, the locus is given by the formulas

$$\begin{aligned} x &= (|a| + |b|)\cos(t - t_0), \\ y &= \big||a| - |b|\big| \sin(t - t_0), \qquad 0 \leq t \leq 2\pi, \end{aligned} \tag{16.14}$$

from which we can read off the result.

If $|a| = |b|$ then the locus (16.13) is a line segment traversed in both directions.

Lemma 1 tells us that the first two terms of (16.11) represent an affine image of a regular n-gon. It is insribed in an ellipse with axes $|a_1| + |a_{-1}|\ ||a_1| - |a_{-1}||$. Thus we have tracked down the reason for the appearance of an elliptical shape in Fig. 16.1.

If by chance we would have had $|a_1| = |a_{-1}| \neq 0$, the ellipse would have degenerated into a line segment.

If $a_1 = a_{-1} = 0$ then $T^h\mathbf{z}$, suitably scaled, will for large h be approximately the affine image of a regular star (n, k)-gon, or possibly a line segment.

We turn now to Douglas and Neumann's generalization of Napoleon's Theorem.

Let c be a given complex number, and $\mathbf{z} = (z_1, z_2, \ldots, z_n)$ an arbitrary n-gon. On side $z_k z_{k+1}$ of $\mathbf{z}$, the triangle $z_k z_{k+1} z_{k,k+1}$ is erected, where $z_{k,k+1}$ is chosen so that the triangle is directly similar to triangle $01c$ (Fig. 16.2). Points $z_{k,k+1}$, $k = 1, 2, \ldots, n$, called the *free vertices* of the similar triangles, form an n-gon. Using the initial of the emperor, we denote this polygon by $N_c\,\mathbf{z}$.

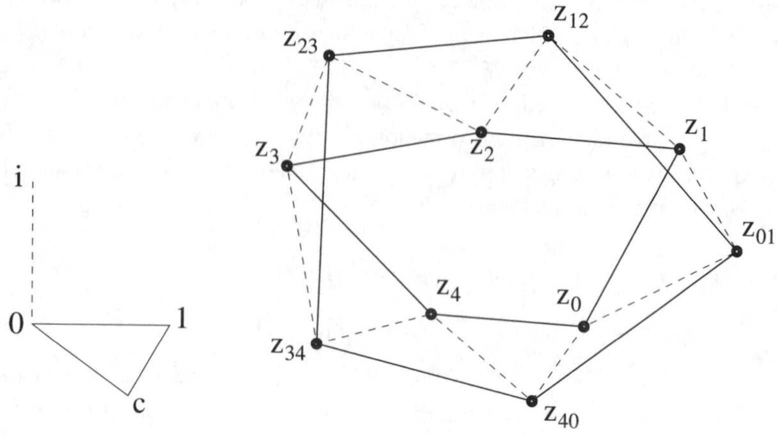

Figure 16.2. Douglas and Neumann construction.

We can express the operator N_c in terms of I and E as follows. From the two vertices z_k, z_{k+1} the free vertex z'_k is found by

$$\frac{z'_k - z_k}{z_{k+1} - z_k} = \frac{c - 0}{1 - 0},$$

from which

$$z'_k = (1 - c)z_k + cz_{k+1}.$$

In terms of the shift operator E,

$$\mathbf{z}' = N_c\,\mathbf{z} = ((1 - c)I + cE)\,\mathbf{z}. \tag{16.15}$$

The vectors $\mathbf{e}^{(k)}$ are eigenvectors of N_c with eigenvalues $1 - c + c\lambda_{c,k} = 1 - c(1 - \omega^k)$. This will be equal to 0, i.e., maps all vertices of the regular star (n, k)-gon $\mathbf{e}^{(k)}$ onto the origin, if

$$c = c_k = \frac{1}{1 - \omega^k}. \tag{16.16}$$

The representation (16.9) gives an arbitrary polygon $\mathbf{z}$ with center of mass at the origin as the sum of $n - 1$ regular star (n, k)-gons. Thus, if we apply to $\mathbf{z}$ successively $n - 2$ Napoleonic operators with $n - 2$ of the $n - 1$ quantities c_k given by (16.16), all but one of the summands in (16.9) are mapped into 0 and we obtain a regular star (n, k')-gon, possibly a single point, where k' is the index which was not used. This is Douglas and Neumann's theorem, except for a simple geometric characterization of the triangles, which we now give.

At this time, it is somewhat simpler to return to $1, 2, \ldots, n - 1$ as the representatives of the nonzero residue classes mod n. The formula $c_k = 1/(1 - \omega^k)$

remains valid. The side $c_k\,1$ of the template triangle $0\,c_k\,1$ is represented by the point $1 - c_k = -\omega^k/(1-\omega^k)$. We have

$$\frac{1-c_k}{c_k} = -\omega^k. \tag{16.17}$$

Since $|\omega| = 1$, the triangle is isosceles. Equation (16.17) also tells us that the exterior angle from side $0\,c_k$ to side $c_k\,1$ is $\frac{k2\pi}{n} - \pi$. Hence the interior angle from side $0\,1$ to side $0\,c_k$ is $\frac{\pi}{2} - \frac{k}{n}\pi$. Thus the triangles to be used in the Douglas-Neumann constructions are any $n-2$ of the $n-1$ isosceles triangles with base angles

$$\frac{\pi}{2} - \frac{k}{n}\pi, \quad k = 1, 2, \ldots, n-1. \tag{16.18}$$

A positive base angle means the triangles are pointing to the left as we go along the polygon in the direction of increasing indices, a negative base angle means they are pointing to the right. If n is even and $k = n/2$, then the angle is 0. Geometrically, there is no triangle but our formula gives $c_{n/2} = 1/2$. This means $N_{n/2}\mathbf{z}$ is obtained by taking the midpoints of the sides of $\mathbf{z}$.

As special cases of the Douglas-Neumann theorem, a great number of geometric facts can be derived.

For $n = 3$, the Douglas and Neumann Theorem reduces to Napoleon's theorem: the two possibilities for the base angle, $\pi/6$ and $-\pi/6$, correspond to the outer Napoleon triangle and the inner Napoleon triangle respectively.

For $n = 4$, the three possible values for the base angle of the isosceles triangles are $\pi/4, 0, -\pi/4$; any one of them can be omitted in the Douglas-Neumann process. When $\pi/4$ (or $-\pi/4$) is omitted, we have the following two conclusions:

(a) If right-angled isosceles triangles are erected on the sides of a quadrilateral, pointing outside (or inside), their vertices form a quadrilateral the midpoints of whose sides are the vertices of a square (Fig. 16.3, left).

(b) If the sides of the parallelogram formed by the midpoints of the sides of an arbitrary quadrilateral are made the bases of right-angled isosceles triangles pointing outside (or inside), their vertices form a square (Fig. 16.3, right).

The fact that the order in which the operations are carried out does not matter, i.e., that (a) and (b) lead to the same square, can be expressed as follows: Suppose we have two squares $ABCD$ and $AB'C'D'$, both described counterclockwise. If their centers E and E' are made vertices of a square $EFE'F'$, then F and F' are the mid-points of the lines $B'D$ and BD', respectively. (If C, A, C' are any three consecutive vertices of the original quadrilateral in (a) or (b) then F' and F are the points of the interior and the exterior square associated with these three points.)

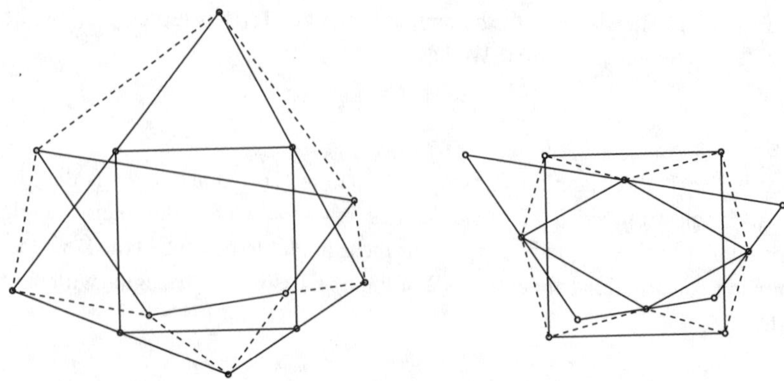

Figure 16.3. Douglas-Neumann for (a) and (b)

When 0 is omitted, we have:

(c) If the sides of an arbitrary quadrilateral are made the bases of right isosceles triangles all outside (all inside), the free vertices form a new quadrilateral. If the sides of the new quadrilateral are made the bases of right isosceles triangles all inside (all outside), then the third free vertex coincides with the first, and the fourth with the second.

Fifty years have passed since the discovery of the Douglas-Neumann Theorem; yet, it has attracted further research in recent years (see[Chang '82] and [Chang and Davis '83]).

Exercises

16.1 If directly similar triangles ABC', BCA', ACB' are erected respectively externally, externally, and internally on the sides of a triangle ABC, show that $A'B'C'B$ is a parallelogram.

16.2 If $ABCD$, $AEBK$ and $CEFG$ are squares with the same orientation, show that B bisects DF. ([Demir '59])

16.3 If four squares are placed externally (or internally) on the four sides of any parallelogram, then their centers are the vertices of another square.

16.4 Let X, Y, Z be the centers of squares placed externally on the sides BC, CA, AB of a triangle ABC. Show that segment AX is congruent and perpendicular to YZ (also BY to ZX and CZ to XY).

CHAPTER SEVENTEEN

Lagrange Interpolation

In this chapter we derive some facts which can be obtained by substituting complex numbers into algebraic identities and looking at the geometric meaning of the results. The most important topic in the chapter is the Lagrange interpolation formula, which has many applications in addition to the somewhat esoteric geometric inequality we are going to derive from it.

In this chapter, n denotes an integer > 1 and $\omega = e^{\frac{2\pi i}{n}}$. The nth roots of unity are $\omega,\ \omega^2, \dots, \omega^{n-1}$ and $\omega^n = 1$. The reader should keep in mind the following fact because it is often used:

$$(17.1)\qquad 1 + \omega^k + \omega^{2k} + \cdots + \omega^{(n-1)k} = \begin{cases} n & \text{if } n|k \\ 0 & \text{otherwise.} \end{cases}$$

Indeed, in the first case all terms in the sum are $= 1$; in the second case $\omega^k \neq 1$, and the sum of the geometric series (17.1) has the value $(1-\omega^{nk})/(1-\omega^k) = 0$.

PROBLEM 17.1. Show that the sum of the squared distances from any point P on the circumcircle of a regular polygon to its vertices is independent of P.

SOLUTION. Without loss of generality, we assume that the circumcircle is the unit circle in the complex plane, and the vertices of the polygon are the nth roots of unity $\omega^j,\ j = 1, 2, \dots, n$. Let z be any point on the unit circle. The desired sum is

$$\begin{aligned}\sum_{j=1}^{n} |z-\omega^j|^2 &= \sum_{j=1}^{n} (z-\omega^j)(\overline{z}-\overline{\omega}_j) \\ &= \sum_{j=1}^{n} \left(|z|^2 + |\omega^j|^2 - z\overline{\omega}^j - \overline{z}\omega^j\right) \\ &= 2n - z\sum_{j=1}^{n} \overline{\omega}^j - \overline{z}\sum_{j=1}^{n} \omega^j = 2n.\end{aligned}$$

Hence the sum of squared distances is $2n$, and is independent of the point z.

PROBLEM 17.2. Given a regular n-gon inscribed in a unit circle, prove that the product of the distances from one vertex to the other vertices is n.

SOLUTION. Let the circle be the unit circle in the complex plane, and let the vertices of the regular n-gon be $1, \omega, \ldots, \omega^{n-1}$.

When we factor the polynomial $z^n - 1$, we obtain, on the one hand,

$$(z-1)\sum_{0}^{n-1} z^h,$$

and if we factor it completely, we get

$$(z-1)(z-\omega)(z-\omega^2)\cdots(z-\omega^{n-1}).$$

Hence, for all $z \neq 1$,

$$\sum_{0}^{n-1} z^h = \prod_{1}^{n-1}(z-\omega^j);$$

Substituting $z = 1$ we get: $n = (1-\omega)(1-\omega^2)\cdots(1-\omega^{n-1})$ from which the assertion follows.

PROBLEM 17.3. In the plane, a finite number of points and a circle C of radius 1 are given. Show that there exists a point on C such that the product of its distances to the given points is at least 1.

SOLUTION. Let C be the unit circle in the complex plane, and denote the given points by $z_1, z_2, \ldots, z_{n-1}$ $(n \geq 2)$. Assume that z varies on the unit circle, i.e., $|z| = 1$. Set $g(z) = (z-z_1)(z-z_2)\cdots(z-z_{n-1})$. Our goal is to show that there exists a point z with $|z| = 1$ such that $|g(z)| > 1$.

Define the polynomial $f(z) = zg(z)$ of degree n and observe that $|f(z)| = |g(z)|$. Expanding $f(z)$, we have

$$f(z) = z^n + a_1 z^{n-1} + a_2 z^{n-2} + \cdots + a_{n-1} z,$$

where $a_1, a_2, \ldots, a_{n-1}$ are constants determined by $z_1, z_2, \ldots, z_{n-1}$. We claim that at least one of the nth roots of unity, ω^j, has the desired property. To see this set $z = \omega^j$ in the expression above and sum over j; making use of (17.1), we get

$$\sum_{j=0}^{n-1} f(\omega^j) = \sum_{j=0}^{n-1} (\omega^j)^n + a_1 \sum_{j=0}^{n-1} (\omega^j)^{n-1} + \cdots + a_{n-1} \sum_{j=0}^{n-1} \omega^j = n.$$

Hence at least one of the summands has modulus ≥ 1.

To address the next problem of this chapter, we need the interpolation formula of Lagrange. Suppose that $z_1, z_2, \ldots, z_n$ are distinct numbers and $w_1, w_2, \ldots, w_n$ are arbitrary numbers. We are going to show that there is exactly one polynomial $p(z)$, of degree at most $n-1$, such that

$$p(z_k) = w_k \quad \text{for} \quad k = 1, \ldots, n. \tag{17.2}$$

A polynomial which assumes assigned values at assigned points is called an *interpolation* polynomial. Different formulas for the interpolation polynomial $p(z)$ were given by several mathematicians. The formulas of Isaac Newton (1642–1727) and Joseph Louis Lagrange (1736–1813) are the most notable.

We prove first that there is at most one polynomial $p(z)$ of degree $< n$ which solves Problem (17.2). If there were two, their difference would be a polynomial of degree $< n$ vanishing at all n points $z_1, \ldots, z_n$. But a theorem of algebra says that a polynomial of degree $< n$ which vanishes at n points is identically 0.

To show that there is a polynomial $p(z)$ which satisfies (17.2), we first consider the special case where $w_j = 1$ and $w_k = 0$ for $k \neq j$. We need a polynomial $p_j(z)$ which vanishes for $z = z_k$, $k \neq j$ and is $= 1$ at $z = z_j$. The first condition is satisfied by any polynomial of the form

$$A \prod_{k \neq j} (z - z_k),$$

and the condition $p_j(z_j) = 1$ gives us the value of the constant A:

$$p_j(z) = \frac{\prod_{k \neq j} (z - z_k)}{\prod_{k \neq j} (z_j - z_k)}.$$

The polynomial which satisfies (17.2) is easily seen to be

$$p(z) = w_1 p_1(z) + w_2 p_2(z) + \cdots + w_n p_n(z), \tag{17.3}$$

and this is Lagrange's interpolation formula.

The argument we gave to demonstrate that there is a polynomial $p(z)$ of degree $\leq n-1$ which takes the given values is very simple and, as we shall see, the formula for $p(z)$ which we obtained is useful. Nevertheless, it is interesting to observe that the existence of $p(z)$ is a consequence of a general theorem of linear algebra and can be deduced without producing a formula for it.

The conditions (17.2) are a system of n linear equations for the n coefficients of the polynomial $p(z)$. We have proved that for any given $w_1, \ldots, w_n$ there is at most one $p(z)$ with these values at the points $z_1, \ldots, z_n$. The theorem of linear algebra that we can apply is this: If the coefficients of a system of n linear equations for n unknowns are such that for any given right-hand side there is at most one solution, then for any given right-hand side there exists a solution.

Next we use the Lagrange formula to derive an identity. If $w_1 = \cdots = w_n = 1$, the polynomial $p(z) \equiv 1$ of degree 0 solves this interpolation problem; so does the polynomial $p(z) = \sum_1^n p_j(z)$ obtained by setting $w_j = 1$, $j = 1, \ldots, n$, in (17.3). Therefore, by uniqueness, we have the identity

$$p_1(z) + \cdots + p_n(z) = 1.$$

A longer way of writing this is

$$\sum_{j=1}^{n} \prod_{k \neq j} \frac{(z - z_k)}{(z_j - z_k)} = 1.$$

Setting $z = 0$, it reduces to

$$\sum_{j=1}^{n} \prod_{k \neq j} \frac{z_k}{(z_k - z_j)} = 1. \tag{17.4}$$

A little manipulation of this identity will give the following geometric theorem due to H. S. Shapiro (see the Editor's comments on Problem E 1023, *Amer. Math. Monthly*, **60** (1953), 119–121).

PROBLEM 17.4. Let $P_1, P_2, \ldots, P_n$ denote n distinct points on a circle of unit radius, and set d_k equal to the product of the distances from P_k to all the other points. Then

$$\sum_{k=1}^{n} \frac{1}{d_k} \geq 1, \tag{17.5}$$

equality being attained if and only if the points form a regular n-gon.

PROOF. Under the hypothesis, the numerators in the identity (17.4) have modulus 1, and the inequality (17.5) follows by taking absolute values in (17.4).

Suppose now that equality holds in (17.5). Let

$$r_j = \prod_{k \neq j} \frac{z_k}{z_k - z_j}.$$

Since $r_1 + \cdots + r_n = 1$, $|r_1| + \cdots + |r_n| = 1$ implies that all the r_j are nonnegative real numbers. Hence

$$\overline{r_j} = r_j. \tag{17.6}$$

The hypothesis $|z_j| = 1$ implies $\overline{z_j} = 1/z_j$. Hence

$$r_1 = \prod_{k=2}^{n} \frac{z_k}{z_k - z_1} = \overline{r_1} = \prod_{k=2}^{n} \frac{\frac{1}{z_k}}{\frac{1}{z_k} - \frac{1}{z_1}} = \prod_{k=2}^{n} \frac{z_1}{z_1 - z_k}. \tag{17.7}$$

The equation simplifies to $z_1^{n-1} = (-1)^{n-1} z_2 \cdots z_n$, or $z_1^n = (-1)^{n-1} z_1 z_2 \cdots z_n$. The right hand side is symmetrical in $z_1,\ z_2, \ldots, z_n$. Hence $z_1^n = z_2^n = \cdots = z_n^n$. Consequently all the ratios z_j/z_1 are nth roots of unity, and since the z_j are all distinct, they form a regular n-gon.

An extension of H. S. Shapiro's result can be found in [Chang '85].

Exercises

17.1 Prove the identity:

$$\sin\frac{\pi}{n}\sin\frac{2\pi}{n}\cdots\sin\frac{(n-1)\pi}{n} = \frac{n}{2^{n-1}}.$$

17.2 Let $f(z)$ be a polynomial of degree $m < n$. Show that

$$f(z) = \frac{1}{n}\sum_{j=1}^{n} f(z_j),$$

where $z_1, z_2, \ldots, z_n$ are the vertices of any regular n-gon with center z.

CHAPTER EIGHTEEN

The Isoperimetric Problem

We have seen that according to several different criteria, the smoothest of all n-gons are the regular ones. Moreover, we showed in Chapter 3 that when $n = 3$, the equilateral triangle has the greatest area of all triangles with given perimeter. It is natural to ask the corresponding questions for arbitrary closed curves. First of all: what are the smoothest figures among the set of all closed curves? It is intuitively clear that the answer should be circles. A circle is symmetric with respect to its center, and also with respect to any diameter. No point on a circle has any special geometric feature not possessed by any other point of the circle.

Next, consider the *isoperimetric problem*: among all closed curves in the plane having a given perimeter, find the one enclosing the greatest area. In this form, the problem was known to the Greeks. They also knew the solution, namely, the circle. From the study of such problems has emerged a branch of mathematics known as the *calculus of variations*. The Swiss mathematician Jakob Steiner (1796–1863) proposed several ingenious arguments to show that the solution to the isoperimetric problem must be a circle. In the following we sketch one of those arguments, see [Pedoe '57].

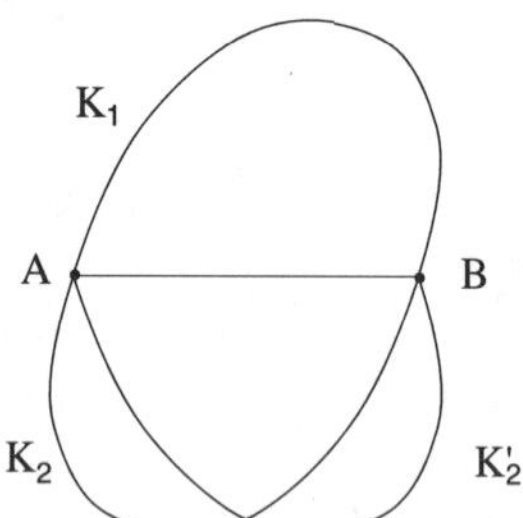

Figure 18.1. Symmetrization.

Given any closed plane curve K which is not a circle, Steiner gave a simple geometric method for constructing a new closed curve K^* in the plane with the same perimeter as K but enclosing a greater area. It follows that K cannot be the solution to the isoperimetric problem.

First, if K is not convex, then taking the convex hull of K gives a figure of shorter perimeter and greater area, so we may assume that K is convex. Choose two points A and B on K which divide K into two arcs K_1 and K_2 of equal length (Fig. 18.1). If K is not a circle we can choose A and B so that neither of the arcs K_1 and K_2 is a semicircle. The area bounded by K_i together with line AB is denoted by F_i $(i = 1, 2)$ so $F = F_1 + F_2$ is the area bounded by K. Assume that $F_1 \geq F_2$. We now wipe out the arc K_2, and substitute in its place the arc K_2' obtained from K_1 by reflection in the line AB. The closed curve K' bounded by K_1 and K_2' evidently has the same perimeter as K. Its area is $F' = 2F_1 \geq F$.

If K' is not convex, then its convex hull has larger area and smaller perimeter. Let us therefore consider a convex curve K' which is symmetric about a line AB and is not a circle (Fig. 18.2, left side).

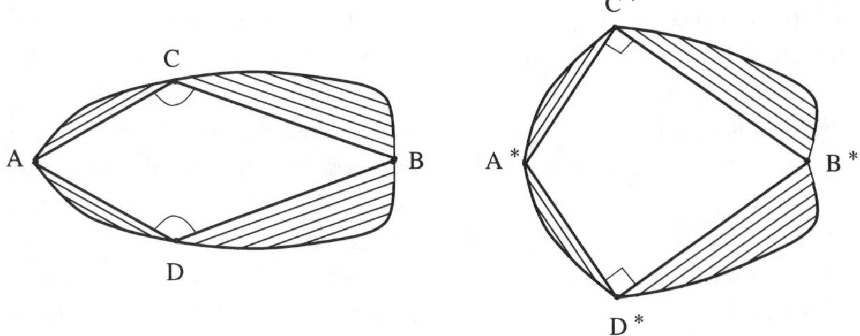

Figure 18.2. Steiner linkage construction.

Since the upper half of K' is not a semicircle, we can find a point C on it so that $\angle ACB$ is not a right angle. Let D be the reflection of C in AB. If we cut out from the area bounded by K' the quadrilateral $ACBD$, four lunes remain. We imagine that the sides of the quadrilateral are bars, hinged at A, B, C and D, and that the lunes are fastened to these bars. Among all triangles with two sides of given lengths, the one with the largest area is the one with a right angle between the given sides. Hence, if we move the bars about the hinges to a position where the angles at C and D are both right angles, (right side of Fig. 18.2), we get a figure with a larger area and the same perimeter as K'.

Steiner's construction shows that a closed curve other than a circle cannot be the solution to the isoperimetric problem. Does this mean that we now have a rigorous proof of the isoperimetric property of the circle? No! Like the earliest proofs, Steiner's reasoning assumed that a solution exists. Ignoring the question of existence may lead to incorrect conclusions. A simple illustration of the logical situation is the following. If n is any positive integer other than 1, $n^2 > n$, hence n is not the largest among natural numbers. If someone thoughtlessly assumed that just because we can write down the phrase "the largest natural number", such a thing exists, he would have to conclude that 1 must be it.

Only in 1870 did Weierstrass point out that one may not assume without proof that a curve bounding a region of largest area with a given perimeter exists, because there are similar problems which do not have solutions. For instance, among the functions which are continuous and nonnegative in $0 \leq x \leq 1$, and which have the value 1 at $x = 1$, there is none whose graph bounds the least area between the x-axis and the lines $x = 0$ and $x = 1$.

Weierstrass gave a complete proof of the existence of a solution to the isoperimetric problem. Although at the time people seemed a bit helpless in dealing with such problems, it is in fact fairly easy to give a rigorous proof based on a construction such as the one we presented. This is done by showing that if a convex curve of length p differs from a circle enough, (e.g., in the Steiner linkage construction, if there is a point C that subtends an angle sufficiently different from a right angle) then the increase in area produced by the linkage construction will exceed some corresponding positive amount. Since the area enclosed by a curve with length p is certainly $< p^2$, it follows that after a finite number of transformations the curve will differ from a circle by only a small amount. The circle which the sequence of curves approaches will have a greater area and a circumference less than or equal to the length of the original curve. For a complete proof of the isoperimetric property without calculus, the reader is referred to [Pedoe '57].

Steiner and others had several other constructions which can be used in a manner similar to the linkage construction. For instance, one can use a different kind of symmetrization followed by forming a convex hull to obtain from a noncircular curve a curve of greater area and smaller perimeter.

Instead of presenting the somewhat tedious details of a proof that suitable sequences of such transformations create a sequence of curves which approach a circle, we give here an elegant proof of the isoperimetric property of the circle due to Peter Lax. It does not depend on iterated transformations and does not require a convergence proof. Instead, it requires some calculus and the inequality between the arithmetic and the geometric mean.

First we express the isoperimetric property of the circle as an inequality. Let L be the length of a closed curve K and A the area bounded by K. A circle with perimeter L has radius $\frac{L}{2\pi}$ and area $\frac{L^2}{4\pi}$. The statement that the area enclosed by a curve of length L is $\leq$ the area of a circle with the same length can be formulated as

The Isoperimetric Inequality. Let L and A be the perimeter and area, respectively, of a figure bounded by a closed curve; we have

$$L^2 \geq 4\pi A, \tag{18.1}$$

and equality holds if and only if the figure is a circle.

PROOF. We consider a closed curve with perimeter 2π. Let $(x(s), y(s))$ be a parametric representation of the curve with clockwise orientation, in which s is the arclength, $0 \leq s \leq 2\pi$. Suppose that we have so positioned the curve that the points $(x(0), y(0))$ and $(x(\pi), y(\pi))$ lie on the x-axis, i.e.,

$$y(0) = y(\pi) = 0. \tag{18.2}$$

The area enclosed by the curve is given by the formula

$$A = \int_0^{2\pi} y\dot{x}\, ds, \tag{18.3}$$

where the dot denotes differentiation with respect to s. We show that $A \leq \pi$.

According to the arithmetic mean-geometric mean inequality for nonnegative numbers a and b, we have $ab \leq (a^2 + b^2)/2$; equality holds only when $a = b$. Applying this to $|y| = a$, $|\dot{x}| = b$, we get

$$A \leq \int_0^{2\pi} |y||\dot{x}|\, ds \leq \frac{1}{2}\int_0^{2\pi} (y^2 + \dot{x}^2)\, ds. \tag{18.4}$$

Since s is arclength, $\dot{x}^2 + \dot{y}^2 = 1$; so we can rewrite (18.4) as

$$A \leq \frac{1}{2}\int_0^{2\pi} (y^2 + 1 - \dot{y}^2)\, ds.$$

Since $y = 0$ at $s = 0$ and $s = \pi$, we can factor y as

$$y(s) = u(s)\sin s, \tag{18.5}$$

where u is bounded and differentiable. Differentiating (18.5), we obtain

$$\dot{y} = \dot{u}\sin s + u\cos s.$$

Therefore

$$\begin{aligned} y^2 + 1 - \dot{y}^2 &= u^2(\sin^2 s - \cos^2 s) - 2u\dot{u}\sin s\cos s + 1 - \dot{u}^2\sin^2 s \\ &= -(u^2\cos 2s + u\dot{u}\sin 2s) + 1 - \dot{u}^2\sin^2 s \\ &= -\frac{1}{2}\frac{d}{ds}(u^2\sin 2s) + 1 - \dot{u}\sin^2 s. \end{aligned}$$

Since

$$\int_0^{2\pi} \frac{d}{ds}(u^2\sin 2s)\, ds = (u^2\sin 2s)|_0^{2\pi} = 0,$$

we get

$$A \le \frac{1}{2}\int_0^{2\pi} (1 - \dot{u}^2 \sin^2 s)\, ds \le \pi. \tag{18.6}$$

This is the isoperimetric inequality for our chosen value $L = 2\pi$.

Now $A = \pi$ only if equality holds in all the inequalities we have used. In (18.4) the first = holds if $\dot{x}y > 0$; the second holds if $\dot{x} \equiv y$. And in (18.6), equality holds if $\dot{u} \equiv 0$. From $\dot{u} \equiv 0$ we see that $u(s) = c$, a constant. From (18.5) we have $y = c \sin s$. Setting $\dot{x} = y = c \sin s$ and $\dot{y} = c \cos s$ into $\dot{x}^2 + \dot{y}^2 = 1$ gives $c = \pm 1$. The clockwise orientation requires $c = -1$. Hence, $y = -\sin s$. By integrating $\dot{x} = -\sin s$, we obtain

$$x = \cos s + \text{constant}.$$

We see that $(x(s), y(s))$ represents a circle of radius 1, with the center on the x-axis. Q.E.D.

Exercise

18.1 Show that a circular disk D can never be covered by two smaller circular disks D_1 and D_2.

CHAPTER NINETEEN

Formulas for Iterates

We have seen a few examples of functional iterations in Chapter 1. In this and the following chapters, we explore in more detail some of the interesting phenomena connected with them.

Let I be a finite or infinite interval. Let f be a function which assigns to each point x of I a point $x' = f(x)$ in I. We say f produces a map $x \to x'$ of I into itself. The $'$ denotes the result of applying the map f once, not a derivative. We are interested in the properties of the nth iterate f^n of the function, and the sequence of images $x', x'', \ldots, x^{(n)}, \ldots$ of a point x under iterations of f.

For a linear function $f(x)$, we found the explicit formula for $f^n(x)$ in Chapter 1. If $f(x)$ is an arbitrarily given function, a single formula valid for all n is unlikely to exist. In case it does exist, the *method of conjugate functions*[1] may help us find it, as we shall see. It must be admitted though that our examples have been constructed backwards, starting from the result.

EXAMPLE 19.1. Let $f(x) = x + 2\sqrt{x} + 1, x \geq 0$. Find $f^n(x)$.

SOLUTION. Set $u = h(x) = \sqrt{x}$ or $x = u^2$. We can think of u as a new label or coordinate for points on the half-line of nonnegative numbers. Then the new label, u', of the image point $x' = f(x)$ of x is

$$u' = \sqrt{x'} = \sqrt{u^2 + 2u + 1} = u + 1.$$

Thus, in terms of the variable u, our transformation has the simple form $u' = g(u) = u + 1$. Iterating this map n times, we get $g^n(u) = u + n$. In terms of the original variable x, our formula is $f^n(x) = (\sqrt{x} + n)^2$.

Now we describe the method of conjugate functions in general terms. Let us denote the relation between x and the new "coordinate" u by $u = h(x)$. Then $x = h^{-1}(u)$. (In Example 1, $h(x) = \sqrt{x},\ h^{-1}(u) = u^2$.) Let the map from u to u' be denoted by g: $u' = g(u)$. (In the example, $g(u) = u + 1$.) A schematic

[1]The method goes back to F.W.K.E. Schröder (1841–1902), a pioneer in the general theory of iterations. See, for example, [Melzak '73]

diagram will help:

$$\begin{array}{ccc} x & \xrightarrow{f} & x' \\ {\scriptstyle h}\downarrow & & \downarrow{\scriptstyle h} \\ u & \xrightarrow{g} & u' \end{array}$$

Here x' is the image of x under f; it is also the image of u' under h^{-1}, where u' is the image of u under g, and u is the image of x under h. Thus we can get from x to $x' = f(x)$ via u as follows:

(19.1) $$u = h(x), \qquad u' = g(u), \qquad x' = h^{-1}(u') \quad \text{or} \quad f = h^{-1} \circ g \circ h.$$

The functions f and g related by $f = h^{-1} \circ g \circ h$ are called *conjugates* of each other.

The second iterate x'' can be similarly computed by means of h and g:

(19.2) $$u = h(x), \qquad u'' = g^2(u), \qquad x' = h^{-1}(u') \quad \text{or} \quad f^2 = h^{-1} \circ g^2 \circ h.$$

We note that this formula can be derived from (19.1) by a formal computation:

$$f^2 = h^{-1} \circ g \circ h \circ h^{-1} \circ g \circ h.$$

By the associative law, the $h \circ h^{-1}$ in the middle can be replaced by the identity map I and the I can then be deleted, leading to (19.2). In the same way we get the formula for the nth iterate

(19.3) $$f^n = h^{-1} \circ g^n \circ h$$

for any positive integer n, and even for negative integers if f has an inverse.

EXAMPLE 19.2. We cook up the map $x' = f(x)$ which maps $x = 1/\sqrt{u}$ into $x' = 1/\sqrt{u+c}$, where c is any positive constant. The nth image point will then be $x^{(n)} = 1/\sqrt{u+nc}$. The result of this cheap trick will be useful in making estimates by comparison of the iterates of a map whose iterates we can not express by a simple formula.

We have

$$u = \frac{1}{x^2}, \qquad u' = u + c, \qquad x' = \frac{1}{\sqrt{u+c}} = \frac{1}{\sqrt{c + 1/x^2}} = \frac{x}{\sqrt{1+cx^2}}.$$

Thus if

$$f(x) = \frac{x}{\sqrt{1+cx^2}},$$

where c is any positive constant, the nth iterate is given by the formula

(19.4) $$f^n(x) = \frac{x}{\sqrt{1+ncx^2}}.$$

As one would expect, the formula is valid even for negative n, except that one may get a negative number under the square root.

EXAMPLE 19.3. Let

$$f(x) = \frac{2x}{1-x^2}. \tag{19.5}$$

Find $f^n(x)$.

SOLUTION. This function does not fit the general discussion above because there is no interval which it maps into itself. For instance, we can not take the whole real line because $f(1)$ and $f(-1)$ are undefined. Leaving out those points does not help much because the second iterate $f^2(x)$ will still be undefined at the points where $f(x) = \pm 1$. We would have to remove an infinite set of points to get a set where all iterates of f are defined.

Let us ignore this difficulty for the moment and note that our map is the one which occurs in the trigonometric formula

$$\tan 2u = \frac{2\tan u}{1-\tan^2 u}. \tag{19.6}$$

We set $u = h(x) = \arctan x$, so that $x = \tan u$. Thus $x' = \tan 2u$, $u' = g(u) = \arctan x' = 2u$, and the nth iterate of g is $g^n(u) = 2^n u$. Finally, $f^n(x) = \tan(2^n u) = \tan(2^n \arctan x)$.

We describe a useful construction to get around the fact that the map (19.5) is undefined for $x = \pm 1$. The construction adds an element which we call ∞ to the domain and range of the map (19.5), which we may consider to be the set of real numbers, or the set of complex numbers. The rules to be used for evaluating formulas containing ∞ are as follows. For $a \neq 0$ we assign the value ∞ to the expression $a/0$. If a is any real number, we assign the value ∞ to $\infty + a$ and the value 0 to a/∞.

We leave $\infty - \infty$ and ∞/∞ undefined. From the point of view of algebraic operations, enlarging our set of numbers in the way we did is very detrimental. The enlarged set of numbers is no longer a field. One of many consequences of this is that we can no longer conclude from $a + c = b + c$ that $a = b$. However, the study of maps such as (19.5) becomes simpler, especially in the complex domain, because now the map (19.5) is defined everywhere. We have $f(1) = f(-1) = \infty$ and if we write $f(x) = -2/(x - 1/x)$ we get $f(\infty) = 0$.

We give a geometric interpretation of ∞. Take a circle S which touches the real number line at the origin 0, as in Fig. 19.1. Let P be the other end of the diameter through 0. We map each point x of the number line into a point ξ by central projection with center P. (The full advantage of this procedure appears

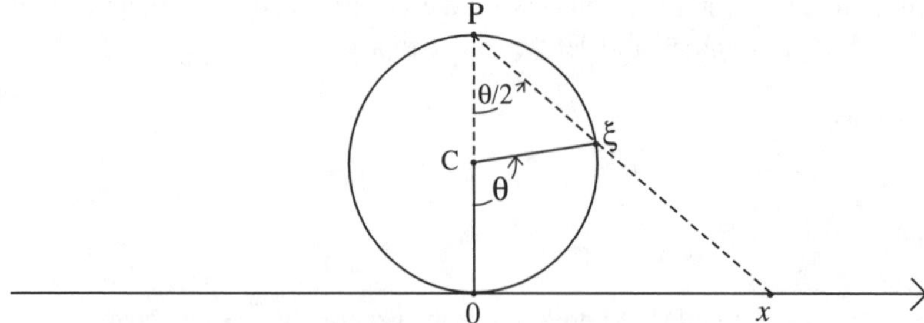

Figure 19.1. Stereographic projection.

when one is using complex numbers; then $\mathcal{S}$ is a sphere and P is its north pole. This is the reason for the notation.) If x is very large, ξ is very close to P, and this is so irrespective of whether x is large and positive or large and negative. Thus P can be regarded as the "projection" of ∞ onto the circle $\mathcal{S}$. This procedure also makes it appear natural that we introduced just a single element ∞. One might think that $+\infty$ and $-\infty$ are poles apart and can never be lumped together but in this context projection does bring them together at the same pole. The reason the circle $\mathcal{S}$ is so convenient for studying the map (19.5) is that the map is not only defined everywhere on $\mathcal{S}$, it is also continuous at every point of $\mathcal{S}$.

If we take the diameter of $\mathcal{S}$ to be 1, the map on $\mathcal{S}$ corresponding to (19.5) becomes especially simple. We have, see Fig. 19.1, $x = \tan(\theta/2)$. The observation we made earlier gives $x' = \tan\theta$. Hence the angle θ' corresponding to x' is just 2θ.

A projection of a plane onto a sphere in the manner we described is called a *stereographic projection.*

EXAMPLE 19.4. The method of conjugate functions enables us to find an explicit formula for the nth approximant x_n in computing $\sqrt{a}$ by means of the iteration $x_{n+1} = \frac{1}{2}(x_n + a/x_n)$, which we discussed in Chapter 1. There we had the recursion formula (1.4) for the error in successive approximants:

$$x_{n+1} - \sqrt{a} = \frac{1}{2x_n}(x_n - \sqrt{a})^2.$$

Let $x_n - \sqrt{a} = 2\sqrt{a}\,u_n$. Then $x_n = \sqrt{a}(2u_n + 1)$ and

$$u_{n+1} = \frac{u_n^2}{(2u_n + 1)}.$$

If we set $u_i = 1/v_i,\ i = 1, 2, \ldots,$ we get

$$v_{n+1} = v_n^2 + 2v_n.$$

Finally, we set $w_i = v_i + 1$ and the recursion simplifies to $w_{n+1} = w_n^2$. Hence

$$w_n = w_0^{2^n}.$$

We express our result in terms of the x_n. The above chain of substitutions gives

$$w_0 = \frac{x_0 + \sqrt{a}}{x_0 - \sqrt{a}} \quad \text{and} \quad x_n = \sqrt{a}\frac{w_n + 1}{w_n - 1}.$$

We note in passing that the function $(x + 1)/(x - 1)$, which connects w_n and $x_n/\sqrt{a}$, is its own inverse. We can now write down the explicit formula for x_n:

$$(19.7) \qquad x_n = \sqrt{a}\,\frac{\left(\frac{x_0+\sqrt{a}}{x_0-\sqrt{a}}\right)^{2^n} + 1}{\left(\frac{x_0+\sqrt{a}}{x_0-\sqrt{a}}\right)^{2^n} - 1} = \sqrt{a}\,\frac{(x_0 + \sqrt{a})^{2^n} + (x_0 - \sqrt{a})^{2^n}}{(x_0 + \sqrt{a})^{2^n} - (x_0 - \sqrt{a})^{2^n}}.$$

The main interest of this formula is that it is an explicit formula for the general term of the recursively defined sequence $x_0, x_1, \ldots$. The formula also tells us whether for any given complex values of a and x_0 the sequence converges, and to what limit. (For definiteness, let $\sqrt{a}$ denote the square root with positive real part if a is not negative or 0, and the square root with nonnegative imaginary part if $a \leq 0$.) If x_0 is closer to $\sqrt{a}$ than to $-\sqrt{a}$ then

$$\left|\frac{x_0 + \sqrt{a}}{x_0 - \sqrt{a}}\right| > 1$$

and, when raised to a high power, goes to infinity; so the middle expression in (19.7) approaches $\sqrt{a}$. If x_0 is closer to $-\sqrt{a}$ than to $\sqrt{a}$, this fraction is <1, and the middle expression in (19.7) tells us that x_n approaches $-\sqrt{a}$. If x_0 is equidistant from the two square roots, the iteration does not converge.

Finally, let us mention a point which may have puzzled the reader. The formula (19.7) contains $\sqrt{a}$ in 5 places, whereas of course we do not need to know $\sqrt{a}$ to compute the x_i; the purpose of the iteration is to find $\sqrt{a}$. To clear up this seeming discrepancy, look at the second form of the expression. In the numerator, odd powers of $\sqrt{a}$ cancel out. In the denominator the even powers cancel out so that $\sqrt{a}$ can be factored out and otherwise only even powers are left. The factor $\sqrt{a}$ in the denominator cancels with the factor in front of the fraction so that our expression contains only powers of a and x_0, as one would expect.

Exercises

19.1 Let

$$f(x) = \frac{x^2}{2x - 1}.$$

Find $f^n(x)$.

19.2 Find a function $\phi(x)$ such that

$$\phi^5(x) = x^2 + 2x.$$

19.3 Let

$$f(x) = \frac{x}{\sqrt{x^2 + c}},$$

where $c \geq 0$. Find $f^n(x)$. (There is a formula valid for $c \neq 1$ and another formula for $c = 1$.)

CHAPTER TWENTY

Convergent Orbits

In this chapter we turn from finding a formula for the nth iterate to investigating the sequence of images of a point. Let the interval I be the domain of a function f and suppose $f(I) \subseteq I$. For any $x \in I$, consider the behavior of the sequence

$$x_0, \ x_1 = f(x_0), \qquad x_2 = f^2(x_0), \ldots.$$

This sequence is called the *orbit* of x_0 under f, and the iterative procedure is an example of a *discrete dynamical system.* For example, input a number x_0 in a scientific calculator and press the SIN key over and over again. We get an orbit and will see that the orbit tends to zero as we strike more and more times. If the calculator is in degree mode, the convergence is very rapid, if it is in radian mode, it is very slow. Later in this chapter we shall obtain very precise information about the rate of convergence of this sequence.

If the calculator is in radian mode and we iterate the function $\tan x$, we get an apparently irregular sequence instead of a convergent one if we start with any number other than 0.

Orbits can be quite complicated sets, even for very simple nonlinear functions. The concept of orbit raises many intriguing questions, such as: Does the sequence have a finite limit as n becomes large? If the limit exists, how can we find it? What is the rate of convergence? If the orbit does not have a finite limit, does it tend to infinity as n gets large? Research into such questions has spawned a major branch of modern mathematics.

The functions discussed in this book are all continuous. Geometrically, a continuous function is one whose graph can be drawn without lifting the pencil. A formal definition of continuous function can be found in Appendix B.

We are specially interested in orbits which converge to a finite number. Suppose $x_0 \in I$ and the orbit of x_0 converges to the point $x^* \in I$. Then it follows from $x_{n+1} = f(x_n)$ and the continuity of f that

$$x^* = \lim_{n\to\infty} x_{n+1} = f\left(\lim_{n\to\infty} x_n\right) = f(x^*). \tag{20.1}$$

This means that x^* is a fixed point of f.

Successive images of a point x_0 can be constructed by means of the graph of $f(x)$ as follows. We first draw the *diagonal line* $y = x$, which makes a 45° angle with the x-axis and y-axis, provided the units on the two axes are equal. The next point on the orbit of x_0 is the number $f(x_0)$. The graph of f allows us to construct this number, since $(x_0, f(x_0))$ is the point on the graph over x_0. A vertical line from (x_0, x_0) to the graph of f meets the graph at $(x_0, f(x_0))$. Then a horizontal line from $(x_0, f(x_0))$ meets the diagonal line at $(f(x_0), f(x_0)) = (x_1, x_1)$. In general, we obtain the points (x_n, x_n) by drawing line segments vertically from the diagonal to the graph and then horizontally to the diagonal. See Fig. 20.1, which illustrates the orbits of two points under the map of the function shown. Some authors call this way of visualizing orbits *graphical analysis*. In some cases it enables one to read off qualitative information about the orbits, see [Devaney '89].

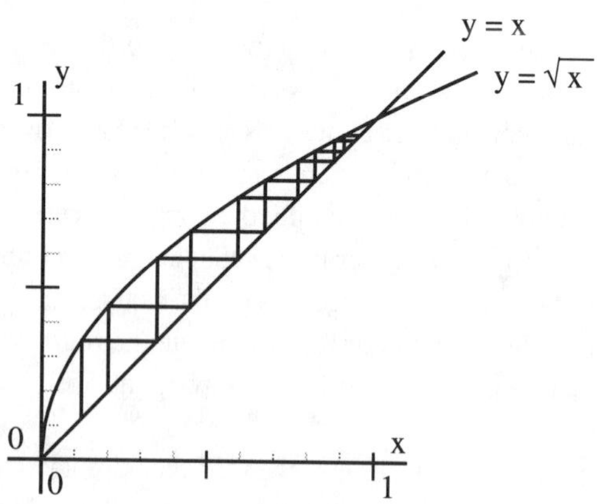

Figure 20.1. Graphical analysis.

If for any initial value $x_0 \in I$ the orbit of x_0 converges to the same fixed point x^*, then x^* is said to be a *globally stable fixed point*. If there is an interval containing a fixed point x^* of f such that for any initial value in this interval the corresponding orbit converges to x^*, then x^* is called a *stable fixed point* of f. A globally stable fixed point is, of course, a stable fixed point.

How can we guarantee the convergence of an orbit? We have the following theorems.

THEOREM 20.1. *Let the function f be continuous on a closed interval I and $f(I) \subseteq I$. If there is a constant $k < 1$ such that for any two points $x_1, x_2 \in I$*

$$|f(x_1) - f(x_2)| \leq k|x_1 - x_2|, \tag{20.2}$$

then there is a globally stable fixed point of f in I.

PROOF. Suppose that $I = [a,b]$. Since $f(I) \subseteq I$ we see that $a \leq f(a)$, $f(b) \leq b$. Thus, $f(a) - a \geq 0$ and $f(b) - b \leq 0$, by the intermediate value theorem, $f(x) - x = 0$ has a root, denoted by x^*, in I. It remains to show that x^* is globally stable. Let x_0 be any point in I, and $x_1, x_2, \ldots$ its successive images under f. We want to show the sequence converges to x^*. We have $|x_{n+1} - x^*| = |f(x_n) - f(x^*)| \leq k|x_n - x^*|$ and hence

$$\left|x_{n+1} - x^*\right| \leq k^{n+1}\left|x_0 - x^*\right|$$

which shows that the sequence converges to x^*.

The above method of showing the existence of a fixed point is called the *principle of contracting maps.* It gives an easy way to compute the fixed point, which we shall explore a little later, and it generalizes easily to maps in two or more dimensions.

With some extra work, we can weaken the hypothesis $k < 1$:

THEOREM 20.2. *Let the function f be continuous on a closed interval I and $f(I) \subseteq I$. If any two distinct points $x_1, x_2 \in I$ satisfy*

$$|f(x_1) - f(x_2)| < |x_1 - x_2|,$$

then there is a globally stable fixed point of f in I.

PROOF. We conclude as before that there is a fixed point x^* in I. Let ϵ be a positive number and let I_ϵ be the interval $(x^* - \epsilon,\ x^* + \epsilon)$. Let $J = I - I_\epsilon$. We have to prove that no matter how small ϵ is, there is an integer $N(\epsilon)$ such that $x_n \in I_\epsilon$ for $n \geq N(\epsilon)$.

J is a closed set; it is also bounded. The function

$$r(x) = \frac{|f(x) - x^*|}{|x - x^*|}$$

is continuous at every point of J. Hence, there is a point $x_{\max} \in J$ such that $r(x) \leq r(x_{\max}) = m$ for all $x \in J$. By hypothesis, $m<1$. Thus, as long as $x_n \in J$, we have

$$\left|x_{n+1} - x^*\right| \leq m\left|x_n - x^*\right|.$$

From this we see that in a finite number of steps, $|x_N - x^*|$ will become smaller than ϵ. While the above inequality need not be valid after that, the hypothesis ensures that $|x_n - x^*| < \epsilon$ does remain valid for all $n \geq N$. This completes the proof of Theorem 20.2.

Since most of the functions we are interested in are differentiable, the following theorem is often more convenient.

THEOREM 20.3. *Let x^* be a fixed point of a function f which is differentiable at x^*.*

1. *If $|f'(x^*)| < 1$, then x^* is a stable fixed point;*
2. *if $|f'(x^*)| > 1$, then x^* is not a stable fixed point.*

PROOF. The definition of the derivative gives

$$\lim_{x \to x^*} \frac{f(x) - f(x^*)}{x - x^*} = f'(x^*).$$

Assume that $|f'(x^*)| < 1$. Pick any positive number $\epsilon < 1 - |f'(x^*)|$. Then $q = \epsilon + |f'(x^*)|$ is a number in $(0, 1)$. By the definition of limit, there is a sufficiently small number δ such that for any $x \in (x^* - \delta, x^* + \delta)$ with $x \neq x^*$, we have

$$\left| \frac{f(x) - f(x^*)}{x - x^*} - f'(x^*) \right| < \epsilon.$$

Hence

$$|f(x) - x^*| = |f(x) - f(x^*)| < (\epsilon + |f'(x^*)|)|x - x^*| = q|x - x^*|.$$

We see from the last inequality that $f(x) \in (x^* - \delta, x^* + \delta)$.We can apply the inequality n times to get $|f^n(x) - x^*| \leq q^n|x - x^*|$. We conclude that $\lim_{n\to\infty} f^n(x) = x^*$ for all $x \in (x^* - \delta, x^* + \delta)$. Hence x^* is a stable fixed point of f. The proof for the first statement is completed.

The second statement can be proved in a similar way.

We next turn our discussion to the rate of convergence of iterations. We have mentioned at the beginning of this chapter that for any real x, the orbit

$$x, \ \sin x, \ \sin\sin x, \ \sin\sin\sin x, \ldots$$

tends to zero as $n \to \infty$. (In all theoretical discussions of the function $\sin x$ it is assumed that the angle is measured in radians.) As in Chapters 12 and 15, we ask: what is the rate of convergence? To answer this question, we first prove the following useful result.

THEOREM 20.4. *Let f, g and h be functions defined in an interval I and $f(I) \subseteq I$, $g(I) \subseteq I$, $h(I) \subseteq I$. Suppose that g and h are increasing on I. If*

$$g(x) \leq f(x) \leq h(x), \quad x \in I \tag{20.3}$$

then

$$g^n(x) \leq f^n(x) \leq h^n(x), \quad x \in I,$$

for $n = 1, 2, 3, \ldots$.

PROOF. We prove the statement by mathematical induction. For $n = 1$, the assertion to be proved is just the hypothesis. Assume that for some positive integer n

$$g^n(x) \leq f^n(x) \leq h^n(x), \quad x \in I.$$

By (20.3) we have

$$g(f^n(x)) \leq f(f^n(x)) \leq h(f^n(x)), \quad x \in I,$$

i.e.,

$$g(f^n(x)) \leq f^{n+1}(x) \leq h(f^n(x)), \quad x \in I.$$

Since g and h are increasing on I, we have by the induction hypothesis

$$g^{n+1}(x) = g(g^n(x)) \leq g(f^n(x)), \qquad h(f^n(x)) \leq h(h^n(x)) = h^{n+1}(x).$$

Thus

$$g^{n+1}(x) \leq f^{n+1}(x) \leq h^{n+1}(x)$$

and the induction is completed.

We are now ready to solve

PROBLEM 20.1. Show that if $0 < x < \pi/2$, then

$$\lim_{n\to\infty} \sqrt{n} \sin \circ \sin \circ \cdots \circ \sin x = \lim_{n\to\infty} \sqrt{n} \sin^{(n)}(x) = \sqrt{3}. \tag{20.4}$$

SOLUTION. In Chapter 19, Example 2 we considered the iterations of the map $g(x) = x/\sqrt{1 + cx^2}$. For positive x, $0 < g(x) < x$ and hence the sequence $x_0, x_1, \ldots$ is a decreasing sequence and bounded from below, so it must converge. (See Theorem 4 in Appendix B.) The continuity of g implies that the limit x^* satisfies $g(x^*) = x^*$ and hence it is 0.

For this particular function we can read off all this and more from the formula (19.4) for the n-th iterate, which we can write as

$$g^n(x) = \frac{1}{\sqrt{\frac{1}{x^2} + cn}}.$$

We see from our formula that

$$g^n(x) \sim \frac{1}{\sqrt{cn}} \quad \text{as } n \to \infty; \tag{20.5}$$

we use the symbol $\sim$ to say that the ratio of the expressions on its two sides approaches 1. The right side is independent of the initial value x. This seems surprising at first but it can be seen to be a consequence of the fact that g is an increasing function and $g(x) \sim x$ as $x \to 0$. Because of this, as x approaches 0, one more application of g or indeed any fixed number k of applications changes x by a smaller and smaller percentage. On the other hand, given any two initial values, some fixed number k of extra iterations on one or the other at the initial stage can compensate for the difference between them. But k additional iterations at the beginning have the same effect on the final result as k additional iterations at the end. That is why the right-hand side of the asymptotic formula (20.5) does not depend on the initial value x.

Consider now the function $\sin x$. One can show using Taylor's formula, or by a geometrical argument (see Appendix A) that for small values of x, $x - \frac{1}{6}x^3$ is a very good approximation to $\sin x$, in the sense that

$$x - \sin x \sim \tfrac{1}{6}x^3 \quad \text{as } x \to 0. \tag{20.6}$$

Formula (2) of Appendix A tells us that for small x, the function $g(x) = x/\sqrt{1+cx^2}$ is very well approximated by $x - \frac{1}{2}cx^3$, again in the sense that

$$x - g(x) \sim \tfrac{1}{2}cx^3 \quad \text{as } x \to 0. \tag{20.7}$$

Now take any $\epsilon > 0$ and assign the value $1/3 + \epsilon$ to c. Comparing (20.7) and (20.6) we see that for sufficiently small positive values of x,

$$g(x) < \sin x. \tag{20.8}$$

Thus, by Theorem 20.4, for such an x

$$g^n(x) = \frac{1}{\sqrt{\frac{1}{x^2} + \left(\frac{1}{3} + \epsilon\right)n}} < \sin^{(n)} x,$$

that is,

$$\frac{\sqrt{3}}{\sqrt{1 + 3\epsilon + \frac{3}{nx^2}}} < \sqrt{n}\,\sin^{(n)} x. \tag{20.9}$$

For large enough values of n

$$\sqrt{\frac{3}{1+4\epsilon}} < \sqrt{n}\,\sin^{(n)} x. \tag{20.10}$$

Our argument required that x be small enough for (20.8) to hold. Since the right-hand side of our inequality is an increasing function of x up to $x = \pi/2$, the conclusion holds for any x in $(0, \pi/2)$. Observe that the smaller ϵ is, and then the smaller x is, the larger n has to be for (20.10) to become valid.

Next, we get an upper bound for $\sin^{(n)} x$ by using the comparison function $g(x)$ with $c = 1/3 - \epsilon$ where $\epsilon < 1/3$. We get that for small enough positive values of x, (how small depends on how small ϵ is), $\sin x < g(x)$. We conclude as before that

$$\sqrt{n}\,\sin^{(n)} x < \sqrt{\frac{3}{1-3\epsilon+3/nx^2}} < \sqrt{\frac{3}{1-3\epsilon}} \tag{20.11}$$

for small enough positive values of x. If x is not small enough, let k be such that $\sin^{(k)} x$ is small enough. Then (20.11) gives

$$\sqrt{n-k}\,\sin^{(n)} x < \sqrt{\frac{3}{1-3\epsilon}}$$

from which it follows that

$$\sqrt{n}\,\sin^{(n)} x < \sqrt{1+\frac{k}{n-k}}\,\sqrt{\frac{3}{1-3\epsilon}}. \tag{20.12}$$

Since the lower and upper bounds (20.10) and (20.12) for $\sqrt{n}\,\sin^{(n)} x$ for large n can be made as close to $\sqrt{3}$ as we want, (20.4) has been proved.

Exercises

20.1 A sequence is defined recursively by $a_0 = 2$ and

$$a_{n+1} = \frac{a_n^2}{2a_n - 1}, \quad n = 0, 1, 2, \ldots.$$

Find $\lim_{n\to\infty} a_n$.

20.2 Let $f(x) = 2^x$ and $g(x) = 3^x$ for all real x. Show that

$$f^n(1) < g^{n-1}(1) < f^{n+1}(1)$$

for all integers $n > 2$. [Propp '83]

CHAPTER TWENTY-ONE

Finding Roots by Iteration

In high school, we learned formulas for the roots of polynomials of degree one and two. But for an arbitrarily given polynomial equation of degree greater than four and for most other kinds of equations there is no analogue of the quadratic formula. In this chapter, we discuss iterative numerical procedures to *approximate roots of equations*.

The hypothesis and proof of Theorem 20.1 of the previous chapter furnishes a method of solving equations. We restate it from this point of view with a tiny modification of the hypothesis:

THEOREM 21.1. *Suppose the function $\phi(x)$ is defined in an interval I and it maps I into itself. Assume also that there is a constant $k < 1$ such that $|\phi'(x)| \le k < 1$ for $x \in I$. Then for any $x_0 \in I$, the sequence $x_1 = \phi(x_0)$, $x_2 = \phi(x_1), \ldots$ converges to a solution x^* of the equation $x^* = \phi(x^*)$. The error decreases at least as fast as a geometric sequence with ratio $k : |x_n - x^*| \le k^n |x_0 - x^*|$.*

Theorem 20.2 in the previous chapter told us that the weaker hypothesis $|\phi'(x)| \le 1$ suffices for convergence. Since the convergence in this case can be very slow, we do not formulate that theorem in the present context.

The iteration method of solving equations goes back to Augustin Cauchy (1789–1857). It is very important because, unlike the even simpler interval halving method, it can easily be extended to systems of equations with several unknowns and to even more general types of equations.

There are many ways in which an equation $f(x) = 0$ can be brought to the form $x = \phi(x)$. To be able to obtain a solution x^* by iterating ϕ, we need to have $|\phi'(x^*)| < 1$ and for the iteration to converge rapidly, we want $|\phi'(x^*)|$ to be small. As an example we use the equation for the golden ratio, $f(x) = 1 - x - x^2 = 0$. This function decreases for $x > 0$; at $x = 0.5$ its value is 0.25, while at $x = 0.75$, its value is -0.3125. Thus the positive root of the equation is somewhat greater than 0.5. The simplest way to bring the equation to the form $x = \phi(x)$ is to add x to both sides. We get $x + f(x) = \phi(x) = 1 - x^2$, and when $f(x) = 0$, $x = \phi(x)$. If we take $x_0 = 0.5$ and iterate ϕ we get the sequence (printed vertically in two columns)

0.500 000 000	0.346 176 147...
0.750 000 000	0.880 162 074...
0.437 500 000	0.225 314 721...
0.808 593 750	0.949 233 276...,

which does not converge. Indeed, we have $f'(x_0) = -2$ and hence $\phi'(x_0) = 1 + f'(x_0) = -1$, and $\phi'(x) < -1$ for $x > x_0$. One way to remedy this is to realize that the solutions of the equation $f(x) = 0$ are the same as those of $g(x) = 0$, if g is any nonzero multiple of f. We set $g(x) = cf(x)$ and first consider a constant multiple c of f. Now $f(x) = 0 \Leftrightarrow g(x) = 0 \Leftrightarrow \phi(x) = g(x) + x = x$. (We take the liberty of using the old name for the new ϕ function. We are trying to strike a balance between being precise and having too much notation.) To make $|\phi'(x)| = |1 + g'(x)| = |1 + cf'(x)|$ small near $x = x_0$, i.e., to make $g'(x)$ close to -1 near $x = x_0$, we set $1 + cf'(x_0) = 0$ so that $c = \frac{1}{f'(x_0)}$ and we obtain

$$g(x) = -\frac{f(x)}{f'(x_0)} = \frac{1}{2}(1 - x - x^2). \tag{21.1}$$

Then $g'(x_0) = -1$ and the derivative ϕ' is 0 at $x = x_0$. The iteration now has the form

$$x_{n+1} = x_n - \frac{f(x_n)}{f'(x_0)} = \frac{1}{2}\left(1 + x_n - x_n^2\right). \tag{21.2}$$

and yields the sequence:

0.500 000 000	0.618 022 232...
0.625 000 000	0.618 035 376...
0.617 187 500	0.618 033 825...
0.618 133 544...	0.618 034 008...

The convergence is quite good. It would be even better if the x where $\phi'(x) = 0$ were closer to the root x^* or, better still, at x^* itself. We cannot use $f'(x^*)$ instead of $f'(x_0)$ in the denominator in (21.1) because we do not know x^*. However, we can at each step use our current approximation to $f'(x^*)$; in other words, we can replace (21.2) by

$$x_{n+1} = x_n - \frac{f(x_n)}{f'(x_n)}. \tag{21.3}$$

We get the sequence

0.500 000 000	0.618 033 989...
0.625 000 000	0.618 033 988...
0.618 055 555...	0.618 033 988...

The convergence is very rapid. After two iterations, four digits are correct and after three, nearly nine.

The iteration (21.3) is based on the ϕ-function

$$\phi(x) = x - \frac{f(x)}{f'(x)}\,. \tag{21.4}$$

We could have arrived at (21.3) by using that x^* is a zero not only of constant multiples of f, but of all functions of the form $g(x) = c(x)f(x)$. Then $\phi(x) = x + c(x)f(x)$ and $\phi'(x) = 1 + c(x)f'(x) + c'(x)f(x)$. For fastest convergence we want $\phi'(x^*) = 0$. Since the last term in the expression for $\phi'(x)$ vanishes at $x = x^*$, such a ϕ is obtained by taking $c(x) = \frac{-1}{f'(x)}$.

The reader may recognize (21.3) as *Newton's method* for solving $f(x) = 0$. We obtained it as one instance of a very general procedure. It can be understood more easily by looking at the geometric meaning of the construction, shown in Fig. 21.1.

There is a minor complication if $f'(x^*) = 0$; we assume this is not the case. Then

$$\begin{aligned}\phi'(x) &= \frac{f(x)f''(x)}{(f'(x))^2}\,,\\ \phi''(x) &= \frac{f''(x)}{f'(x)} + \frac{f(x)f'''(x)}{(f'(x))^2} - 2\frac{f(x)(f''(x))^2}{(f'(x))^3}\,.\end{aligned} \tag{21.5}$$

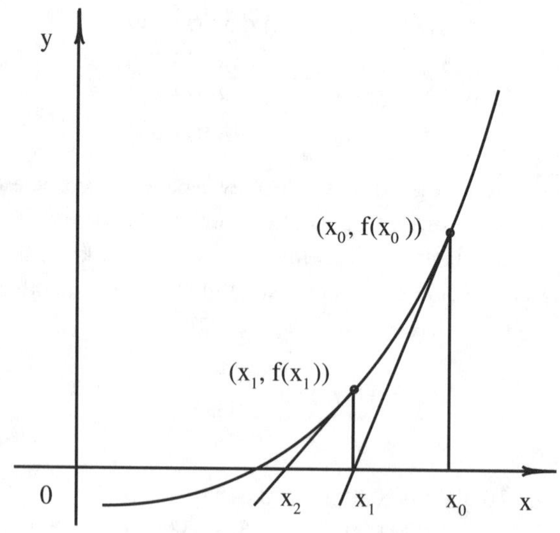

Figure 21.1. Newton's iteration.

The values at the root are

$$\phi'(x^*) = 0, \qquad \phi''(x^*) = \frac{f''(x^*)}{f'(x^*)}. \tag{21.6}$$

We see that, without knowing x^*, we have constructed a ϕ whose derivative vanishes at $x = x^*$. The very rapid convergence of Newton's iteration, once we are close to the solution, can be proved from this property.

THEOREM 21.2. *Suppose*

$$\begin{aligned} \phi(x^*) &= x^*, \qquad \phi'(x^*) = 0, \quad \textit{and} \\ |\phi''(x)| &\le M \quad \textit{for}\ |x - x^*| \le \tfrac{1}{M}. \end{aligned} \tag{21.7}$$

Then if $|x_n - x^*| \le 1/M$, *we have*

$$\left|x_{n+1} - x^*\right| \le \tfrac{1}{2}M\left|x_n - x^*\right|^2 \le \tfrac{1}{2}\left|x_n - x^*\right|. \tag{21.8}$$

PROOF. By Taylor's formula with remainder we have $|\phi(x) - x^*| = |\phi(x) - \phi(x^*)| \le \frac{1}{2}M(x - x^*)^2$ in the interval in question. Substituting x_n for x, we get (21.8).

The first inequality in (21.8) says that, apart from a constant factor, the distance of the approximant from the solution is squared with each iteration. This is called *quadratic convergence.* When we are close to the solution, it means that the number of correct digits approximately doubles with each iteration.

EXAMPLE 21.1. Let $m > 1$ be a positive integer and $a > 0$ a real number. To find the positive mth root of a, apply Newton's method with $f(x) = x^m - a$. We get

$$\phi(x) = x - \frac{x^m - a}{mx^{m-1}} = \frac{1}{m}\left((m-1)x + \frac{a}{x^{m-1}}\right). \tag{21.9}$$

We show that for any positive initial value x_0, the orbit converges to $a^{1/m}$.

The right side of (21.9) is the arithmetic mean of the m positive numbers: $x, x, \ldots, x, a/x^{m-1}$. The geometric mean of these numbers is $a^{1/m}$. Thus the arithmetic-geometric mean inequality gives $\phi(x) \ge a^{1/m}$. This implies that no matter where the initial value $x_0 > 0$ lies, all the iterates $x_1, x_2, x_3, \ldots$ are $\ge a^{1/m}$.

If $x_n \ge a^{1/m}$, then $a/x_n^{m-1} \le x_n$. The next approximant x_{n+1} is a weighted mean of $a/x_n^{m-1} \le x_n$ and x_n:

$$x_{n+1} = \frac{m-1}{m}x_n + \frac{1}{m}\frac{a}{x_n^{m-1}}.$$

Thus $x_{n+1} \le x_n$, i.e., the sequence $x_1, x_2, x_3, \ldots$ is decreasing. Since it is also bounded below, it converges. Its limit is a fixed point of ϕ, hence it is $a^{1/m}$.

This algorithm generalizes the iteration $x_{n+1} = \frac{1}{2}(x_n + a/x_n)$ for finding $\sqrt{a}$ which we discussed in the first chapter.

EXAMPLE 21.2. Find $70^{1/5}$ using Newton iteration with initial value $x_0 = 2$. We find

2	2.338 942 862 . . .
2.475	2.338 942 837 . . .
2.353 101 695 . . .	2.338 942 837 . . .
2.339 112 206 . . .	. . . .

We can observe the doubling of the number of correct digits at each step once we get close to the root.

Exercise

21.1 Someone justified the iteration $x_{n+1} = \frac{1}{2}(x_n + a/x_n)$ for finding $\sqrt{a}$ as follows. No matter on which side of $\sqrt{a}$ the number x_n is, a/x_n is on the opposite side. Hence, taking the arithmetic mean of x_n and a/x_n should give a better approximant x_{n+1}. For mth roots this reasoning suggests a sequence with $x_{n+1} = \frac{1}{2}(x_n + a/x_n^{m-1})$. Check how this sequence works for fifth roots.

CHAPTER TWENTY-TWO

Chebyshev Polynomials

This chapter considers another problem of the 1976 IMO whose solution is iterative. Even though this problem is quite elementary, it serves nicely as an introduction to Chebyshev polynomials, a topic of fundamental significance in approximation theory.

PROBLEM 22.1. Let $P_1(x) = x^2 - 2$ and let $P_j(x) = P_1(P_{j-1}(x))$ for $j = 2, 3, \ldots$. Show that, for any positive integer n, the roots of the equation $P_n(x) = x$ are real and distinct.

SOLUTION 1. Note that $P_1(x) = x^2 - 2$ is an even function, that is,

$$P_1(-x) = P_1(x).$$

By mathematical induction we see that $P_n(x)$ is even and of degree 2^n, and that $P_n(-2) = P_n(2) = 2$.

We shall show that the equation $P_n(x) = x$ has 2^n roots in the interval $-2 \leq x \leq 2$. Since the polynomial $P_n(x) - x$ has degree 2^n, it follows that there are no other roots of $P_n(x) = x$.

Figure 22.1 shows the graphs of P_1 and P_2. As x goes from -2 to 2, the value of $P_1(x)$ goes monotonically down from 2 to -2 and then monotonically back up to 2. We shall call this a *wiggle*. The graph of P_2 consists of two wiggles, one on the interval $[-2, 0]$, and the other on the interval $[0, 2]$. We assert that the graph of P_n consists of 2^{n-1} wiggles. We have already seen this for $n = 1$ and $n = 2$. Suppose it holds for some integer n, and consider the graph of P_{n+1}. On the interval $[0, 2]$, P_1 increases monotonically from -2 to 2, so by the induction hypothesis $P_n(P_1(x)) = P_{n+1}(x)$ wiggles 2^{n-1} times there. Since $P_{n+1}(x)$ is even, it also wiggles 2^{n-1} times on the interval $[-2, 0]$. Thus it has altogether 2^n wiggles in $[-2, 2]$, and the induction is complete.

We show next that each wiggle of $y = P_n(x)$ intersects the line $y = x$ at least twice; since there are 2^{n-1} wiggles, this gives at least $2^{n-1} \cdot 2 = 2^n$ roots of $P_n(x) = x$; and since the degree of $P_n - x$ is 2^n, it has exactly 2^n zeros.

Let a typical wiggle of P_n occur on the interval $[a, c]$; then $P_n(a) = P_n(c) = 2$ and $P_n(b) = -2$ for some b with $a < b < c$. Hence the polynomial

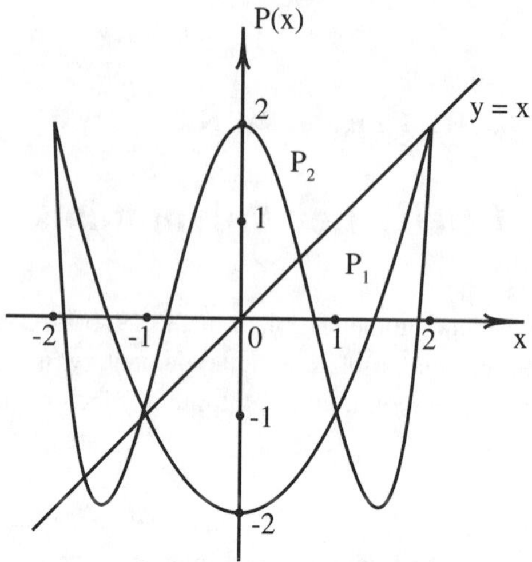

Figure 22.1. Graphs of P_1 and its iterate.

$Q_n(x) = P_n(x) - x$ satisfies $Q_n(a) > 0$, $Q_n(b) < 0$ and $Q_n(c) \geq 0$. (The equality $Q_n(c) = 0$ holds only for $c = 2$, that is, for the rightmost wiggle.) By the intermediate value theorem for continuous functions, it follows that $Q_n(x) = 0$ for at least two values of x, at least one in the interval (a, b) and one in (b, c). This completes the proof.

The second solution is analytical and leads to some important observations.

SOLUTION 2. The substitution $x = 2t$ carries the interval $[-2, 2]$ onto the standard symmetric interval $[-1, 1]$. We use this substitution te rescale our problem; that is, we set

$$Q_1(t) \stackrel{\text{def}}{=} \tfrac{1}{2} P_1(2t) = 2t^2 - 1,$$

and $Q_j = Q_1 \circ Q_{j-1}$, that is,

$$Q_j(t) = Q_1[Q_{j-1}(t)].$$

Clearly $Q_j(t) = P_j(2t)$, and each Q_j maps the interval $[-1, 1]$ into itself.

Next, we make a less trivial change of variable. For $-1 \leq t \leq 1$, we set

$$\theta = \arccos t, \qquad t = \cos\theta,$$

where θ is taken to lie in $[0, \pi]$. Now we calculate

$$Q_1(t) = Q_1(\cos\theta) = 2\cos^2\theta - 1 = \cos 2\theta.$$

Repeating this over and over again, we get

$$Q_2(t) = Q_1(Q_1(t)) = \cos 4\theta,$$

$$\dots = \dots,$$

$$Q_j(t) = Q_1(Q_{j-1}(t)) = \cos 2^j\theta. \tag{22.1}$$

Hence the equation $P_n(x) = x$ is transformed into

$$\cos 2^n\theta = \cos\theta. \tag{22.2}$$

Here θ is in the interval $[0, \pi]$. Since $\cos x$ is monotone decreasing in the interval $[0, \pi]$ and even, the values x in $[-\pi, \pi]$ for which

$$\cos x = \cos\theta \tag{22.3}$$

are $x = \theta$ and $x = -\theta$. The cosine function has period 2π, hence the set of all solutions x of (22.3) consists of the numbers which differ from θ or $-\theta$ by a multiple of 2π. Consequently, (22.2) is equivalent to

$$2^n\theta = \theta + 2k\pi \quad \text{or} \quad 2^n\theta = -\theta + 2k\pi,$$

where k is an integer. Solving for θ we get

$$\theta = \frac{2k\pi}{2^n - 1} \quad \text{or} \quad \theta = \frac{2k\pi}{2^n + 1}.$$

Since θ is in $[0, \pi]$, the first expression yields 2^{n-1} distinct roots for $k = 1, 2, \dots, 2^{n-1} - 1$, the second yields another 2^{n-1} distinct roots for $k = 0, 1, \dots, 2^{n-1}$. It follows that there are altogether 2^n distinct roots in $[0, \pi]$. Hence there are altogether 2^n distinct values $x = 2\cos\theta \in [-2, 2]$ satisfying the equation $P_n(x) = x$.

The last of equations (22.1) is the defining equation of the *Chebyshev polynomial* of degree 2^j. The second solution of the problem tells us that the Chebyshev polynomials with degree of powers of 2 can be generated by iterating the quadratic $2t^2 - 1$.

Since Chebyshev polynomials are so important in interpolation and approximation of functions, we discuss them a little further.

For $t \in [-1, 1]$, define $\theta = \arccos t$ and

$$T_n(t) \stackrel{\text{def}}{=} \cos n\theta. \tag{22.4}$$

We show that $T_n(t)$ is a polynomial of degree n in t. From the definition we have immediately that $T_0 = 1,\ T_1 = t$; From the addition formula for the cosine, we obtain the trigonometric identity

$$\cos(n+1)\theta = 2\cos\theta\cos n\theta - \cos(n-1)\theta.$$

There follows a three term recurrence relation

$$T_{n+1}(t) = 2tT_n(t) - T_{n-1}(t), \quad n = 1, 2, \ldots.$$

Using it, we can compute the first few Chebyshev polynomials explicitly. We find:

$$\begin{aligned}
T_0(t) &= 1 \\
T_1(t) &= t \\
T_2(t) &= 2t^2 - 1 \\
T_3(t) &= 4t^3 - 3t \\
T_4(t) &= 8t^4 - 8t^2 + 1 \\
T_5(t) &= 16t^5 - 20t^3 + 5t \\
T_6(t) &= 32t^6 - 48t^4 + 18t^2 - 1.
\end{aligned}$$

By using the recurrence relation and mathematical induction, we can show that $T_n(t)$ is even when n is even and odd when n is odd; furthermore, $T_n(t)$ is of degree n with leading coefficient 2^{n-1}.

We see that $|T_n(t)| = |\cos n\theta| \leq 1$ for all t in [-1,1], and that $T_n(t)$ has extreme values at the $n + 1$ points

$$t_k = \cos \frac{m}{n}\pi, \quad m = 0, 1, \ldots, n,$$

where it assumes the alternating values $(-1)^m$, as shown in Fig. 22.2. Thus $T_n(t)$ has n distinct roots all in $(-1, 1)$. For any function $f(t)$, continuous in $[-1, 1]$, let $\|f\|$ denote the maximum of $|f(t)|$ for $t \in [-1, 1]$. The nonnegative number $\|f\|$ is called the *uniform norm* of f. Then we have $\|T_n\| = 1$ for $n = 0, 1, 2, \ldots$. Define

$$\tau_n(t) = \frac{1}{2^{n-1}} T_n(t);$$

the leading coefficient of $\tau_n(t)$ is 1 and $\|\tau_n\| = 1/2^{n-1}$. We shall see that among all polynomials of degree n with leading coefficient 1, τ_n has the minimal uniform norm.

THEOREM 22.1 (CHEBYSHEV). *Let $p(t)$ be any polynomial of degree n with leading coefficient* 1. *Then*

$$\frac{1}{2^{n-1}} \leq \|p\|.$$

PROOF. On $[-1, 1]$, $|\tau_n|$ assumes its maximum value, $1/2^{n-1}$, $n + 1$ times at the points $t_k = \cos \frac{k}{n}\pi$, $k = 0, 1, \ldots, n$.

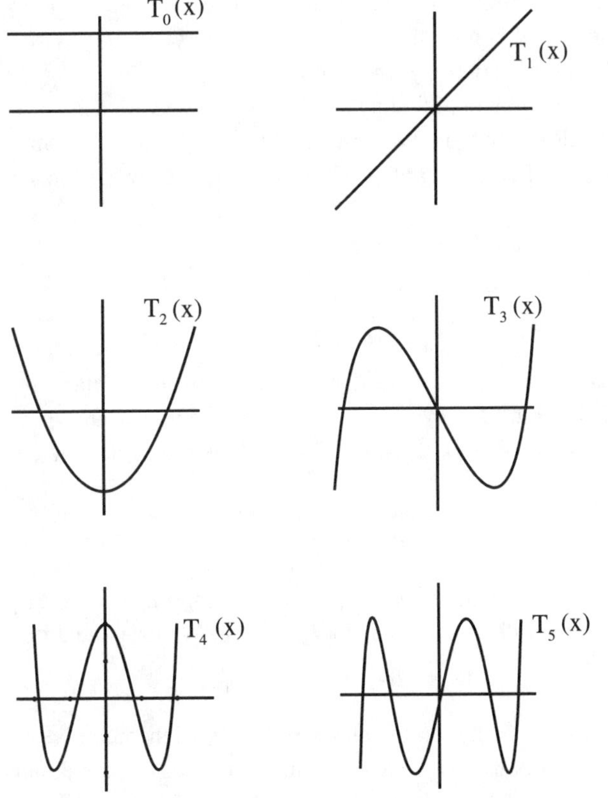

Figure 22.2. The Chebyshev polynomials.

Suppose there were a $p(t)$, a polynomial of degree n with leading coefficient 1, such that $\|p\| < 2^{-(n-1)}$. Form the difference

$$\delta(t) = \tau_n(t) - p(t).$$

Clearly $\delta(t)$ is a polynomial of degree at most $n - 1$. Now

$$\delta(t_k) = \frac{(-1)^k}{2^{n-1}} - p(t_k), \quad k = 0, 1, \ldots, n.$$

These quantities are alternately positive and negative inasmuch as $|p(t_k)| < 1/2^{n-1}$. Thus there are $n + 1$ points where $\delta(t)$ takes values with alternating signs and therefore $\delta(t_n)$ has at least n distinct zeros. Since $\delta(t)$ is of degree at most $n - 1$, it must vanish identically. Thus, $p(t) \equiv \tau_n(t)$. This yields

$$\frac{1}{2^{n-1}} = \|\tau_n\| = \|p\| < \frac{1}{2^{n-1}},$$

a contradiction.

In practice it is often important to approximate a polynomial by polynomials of lower degree. Suppose that $f(t)$ and $g(t)$ are continuous on $[-1, 1]$. We define the distance between f and g to be $\|f - g\|$.

Consider the problem of finding a polynomial $p(t)$ with degree at most $n - 1$ which is the closest to t^n in the sense that $\|t^n - p\|$ is minimal. Since $t^n - p(t)$ is a polynomial of degree n with leading coefficient 1, by the above theorem we have

$$t^n - p(t) = \tau_n(t),$$

that is

$$p(t) = t^n - \tau_n(t).$$

As a second application, consider a point (a, b) in the plane such that a is not in $[-1, 1]$. Find a polynomial $p(t)$ of degree n such that $p(a) = b$, and the uniform norm of p on the interval $[-1, 1]$ is minimal. The answer to the problem is

$$p(t) = \frac{b}{T_n(a)} T_n(t).$$

It is clear that $p(t)$ is of degree n and $p(a) = b$. Suppose there were a polynomial $f(t)$ of degree n with $f(a) = b$ and $\|f\| < \|p\|$. We form the difference

$$d(t) = p(t) - f(t).$$

By the same reasoning as in the proof of the above theorem, we see that $d(t)$ has n distinct roots in $(-1, 1)$ and one more root, a, outside the interval. We therefore must have $d(t) \equiv 0$, that is, $p(t) \equiv f(t)$. This contradicts $\|f\| < \|p\|$. For applications of this result, the reader is referred to [Sederberg & Chang '90].

More material on Chebyshev polynomials can be found in [Davis '75].

Exercises

22.1 [CMO 1986]. Let $f(x) = |1 - 2x|$ for $x \in [0, 1]$. How many solutions of the equation $f^3(x) = x/2$ are there in $[0, 1]$?

22.2 Let $I = [a, b]$ and suppose $f(x)$ is continuous in I. Let M and m be the maximum and the minimum of $f(x)$ on I. Show that, for any constant c,

$$\left\| f(x) - \frac{m + M}{2} \right\| \le \|f(x) - c\|.$$

22.3 [CMO 1983]. Define the function

$$f(x) = |\cos^2 x + 2\cos x \sin x - \sin^2 x + Ax + B|.$$

The maximum of $f(x)$ in the interval $[0, 3\pi/2]$, denoted by M, depends on the parameters A and B. Find A and B for which M is minimal. Justify your conclusion.

22.4 Show that in Theorem 22.1, equality holds only if $p = \tau_n$.

22.5 Show that

$$T_m(T_n(t)) = T_n(T_m(t)) = T_{mn}(t).$$

22.6 Show that

$$T_n(2t^2 - 1) = 2\,(T_n(t))^2 - 1 = T_{2n}(t).$$

22.7 Show that

$$T_m(t)T_n(t) = \tfrac{1}{2}(T_{m+n}(t) + T_{|m-n|}(t)).$$

CHAPTER TWENTY-THREE

Sharkovskii's Theorem

Let I be a closed interval $[a, b]$ on the real line and let $f: I \to I$ be a continuous map. In earlier chapters we considered additional conditions which ensure that the orbit of every point $x \in I$ converges to a fixed point of f. We start here with the observation that continuity alone assures that f has a fixed point in I. Indeed, $f(a) \geq a$ and $f(b) \leq b$, since f maps $[a, b]$ into itself. Hence $f(x) - x$ is ≥ 0 at a and ≤ 0 at b, so it vanishes somewhere on I.[1]

A generalization of the notion of fixed points is the concept of *periodic points*. If n is a positive integer, a point $x \in I$ is called a *periodic point of period n* if $f^n(x) = x$. A positive integral multiple of any period of x is also a period of x. If f has an inverse and n is a period of x then $-n$ is also a period of x, but the interesting things discussed in this chapter occur only when f does not have an inverse, so we do not assume that it has one.

LEMMA 23.1. *Let p be the least positive period of the periodic point x. Every period n of x is an integral multiple of p.*

PROOF. Suppose that n is a larger period which is not a multiple of p. Then $n = qp + r$, $0 < r < p$, where q, r are positive integers. Hence $f^n(x) = f^r(f^{qp}(x))$ and thus $x = f^r(x)$. Hence r is also a period of x, contrary to the assumption that p is the smallest positive period.

We shall call the least positive period of a point the *least period,* which is slightly inaccurate but self-explanatory. Other terms used for the same thing are *prime period* and *fundamental period*. Some authors consider it understood that when they refer to the period of a point, it is the least period.

If n is a period of the point x then the orbit of x is an infinite repetition of the pattern x, $f(x)$, $f^2(x), \ldots, f^{n-1}(x)$.

[1]This is almost embarrassingly obvious. However, the extension to more dimensions by L.E.J. Brouwer created a sensation in 1910 and opened an important new area of mathematics. Brouwer's fixed point theorem says that a continuous map into itself of a square, a cube etc., or any region which can be deformed into these, has at least one fixed point.

The points of period 1 are the fixed points of f and the points of period n are the fixed points of f^n. By Lemma 23.1 they are the periodic points of f whose least periods are divisors of n.

EXAMPLE 23.1. The function $f(x) = x^2 - 1$ has fixed points at $(1 \pm \sqrt{5})/2$ while the points 0 and -1 lie on an orbit of period 2.

Suppose a has least period 2. Then $f(a) = b$, $f(b) = a$ and $a \neq b$. This means that points (a, b) and (b, a) are both on the graph of $y = f(x)$; they are mirror images of each other with respect to the diagonal $y = x$. We conclude that $f(x)$ has a periodic point of least period 2 if and only if there are two distinct points on the graph of $y = f(x)$ which are symmetric with respect to the line $y = x$. Note that if the graph of f is symmetric with respect to the diagonal then every point is periodic with least period 2 or 1.

EXAMPLE 23.2. Unlike points of periods 1 and 2, points of least period 3 and higher do not have an obvious interpretation in terms of the graph.

For example, the function defined on $I = [0, 1]$ as

$$f(x) = \begin{cases} x + \frac{1}{2} & \text{if } x \in \left[0, \frac{1}{2}\right]; \\ 2(1-x) & \text{if } x \in \left[\frac{1}{2}, 1\right] \end{cases}$$

shown in Fig. 23.1, has 0, $\frac{1}{2}$ and 1 as periodic points of least period 3, since $f(0) = \frac{1}{2}$, $f(\frac{1}{2}) = 1$, $f(1) = 0$. This is not obvious from looking at Fig. 23.1. In fact, the deceptive simplicity of the figure gives no hint of the fact that *every* positive integer is the least period of some point! This fact is guaranteed by an amazing theorem.

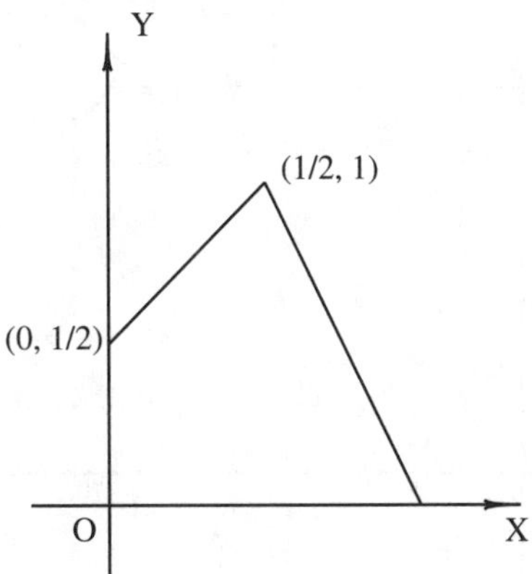

Figure 23.1. Periodic points of least period 3.

In 1975, Tien-Yien Li and J.A. Yorke published a joint paper titled "Period Three Implies Chaos" in *The American Mathematical Monthly*, [Li & Yorke '75]. In this paper they prove the following theorem:

THEOREM 23.1. *Let $f: I \to I$ be continuous. If f has a periodic point of least period three, then f has periodic points of all other least periods.*

The proof will depend on three elementary lemmas.

LEMMA 23.2. *Let $f: I \to I$ be continuous, and let $J = [a, b] \subseteq I$. If $f(J)$ contains J, then f has at least one fixed point in J.*

PROOF. Since $f([a, b]) \supseteq [a, b]$, we can find two numbers c and d in $[a, b]$ such that $f(c) = a$ and $f(d) = b$. If $c = a$ or $d = b$ then that point is a fixed point. Otherwise the points (c, a) and (d, b) on the graph of f are on opposite horizontal sides of the square $J \times J$. Thus the graph of f lying between (c, a) and (d, b) must intersect the diagonal $y = x$ of the square. The x-coordinate of any such intersection is a fixed point of f. (Fig. 23.2).

LEMMA 23.3. *Let $f: I \to I$ be continuous and J a closed subinterval such that $f(I) \supseteq J$. Then there is a closed subinterval K of I such that $f(K) = J$.*

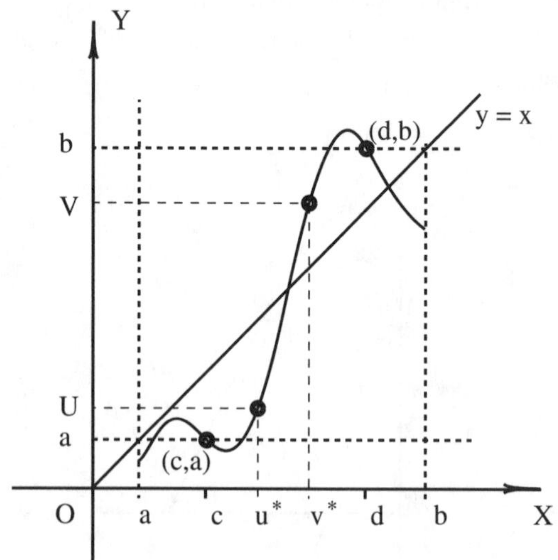

Figure 23.2. Figure for Lemmas 2 and 3.

PROOF. Let $I = [a, b]$ and $J = [U, V]$. The continuity of f implies that there are two points u and v in I such that $f(u) = U$ and $f(v) = V$. Consider the case $u < v$. Define

$$u^* = \text{l.u.b.}\{s \mid f(s) = U, u \leq s < v\};$$

here l.u.b. stands for the least upper bound of the set in the braces. The term "least upper bound" is nearly self-explanatory, but there are some subtleties connected with it, discussed in the Appendix on limits and continuity. Next we define

$$v^* = \text{g.l.b.}\{t \mid f(t) = V,\ u^* < t \leq v\};$$

here g.l.b. stands for greatest lower bound. By the definition of u^*, there are points x arbitrarily close to u^* such that $f(x) = U$ and hence the continuity of f implies that $f(u^*) = U$. Similarly, $f(v^*) = V$. We have made sure that inside the interval $K = [u^*,\ v^*]$ the function $f(x)$ is not equal to U or V. Hence the values of $f(x)$ do not get out of the interval $[U,\ V]$ anywhere on K, and the lemma is proved for the case $u < v$. The case $u > v$ can be handled by an appropriate modification of the above argument which we leave as an exercise, or by applying the above argument to $g(x) = f(a + b - x)$ with $u' = a + b - u,\ v' = a + b - v$.

LEMMA 23.4. *Let $J_0, J_1, \ldots, J_{n-1}$ be closed subintervals of I with $f(J_j) \supseteq J_{j+1},\ j = 0, 1, \ldots, n-2$ and $f(J_{n-1}) \supseteq J_0$. There exists at least one point $x_0 \in J_0$ such that $f^n(x_0) = x_0$ and $f^j(x_0) \in J_j$ for $j = 0, 1, \ldots, n-1$.*

This means that as j runs from 0 to n, $f^j(x_0)$ visits $J_0, J_1, \ldots, J_{n-1}$ in order and eventually returns to x_0.

PROOF. Since $f(J_{n-1}) \supseteq J_0$, by Lemma 23.3 we can find a closed interval $K_{n-1} \subseteq J_{n-1}$ such that $f(K_{n-1}) = J_0$. Similarly, from

$$f(J_{n-2}) \supseteq J_{n-1} \supseteq K_{n-1},$$

we can find a closed subinterval K_{n-2} of J_{n-2} such that $f(K_{n-2}) = K_{n-1}$. Proceeding in this manner, we can find $K_1 \subseteq J_1$ such that $f(K_1) = K_2$ where $K_2 \subseteq J_2$. Finally there is a subinterval K_0 of J_0 such that $f(K_0) = K_1$. We see that

$$f(K_0) = K_1, \qquad f^2(K_0) = K_2, \qquad f^3(K_0) = K_3,$$

$$\ldots,$$

$$f^{n-1}(K_0) = K_{n-1}, \qquad f^n(K_0) = J_0 \supseteq K_0.$$

Hence, by Lemma 23.2, there is a point $x_0 \in K_0 \subseteq J_0$ such that $f^n(x_0) = x_0$. It is clear that $f^k(x_0) \in K_k \subseteq J_k$ for $k = 0, 1, \ldots, n-1$. Lemma 23.4 is proved.

Now we return to the

PROOF OF THEOREM 23.1 Let $\alpha < \beta < \gamma$, all in the interval I, form an orbit of period three. There are only two possibilities: $f(\alpha) = \beta$ and $f(\alpha) = \gamma$. We consider the first case in which $f(\alpha) = \beta$, $f(\beta) = \gamma$ and $f(\gamma) = \alpha$. The other possibility is handled similarly. Let $J = [\alpha, \beta]$ and $K = [\beta, \gamma]$. Note that $f(J)$ contains the two points $f(\alpha) = \beta$ and $f(\beta) = \gamma$, thus $f(J) \supseteq K$. Since $f(K)$ contains two points $f(\beta) = \gamma$ and $f(\gamma) = \alpha$, $f(K) \supseteq [\alpha, \gamma] = J \cup K$.

We now show that there exists a point of least period n, where n denotes any positive integer. For $n = 1$, from $f(K) \supseteq K$ and by Lemma 23.2, there is a fixed point of f in K. Next let $n = 2$. Consider the two closed intervals $I_0 = K$ and $I_1 = J$. These two intervals meet all the requirements of Lemma 23.4. We conclude by the lemma that there is a point x_0 in $[\beta, \gamma]$ such that $f(x_0) \in [\alpha, \beta]$ and $f^2(x_0) = x_0$. We claim that x_0 is a point of least period 2. For otherwise, since $[\alpha, \beta] \cap [\beta, \gamma] = \{\beta\}$, we must have $x_0 = \beta$. This implies that $\beta = f(\beta) = \gamma > \beta$, a contradiction.

Of course any one of the points α, β, and γ is periodic of period 3.

Now assume that $n > 3$. Let

$$I_0 = I_1 = I_2 = \cdots = I_{n-2} = K, \qquad I_{n-1} = J.$$

All the requirements of Lemma 23.4 are fulfilled. Hence there is a point x_0 in $K = [\beta, \gamma]$ such that $f^n(x_0) = x_0$. We claim that x_0 is a point of least period n. If there were a positive integer $k < n$ such that $f^k(x_0) = x_0$, then $f^{n-1}(x_0)$ would be one of the following numbers:

$$x_0,\ f(x_0),\ f^2(x_0), \ldots,\ f^{n-2}(x_0);$$

all of these numbers are in K. Since also $f^{n-1}(x_0) \in J$, we have $f^{n-1}(x_0) \in J \cap K$, that is,

$$f^{n-1}(x_0) = \beta.$$

Thus $x_0 = f^n(x_0) = f(\beta) = \gamma$. By Lemma 23.4,

$$f(x_0) = f(\gamma) = \alpha \in I_1 = K = [\beta, \gamma].$$

Since this is impossible, Theorem 23.1 is proved.

By using this theorem we see that the function illustrated in Fig. 23.1 has periodic points of every period.

The following is an example showing that a function can have a point of least period 5 but no point of least period 3.

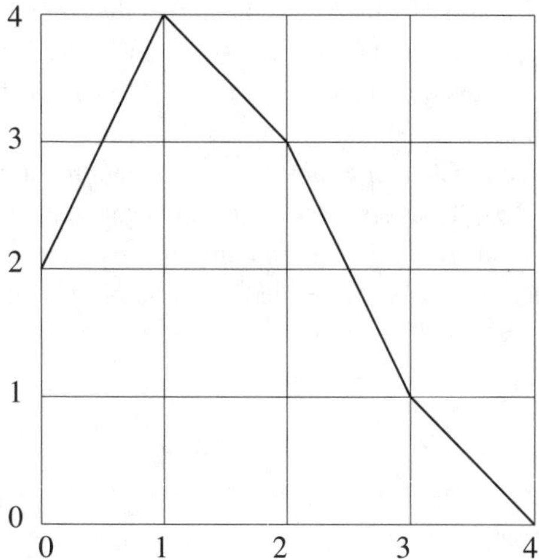

Figure 23.3. Period 5 does not imply period 3.

EXAMPLE 23.3. Consider the continuous piecewise linear function $f : [0, 4] \to [0, 4]$ graphed in Fig. 23.3 with

$$f(0) = 2, \quad f(2) = 3, \quad f(3) = 1, \quad f(1) = 4, \quad f(4) = 0.$$

It is clear that points 0, 1, 2, 3, and 4 are points of least period 5. We show that the function f does not have any point of least period 3. From the figure we see that intervals are mapped as follows:

$$f([0, 1]) = [2, 4], \qquad f^2([0, 1]) = f([2, 4]) = [0, 3],$$
$$f^3([0, 1]) = f([0, 3]) = [1, 4].$$

Similarly

$$f^3([1, 2]) = [2, 4]; \quad f^3([3, 4]) = [0, 3],$$

so f^3 has no fixed points in any of these intervals. Since $f^3([2, 3]) = [0, 4] \supseteq [2, 3]$, f^3 has at least one fixed point in $[2, 3]$. We claim that this point is unique. Indeed, $f : [2, 3] \to [1, 3]$ is monotonically decreasing, as is $f : [1, 3] \to [1, 4]$ and $f : [1, 4] \to [0, 4]$. Therefore f^3 is monotonically decreasing on $[2, 3]$ and the fixed point, denoted by x_0, is unique. From the figure we see that f has a unique fixed point, denoted by x_1. Since $f^3(x_1) = x_1$, by uniqueness we have $x_0 = x_1$. Hence x_0 is just the fixed point of f, rather than a period 3 point.

This theorem is just the beginning of the story. Shortly after the publication of Li and Yorke's paper, it was pointed out that their result is a special case of a result published in 1964 by the Russian mathematician A. N. Sharkovskii (1936–) who gave a complete account of which periods imply other periods for a continuous function. The paper was written in Russian. In 1977, it was introduced to Western mathematicians by Stefan, see [Stefan '77], with a simplified proof for the theorem and other extensions.

As we mentioned in Chapter 8, any positive integer n can be uniquely represented by $n = 2^s(2p+1)$ where s and p are nonnegative integers. Clearly $s = 0$ if and only if n is odd, and $p = 0$ if and only if n is a nonnegative power of 2. Sharkovskii's theorem can best be stated in terms of the following ordering of the positive integers:

$$\begin{gathered} 3,\ 5,\ 7,\ 9,\ldots, \\ 2\cdot 3,\ 2\cdot 5,\ 2\cdot 7,\ 2\cdot 9,\ldots, \\ 2^2\cdot 3,\ 2^2\cdot 5,\ 2^2\cdot 7,\ 2^2\cdot 9,\ldots, \\ \ldots\quad \ldots\quad \ldots\quad \ldots, \\ \ldots 2^5,\ 2^4,\ 2^3,\ 2^2,\ 2,\ 1. \end{gathered}$$

THEOREM 23.2 (SHARKOVSKII). *Suppose $f : I \to I$ is continuous. Suppose f has a point of period m. If m precedes n in the above ordering, then f also has a point of least period n. Moreover, there are functions which have points of least period n but no points of period m.*

The Sharkovskii Theorem reveals a beautiful and amazing relation between periodic points of a continuous function. By the theorem we see that the function f defined in Example 23.3 has periodic points of all periods other than 3. If a continuous function f has a point of least period 7, then f has points of all least periods except possibly 3 and 5. If f has a point of least period 100, then f must have points of least periods 108, 116, and 124.

The proof is a more elaborate version of the proof of Theorem 23.1. The reader may find it in [Devaney '89].

Even though Sharkovskii anticipated Li and Yorke by 11 years, the Li and Yorke paper is important because it popularized the concept of *chaos*. In addition to what we mentioned in Theorem 23.1, they showed that if f has a periodic point of period 3, then there exists an uncountable set S in I such that no point in S is periodic and

1. for any $x \neq y \in S$
$$f^n(x) - f^n(y) \not\to 0 \quad \text{as } n \to \infty;$$
2. for any $x \neq y \in S$ there exists a subsequence of natural numbers $k_1 < k_2 < \cdots < k_n < \cdots$ such that
$$\lim_{n\to\infty} \left(f^{k_n}(x) - f^{k_n}(y)\right) = 0;$$

3. for any $x \in S$ and any periodic point p

$$f^n(x) - f^n(p) \not\to 0 \quad \text{as } n \to \infty.$$

The first and the second statements indicate that for any two different points in S, the distances between corresponding terms in their orbits do not converge to zero as $n \to \infty$, but a subsequence of these distances does converge to 0. We see that the orbits of points in S behave in a complicated, chaotic, almost random fashion. The mathematical formulation of what constitutes chaotic behavior is one of the most important recent advances in mathematics.

Examples of chaos are commonplace in our everyday world: the uniqueness of each snowflake, the dance of flames in a fireplace, the whirlpools in a river, the rise and fall of the stock market. Even quadratic maps can behave unpredictably when iterated. All the ingredients of chaos are present in this simple dynamical system. The reader is encouraged to pursue this topic in books devoted to it, such as the ones by Robert Devaney.

Functions have been studied for hundreds of years but as long as the main objective of research was to find formulas for solutions of equations, there was not much that could be done with iterations. Only in the second half of the 19th century were mathematical concepts and questions of a general nature formulated for the description of iterations. These had their origin in attempts to derive the properties of matter in bulk from the interactions of molecules, pioneered by Ludwig Boltzmann (1844–1906), and in the analysis of the motions of three stars moving under the influence of each other's gravitational attraction (the "three body problem") by Henri Poincaré (1854–1912).

Proving theorems turned out to be even harder than developing the concepts for describing iterations, but by the 1960's some extremely complicated properties of iterations had been established with mathematical rigor. It is surprising that Sharkovskii's relatively simple theorem was not found earlier. It is even more surprising that this splendid result languished in obscurity until Li and Yorke rediscovered part of it. Since then the theorem has generated a ground swell of related publications. This suggests that there is still much to be discovered in this subject.

Periodic points have recently appeared in Mathematical Olympiad problems. In China, there is a nationwide Mathematical Olympiad held on the third Sunday of each October; more than 100,000 senior high-school students participate. The roughly 80 regional winners, representing every province in the mainland, then compete in a special Mathematical Olympiad held the following January. Like the IMO, three problems are given on each of two consecutive days. The purpose of this test is to select twenty students to form an elite class to be trained by experienced mathematicians. From this class, six students are eventually chosen to form the Chinese IMO team. Thus, the special Mathematics Olympiad mentioned above can be regarded as the first-round selection test for the Chinese IMO team.

On January 17, 1989, the first selection tests were held on the campus of the University of Science and Technology of China, Hefei, Anhui Province. The hardest of the six problems was perhaps the following, proposed by Professor Zhang Zhu-Sheng from Beijing University. We note that this was the first year that the Chinese team placed first in the IMO competition.

PROBLEM 23.1. The complex function $f(z)=z^m$ is defined on the unit circle $S=\{z : |z|=1\}$, where $m>1$ is an integer. Determine the number N of periodic points of least period 1989.

SOLUTION. For the purpose of this discussion, let us normalize the arguments of complex numbers to lie in the interval $[0,\ 2\pi)$. Also, let $\operatorname{frac}(x)$ denote the fractional part of x, i.e., $\operatorname{frac}(x)=x-\lfloor x\rfloor$. We have $\arg(f(z))/2\pi = \operatorname{frac}(m\arg(z)/2\pi)$. Thus, in terms of the variable $x=\arg(z)/2\pi$, the map whose points of least period 1989 we have to count is $g(x)=\operatorname{frac}(mx)$.

We have $g(g(x))=\operatorname{frac}(m\operatorname{frac}(mx))=\operatorname{frac}(m^2x)$, because the argument of the outer frac on the left side differs from the argument of the frac on the right side by an integer. Applying this repeatedly we get

$$g^{(1989)}(x)=\operatorname{frac}(m^{1989}x); \tag{23.1}$$

here we used the notation $g^{(n)}$ for the iterate because in the present discussion ordinary exponents also figure prominently.

For $n=1,2,\ldots$, let B_n be the set of points of period n, i.e.

$$B_n=\{x\in[0,\ 1) : \operatorname{frac}(m^n x)=\operatorname{frac}(x)\}. \tag{23.2}$$

By Lemma 23.1, the set of points of least period 1989 is obtained by removing from B_{1989} the points which have 1 or a proper divisor of 1989 as a period. Since $1989=3^2\cdot 13\cdot 17$, each of these numbers is a divisor of at least one of $1989/3=663$, $1989/13=153$, and $1989/17=117$. Hence the set of points of least period 1989 is what remains of B_{1989} when the union of those three has been removed:

$$B_{1989}\setminus(B_{663}\cup B_{153}\cup B_{117}).$$

Let $|B|$ denote the number of elements in a set B. By the principle of inclusion and exclusion we see that

$$\begin{aligned}|B_{663}\cup B_{153}\cup B_{117}| &= |B_{663}|+|B_{153}|+|B_{117}|-|B_{663}\cap B_{153}|\\ &\quad -|B_{663}\cap B_{117}|-|B_{153}\cap B_{117}|+|B_{663}\cap B_{153}\cap B_{117}|.\end{aligned}$$

By Lemma 23.1, the set $B_s\cap B_t$ consists of the points whose least periods are divisors of both s and t, i.e, the points whose least periods are divisors of the greatest common divisor of s and t: $B_s \cap B_t = B_{\text{g.c.d.}(s,t)}$. Similarly, $B_s \cap B_t \cap B_u = B_{\text{g.c.d.}(s,t,u)}$. Thus the number of points of least period 1989 is

$$N = |B_{1989}| - |B_{663}| - |B_{153}| - |B_{117}| + |B_{51}| + |B_{39}| + |B_9| - |B_3|.$$

The equation $g^k(x) = \text{frac}(m^k)x = x$ is equivalent to $\text{frac}((m^k - 1)x) = 0$, $0 \le x < 1$. The solutions of this equation are the fractions $j/(m^k - 1)$, $j = 0, 1, \ldots, m^k - 2$. Thus $|B_j| = m^j - 1$. This gives us the result

$$N = m^{1989} - m^{663} - m^{153} - m^{117} + m^{51} + m^{39} + m^9 - m^3.$$

Problem 23.1 deals with a map of a circle into itself. We must stress that Sharkovskii's theorem does not hold for such maps; a rotation of a circle by one third is a continuous map for which every point is periodic with least period 3, but no other period occurs.

Exercises

23.1 If f is the function in Example 23.2, find five distinct points x_0, x_1, x_2, x_3, x_4 in the interval $[0, 1]$ such that

$$f(x_0) = x_1, \qquad f(x_1) = x_2, \qquad f(x_2) = x_3, \qquad f(x_3) = x_4, \qquad f(x_4) = x_5.$$

23.2 Let $f(x) = 4(x - 1/2)^2$, $x \in [0, 1]$. Show that for any positive integer n, f has at least one point of least period n.

23.3 Show that the function with the following graph has an orbit of period 7 but not one of period 5.

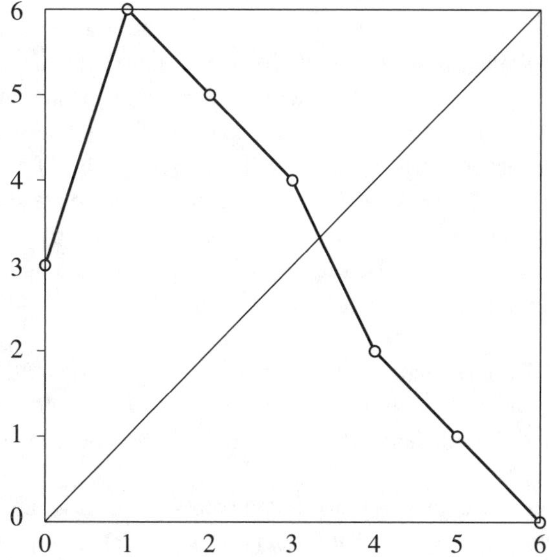

Figure 23.4. Period 7 but not period 5.

CHAPTER TWENTY-FOUR

Variation Diminishing Matrices

Let $p(x) = c_0 + c_1x + c_2x^2 + \cdots + c_nx^n$ be a polynomial with real coefficients, not all of which are 0. Mathematicians have long been interested in easy ways to get information about the number of positive roots of the equation

$$p(x) = 0. \tag{24.1}$$

We shall present here a bound discovered by René Descartes (1596–1650).

If all nonzero coefficients c_i have the same sign, then (24.1) has no positive roots.

If there is one change of sign in the sequence

$$c_0, c_1, \ldots, c_n, \tag{24.2}$$

then there is exactly one positive root, see Exercise 24.2. (We ignore coefficients which are 0 when we count changes of sign.)

When there are two sign changes, there may be no positive roots at all, as in the case $p(x) = 1 - x + x^2 = (1 - x)^2 + x$, or two positive roots as for $p(x) = 1 - 3x + x^2$. The following argument makes it somewhat plausible that when there are two sign changes there will not be more than two roots. Suppose for definiteness that the coefficients are positive, then negative and then positive again. When x is near 0, the lowest degree terms predominate and $p(x)$ is positive. As x increases, the negative terms in the middle may produce a large enough contribution to make the entire sum negative. After some further increase in x the highest degree terms, which at first grow more slowly than the lower degree terms but ultimately grow fastest and are positive, will make the sum positive and keep it positive thereafter. Although this argument is too vague to be convincing, it points us in the direction of

Descartes' Rule of Signs: The number of positive roots of (24.1), with multiple roots counted multiply, is at most equal to the number of sign changes in the sequence of coefficients. If it is less, it is less by an even number.

PROOF. When we say we count multiple roots multiply, we are really saying that we are counting not roots but linear factors of $p(x)$. Our proof will be an induction on the number of factors $x - a,\ \ a > 0$.

If the number of positive roots is 0, $p(x)$ does not change sign on the half-line $0 < x < \infty$. When x is small and positive, the sign of $p(x)$ is the same as the sign of the coefficient of the lowest degree nonzero term, and when x is large enough the sign is that of the highest degree nonzero term. Thus the first and the last nonzero terms have the same sign, and the number of sign changes in (24.2) is even. Thus Descartes' Rule is valid when the number of positive roots is 0.

For the rest of the proof, we may assume that the coefficient of the highest power of x in $p(x)$ is 1. Let $\mathcal{V}\{a_0, \ldots, a_n\}$ denote the number of sign changes in the sequence, ignoring 0's. (The calligraphic V stands for *variation*.) We complete the induction by showing that if

$$\begin{aligned} p(x) &= c_0 + c_1 x + \cdots + x^n \\ &= \left(b_0 + b_1 x + \cdots + x^{n-1}\right)(x - a) = q(x)(x - a), \quad a > 0, \end{aligned}$$

then

$$\mathcal{V}\{c_0, c_1, \ldots, 1\} - \mathcal{V}\{b_0, b_1, \ldots, 1\} \tag{24.3}$$

is positive and odd.

Set $b_i = 0$ and $c_j = 0$ when the indexes fall outside the ranges indicated above; then for all i,

$$c_i = b_{i-1} - ab_i \quad \text{or} \quad -a^i c_i = a^{i+1} b_i - a^i b_{i-1}. \tag{24.4}$$

Summing the above equations we get

$$a^{j+1} b_j = -\left(c_0 + ac_1 + \cdots + a^j c_j\right).$$

As j increases, each sign change in the sum on the right-hand side must be preceded by or coincide with a sign change in the sequence of summands. Thus the expression (24.3) is nonnegative. Moreover, the signs of the lowest-degree nonzero coefficients in $p(x)$ and $q(x)$ are opposite, but the signs of the highest degree nonzero coefficients are the same. So the expression (24.3) is odd. This completes the proof of Descartes' Rule.

We note in passing that it was not until 200 years after Descartes' discovery that a more elaborate algorithm was invented, by Jacques Charles François Sturm (1803–1855), which gives the number of positive roots exactly by counting sign changes in a certain sequence which can be computed by integer arithmetic if the coefficients are integers.

One can represent the value of a polynomial as a matrix product. We have

$$p(x) = \begin{bmatrix} 1 & x & \cdots & x^n \end{bmatrix} \begin{bmatrix} c_0 \\ c_1 \\ \vdots \\ c_n \end{bmatrix}. \tag{24.5}$$

Consequently,

$$\begin{bmatrix} p(a_1) \\ p(a_2) \\ \vdots \\ p(a_m) \end{bmatrix} = \begin{bmatrix} 1 & a_1 & a_1^2 & \cdots & a_1^n \\ 1 & a_2 & a_2^2 & \cdots & a_2^n \\ \vdots & \vdots & \vdots & \ddots & \vdots \\ 1 & a_m & a_m^2 & \cdots & a_m^n \end{bmatrix} \begin{bmatrix} c_0 \\ c_1 \\ \vdots \\ c_n \end{bmatrix} = V(a_1, \ldots, a_m)\mathbf{c} \tag{24.6}$$

where $\mathbf{c}$ denotes the column vector of coefficients. We call the matrix V an $m \times (n+1)$ *Vandermonde matrix.*

Since polynomials are continuous functions, a sign change from $p(a_j)$ to $p(a_{j+1})$ implies that p has a zero between a_j and a_{j+1}. Therefore, $\mathcal{V}\{p(a_1), \ldots, p(a_m)\} \leq$ the number of positive zeros of $p(x)$. Now Descartes' Rule can be formulated as follows:

Let V be the $m \times (n+1)$ Vandermonde matrix formed with the increasing sequence of positive parameters $a_1, a_2, \ldots, a_m$. If $\mathbf{c}$ is any column vector with $n+1$ entries, then $\mathcal{V}(V\mathbf{c}) \leq \mathcal{V}\mathbf{c}$.

We call a transformation of a sequence with this property *variation diminishing.* We shall abbreviate this as v.d.

THEOREM 24.1. *If A is v.d., the following operations on A preserve the property:*

(i) *Multiplying a row or column by a nonnegative constant.*

(ii) *Deleting a row or column.*

(iii) *Inserting a row (column) of* 0's.

(iv) *Adding a positive multiple of a row (column) to an adjacent row (column).*

PROOF. We prove (iv) and leave the rest as exercises.

Let A' be obtained from A by adding α times the ith row to a neighboring row, say the $i+1$st, where $\alpha > 0$. Then for any vector $\mathbf{x}$, the vector $\mathbf{y}' = A'\mathbf{x}$ is obtained from the vector $\mathbf{y} = A\mathbf{x}$ by the same operation. We want to show that $\mathbf{y}'$ has no more sign changes than $\mathbf{y}$. If adding αy_i to y_{i+1} changes the sign of y_{i+1}, it makes the sign of y_{i+1} the same as that of y_i. Thus, even if a new sign change is created between y_{i+1} and y_{i+2}, the total number of sign changes is not increased.

Next, let us consider when A' is obtained by adding α times the ith column to, say, the $i+1$st column, where $\alpha > 0$. Then $\mathbf{y} = A'\mathbf{x} = A\mathbf{x}'$, where $x_j' = x_j$ for $j \neq i$ and $x_i' = x_i + \alpha x_{i+1}$. Thus $\mathcal{V}(\mathbf{y}) \leq \mathcal{V}(\mathbf{x}') \leq \mathcal{V}(\mathbf{x})$; the last inequality is the one we established in the previous paragraph.

THEOREM 24.2. *If the matrices A, B are v.d. and the product AB exists, then it is also v.d.*

PROOF. This follows immediately from the definition of v.d. and the associative property of matrix multiplication.

Next we show that, with the exception of a trivial type, the nonzero entries in a v.d. matrix all have the same sign. First, take the case of matrices of rank 1. (A matrix A has rank 1 if not all its elements are 0 and all its nonzero columns are multiples of the same vector, which we call the range vector. It is easily seen and important to note that the nonzero rows in such a matrix are also proportional to each other.) The following fact is an immediate consequence of the definitions:

LEMMA. *A matrix A of rank 1 is v.d. if and only if all components of its range vector have the same sign.*

We turn now to matrices with rank greater than 1. The next theorem is accessible even to readers who have not yet learned the definition of "rank" of a matrix.

THEOREM 24.3. *If A is a v.d. matrix of rank > 1 then all its elements have the same sign.*

PROOF. Recall that the ith column $\mathbf{c}_i$ of a matrix A is the image under A of the ith unit vector $\mathbf{e}_i$ (all components are 0 except the ith, which is 1):

$$A\mathbf{e}_i = \mathbf{c}_i. \tag{24.7}$$

Since $\mathbf{e}_i$ has no sign change, all components of each column $\mathbf{c}_i$ of a v.d. matrix have the same sign.

Suppose now that A is v.d. and has entries of both signs. In that case, some of the columns of A have only nonnegative entries and some have only nonpositive entries. We shall show that then A is of rank 1.

Let $\mathbf{c}_i$ be a column with at least one positive entry and let $\mathbf{c}_j$ be a column with at least one negative entry. It follows from (24.7) that for an arbitrary number x,

$$A(x\mathbf{e}_i + (1-x)\mathbf{e}_j) = x\mathbf{c}_i + (1-x)\mathbf{c}_j. \tag{24.8}$$

Take $0 \le x \le 1$; then $x\mathbf{e}_i + (1-x)\mathbf{e}_j$ has no sign change, and since A is v.d., neither has $x\mathbf{c}_i + (1-x)\mathbf{c}_j$. Each component of $x\mathbf{c}_i + (1-x)\mathbf{c}_j$ is a linear function of x which is ≤ 0 at $x = 0$ and ≥ 0 at $x = 1$. Such a function either vanishes identically or it has exactly one zero in the interval $[0, 1]$. We claim that all components have a common zero. For, suppose that some component vanishes only at x' and another only at $x'' \ne x'$, then for all values of x between x' and x'', these two components have opposite signs, contradicting that all components of $x\mathbf{c}_i + (1-x)\mathbf{c}_j$ have the same sign. So we conclude that $x'\mathbf{c}_i + (1-x')\mathbf{c}_j = 0$. It follows that $\mathbf{c}_i$ and $\mathbf{c}_j$ are proportional. Since $\mathbf{c}_i$ is any column of A with at least one positive entry and $\mathbf{c}_j$ any column with at least one negative entry, all columns of A are proportional. The nonzero rows of such a matrix A are also proportional; it has rank 1.

How can we tell whether a matrix A is v.d.? In some cases it is obvious and we have discussed various facts which enable one to obtain v.d. matrices. But if we are given a matrix A, it is not clear whether it can be obtained from an obviously v.d. matrix by the given operations, and it could perhaps be v.d. even if it can not be so obtained.

A crude method for checking whether A is v.d. is to list all possible sequences of signs in $\mathbf{x}$ and $\mathbf{y} = A\mathbf{x}$ with more sign changes in $\mathbf{y}$ than in $\mathbf{x}$, and check for each of them if it can occur. We give an example of this procedure.

Consider the matrix

$$A = \begin{bmatrix} 1 & 1 & 1 \\ 1 & 2 & 4 \\ 1 & 3 & 6 \end{bmatrix}. \tag{24.9}$$

As an example of the cases to be checked, let us consider whether it is possible for $A\mathbf{x}$ to have nonzero components with the sign sequence $+,\ -,\ +$ while $x_1 \le 0$ and $x_3 \ge 0$. (Such an $\mathbf{x}$ has at most one sign change.)

To simplify the notation, we shall use $x,\ y,\ z$ instead of $x_1,\ x_2,\ x_3$ to denote the components of $\mathbf{x}$. The question then is, can we satisfy the inequalities

$$\begin{gathered} x \le 0, \quad z \ge 0, \quad x + y + z > 0, \\ x + 2y + 4z < 0, \quad x + 3y + 6z > 0? \end{gathered} \tag{24.10}$$

Finding the solution set of systems of linear inequalities is part of the subject of linear programming. Very efficient but somewhat complicated solution methods are now known. Here we show how one can solve inequalities by simple elimination, a method first described by Jean Baptiste Joseph Fourier in 1824. For little paper and pencil exercises elimination can be faster than the newer methods.

The first step is to solve for x all the inequalities in which x occurs:

$$-y - z < x, \quad -3y - 6z < x, \quad x \le 0, \quad x \le -2y - 4z, \quad 0 \le z.$$

We wrote the inequalities that bound x from below first, and then the inequalities that bound x from above. There is a common solution x to all these inequalities if and only if each of the lower bounds is $<$ each of the upper bounds. (If we had a lower bound with a $\le$ and an upper bound with a $\le$, then this pair of bounds would have to satisfy only the $\le$ relation.) So we can replace the inequalities involving x by the set of inequalities we just mentioned, which eliminates x from our system:

$$\begin{gathered} -y-z<0, \quad -y-z<-2y-4z, \\ -3y-6z<0, \quad -3y-6z<-2y-4z, \quad 0 \le z. \end{gathered} \tag{24.11}$$

We proceed to eliminate y the same way. We solve all our inequalities for y. Write all the lower bounds for y before all the upper bounds:

$$\begin{gathered} -z<y, \quad -2z<y \\ -2z<y, \quad y<-3z, \quad 0 \le z. \end{gathered} \tag{24.12}$$

Eliminating y from this system of inequalities gives

$$-z<-3z, \quad -2z<-3z, \quad -2z<-3z, \quad 0 \le z.$$

The first three inequalities simplify to $z<0$ which is incompatible with the last one, hence there is no solution. So we have eliminated one of the sign change patterns which could produce an increase in variation. For a 3×3 matrix with nonnegative elements one more such computation would cover all possibilities, see Exercise 24.8.

For slightly larger systems the elimination method is no longer practical. If we have q inequalities at a certain stage in the elimination, we could have as many as $q^2/4$ inequalities after eliminating the next variable.

The proliferation of inequalities in the course of checking whether one combination of sign patterns is possible can be avoided by using a better method for solving inequalities. However, the number of combinations of sign patterns to be checked doubles each time one of the dimensions of A is increased by 1, so checking whether a matrix is v.d. by examining all possible sign change sequences for $\mathbf{x}$ and $A\mathbf{x}$ is not practicable much beyond size 10×10 even if we are using good methods for solving inequalities. An alternative method for checking whether a matrix is v.d. is provided by

THEOREM 24.4. *A necessary and sufficient condition that a matrix A of rank r be v.d. is:*

a) *for each $k<r$ all minors of order k have the same sign and*
b) *any two minors of order r which have the same column indices have the same sign.*

The condition is due essentially to Isaac J. Schoenberg. The proof is laborious. It is best presented in *The Convolution Transform,* by I.I. Hirschman and

D.V. Widder, Princeton 1955, where far-reaching ramifications of the variation-diminishing property can also be found. A more recent survey of the subject is the article "Totally Positive Matrices" by Tsuyoshi Ando in *Linear Algebra and Applications* 90, (1987), 165–219.

Schoenberg's necessary and sufficient condition is a neat theorem but as an algorithm for checking whether a matrix A is v.d., Theorem 24.4 is only minimally better than trying all possible sign sequence combinations. The number of minors of an $n \times n$ matrix approximately quadruples each time n is increased by 1. Consequently, the amount of work increases exponentially with n and at about the same rate as in trying all sign combinations.

Exercises

24.1 Show that every $2 \times n$ marix with nonnegative entries is v.d. but there are 3×2 matrices with positive entries which are not v.d.

24.2 Prove that if there is exactly one sign change in the sequence of coefficients of the polynomial $c_0 + c_1x + \cdots + c_nx^n$ then the polynomial has exactly one positive root.

24.3 Prove that if $p(x)$ has n real roots then the number of positive roots is equal to $\mathcal{V}\{c_0, c_1, \ldots, c_{n-1}, 1\}$.

24.4 Show that if $p(x)$ has n real roots and for some k, $c_k = c_{k+1} = 0$ then $c_i = 0$ for $i = 0, 1, \ldots, k-1$.

24.5 Prove that the transformation defined by the matrix

$$S = \begin{bmatrix} 1 & 0 & 0 & \cdots & 0 \\ 1 & 1 & 0 & \cdots & 0 \\ 1 & 1 & 1 & \cdots & 0 \\ \vdots & \vdots & \vdots & \ddots & \vdots \\ 1 & 1 & 1 & \cdots & 1 \end{bmatrix}$$

is v.d.

24.6 Prove the unproved parts of Theorem 24.1.

24.7 Show that the sum of two v.d. matrices need not be v.d.

24.8 List all the sign change sequences which would have to be checked to ascertain that the matrix (24.9) is v.d.

24.9 Use elimination to solve the linear programming problem: Find the smallest value $u = x + 2y$ can have, subject to the inequalities

$$x \geq 0, \quad y \geq 0, \quad 2x + y \geq 3, \quad x + 3y \geq 4. \tag{24.13}$$

Hint: Write the inequalities in terms of the variables y and u and then eliminate y.

24.10 Show that the number of minors of an $m \times n$ matrix A is $\binom{m+n}{n} - 1$. (If $m = n$, the determinant of A is included in the count.)

24.11 Let A be a $(k+1) \times k$ matrix of rank k. Let M_i be the minor obtained by deleting the ith row of A. Show that the orthogonal complement of the column space of

A is generated by the vector

$$\mathbf{v} = [M_1,\ -M_2,\ M_3,\ -M_4, \ldots]^T.$$

24.12 Suppose the vector $\mathbf{b}$ is not alternating. Show there is a strictly alternating vector $\mathbf{a}$ which is orthogonal to $\mathbf{b}$.

24.13 Let A be a v.d. matrix of size $(k+1) \times k$ and rank k. Show that all minors of A of rank k are different from 0 and have the same sign. (The proof of this is short and illustrates the arguments used in the proof of Schoenberg's necessary and sufficient condition.)

CHAPTER TWENTY-FIVE

Approximation by Bernstein Polynomials

The remainder of this book presents some basic concepts from the field of computer-aided geometric design. Bernstein polynomials play a central role in the theory of free-form curves and surfaces.

We start with the identity

$$\begin{aligned}1 &= ((1-x)+x)^n \\ &= (1-x)^n + \binom{n}{1} x(1-x)^{n-1} + \binom{n}{2} x^2(1-x)^{n-2} + \cdots + x^n.\end{aligned}$$

The terms in this identity,

$$\begin{aligned}B_k^n(x) &= \binom{n}{k} x^k (1-x)^{n-k} \\ &= \frac{n}{1}\frac{(n-1)}{2}\cdots\frac{n-(k-1)}{k}\, x^k(1-x)^{n-k}; \qquad k = 0, 1, 2, \ldots, n\end{aligned} \tag{25.1}$$

are called the *Bernstein basis polynomials* of degree n, after Sergei Natanovich Bernstein (1880–1968). (In the current transliteration of the Cyrillic alphabet, the name would be spelled Bernshtein.) It is useful to define binomial coefficients and Bernstein polynomials for integer values of k outside the range $[0,\ n]$ to be 0. This helps avoid having to say something special about certain values of the index in identities.

The two most obvious properties of the $B_k^n(x)$, which the reader should always keep in mind, are

$$B_k^n(x) \geq 0 \quad \text{for } x \in [0, 1] \quad \text{and} \quad B_0^n(x) + B_1^n(x) + \cdots + B_n^n(x) = 1. \tag{25.2}$$

Long before Bernstein utilized them to approximate functions, the $B_k^n(x)$ had come up in probability theory. This is a very interesting topic in its own right, and it helps to understand the properties which Bernstein used to approximate functions, so we shall go into it in some detail.

Consider a random experiment, i.e. an experiment whose outcome depends on chance, such as tossing a die. To facilitate the discussion, we arbitrarily designate a certain outcome or type of outcome as *success*. When tossing a die, getting 6 or getting an odd number could be examples of success.

The basic phenomenon of probability is that although the outcome of any one trial is unpredictable, *if we continue long enough*, the following regularities are observed:

(i) The proportion of successes comes close to a number x which is called the *probability* of success.

(ii) More generally, if we count the outcomes of only those trials which have been preceded by some given pattern of outcomes, for instance, only those which follow a failure, the proportion of successes still approaches the same value x as the number of trials increases.

The mathematical theory of probability uses these basic facts to calculate probabilities of various combinations of outcomes. One very important consequence of (ii) is that in a sequence of two trials, the probability of any given pattern of outcomes, say success in trial 1 and failure in trial 2, is the product of the two individual probabilities, $x(1-x)$ in this case. To see this, consider what happens if we repeat a sequence of two trials N times. If N is large, the number of times the first of the two trials is a success is approximately Nx. By (ii), the proportion among these in which the second trial is a failure is approximately $1-x$. Hence we get approximately $Nx(1-x)$ pairs of trials in which the first trial is a success and the second a failure, i.e., the probability of success followed by failure is $x(1-x)$. This *product rule* is valid for patterns consisting of any number of outcomes.

Let us consider the probability that in a sequence of n trials we get exactly k successes. The product rule tells us that the probability of getting any particular sequence of successes and failures with a total of k successes and $n-k$ failures is $x^k(1-x)^{n-k}$. The number of different sequences of n outcomes with k successes and $n-k$ failures is $\binom{n}{k}$. Hence the probability of exactly k successes in n trials is $\binom{n}{k}x^k(1-x)^{n-k} = B_k^n(x)$. This interpretation suggests properties of the $B_k^n(x)$ which make them useful for approximating arbitrary functions by polynomials.

Let us look at the graphs of the Bernstein basis polynomials for $n = 10$. The phenomena of interest to us would be more pronounced for larger values of n, but the figure would get too crowded.

If the value of n is understood, we set $x_k = \frac{k}{n}$.

The figure suggests that $B_k^n(x_k) > B_j^n(x_k)$ for $j \neq k$, which expresses the plausible statement that if the probability of success is k/n, we are more likely to have k successes in n trials than any other number. It also suggests that if we vary x, the probability of getting exactly k successes in n trials is largest when $x = x_k$. It requires proof that these statements are indeed correct. We leave the proofs as exercises; it helps to be aware of these properties of the Bernstein polynomials but our results about approximation will not depend on them.

We also observe that $B_k^n(x)$ has a fairly narrow peak at $x = x_k$; its values are much smaller when x is some distance away from x_k. In fact, even the sum of *all* the $B_k^n(x)$ whose peaks are some distance away from x is small when n is large.

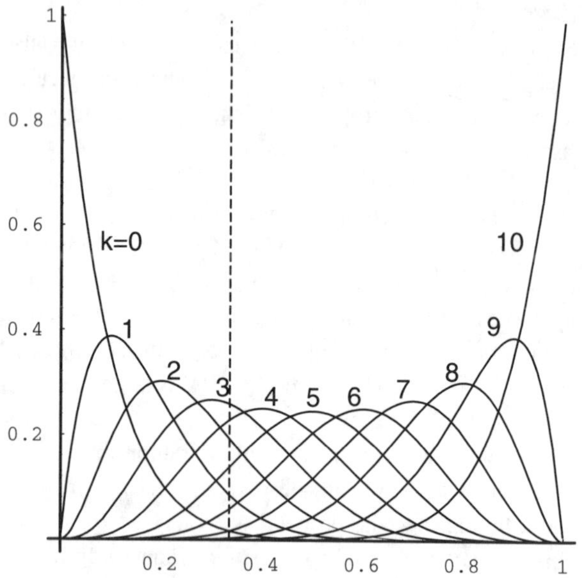

Figure 25.1. The basis polynomials $B_k^{10}(x)$.

In Fig. 25.1 we have drawn a dashed line at $x = 0.335$ to help the reader see how $B_k^n(x)$ varies with k while x and n remain fixed. The same observations could be made at any other x-value.

Our x is between x_3 and x_4. Moving our eyes up and down the dotted line we see the following. The largest of the $B_k^n(x)$ are $B_3^n(x)$ and $B_4^n(x)$. As k moves away from $nx = 3.35$ in either direction, $B_k^n(x)$ gets smaller. Looking at the figure closely we also see that even the ratios $B_{k+1}^n(x)/B_k^n(x)$ get smaller with each step away from nx if $k \geq nx$, and so do the ratios $B_{k-1}^n(x)/B_k^n(x)$ if $k \leq nx$.

Next we look at another fact which the probability interpretation makes plausible, and which will suggest how to use Bernstein polynomials to approximate any continuous function $f(x)$. Let $\delta > 0$ be a given positive number. Think of it as small. The probability that the proportion of successes in n trials differs from x by δ or more is

$$W_{\text{far}}(x, \delta, n) = \sum_{|\frac{k}{n} - x| \geq \delta} B_k^n(x). \tag{25.3}$$

(The inequality $\left|\frac{k}{n} - x\right| \geq \delta$ below the $\sum$ means that the summation is over all values of k satisfying the inequality.)

From the probability interpretation we expect that

$$\text{for } x \text{ and } \delta \text{ fixed,} \quad W_{\text{far}} \to 0 \quad \text{as } n \to \infty. \tag{25.4}$$

We shall give an estimate for W_{far} from which (25.4) will follow. At this stage we are not going to use (25.4); we stated it only to give the reader the motivation for a definition to be made later.

We are ready to discuss approximation of a given continuous function $f(x)$ by polynomials. In theory, specifying an arbitrary function requires giving an infinite number of function values. In practice, a large number of values have to be given unless the function can be represented by a formula. Among formulas representing functions, polynomials are particularly simple. (For periodic functions, trigonometric polynomials are the natural choice.) While we cannot represent an arbitrary function exactly by a polynomial, we shall see that if f is continuous on a closed interval I, we can approximate it by polynomials as closely as we wish.

Earlier, we discussed Lagrange interpolation polynomials of degree n. These are nth degree polynomials which are equal to a given function $f(x)$ at $n+1$ points, but if the points are not close together these polynomials tend to wiggle in between these points and may not even be close to $f(x)$. The advantage of the Bernstein approximating polynomials we are about to introduce is that Bernstein polynomials are as smooth as or even smoother than the function they approximate. This is why they are suitable for creating formulas which closely represent a designer's intentions and have a pleasing appearance.

Let us look again at the equation (25.2), in the light of (25.4). The sum of all the terms in (25.2) is 1, and for large n, the sum of the terms with $|k/n - x| \geq \delta$ is small, so most of the sum is contributed by the terms with $|k/n - x| < \delta$. With this in mind, consider the sum

$$(B^n f)(x) = \sum_{k=0}^{n} f\left(\frac{k}{n}\right) B_k^n(x). \tag{25.5}$$

Here we have deviated from our previous convention; $B^n f$ does not mean the nth iteration of a transformation B. Instead, B^n as a whole represents an operator whose input is a function f and whose output is a polynomial of degree at most n. From here on we will omit the parentheses around $(B^n f)$. The sum (25.5) is a weighted mean of the function values $y_0 = f(x_0), y_1 = f(x_1), \ldots, y_n = f(x_n)$, and as n gets large, most of the weight becomes concentrated near x. If f is continuous, the function values near x are close to $f(x)$. Hence the weighted mean (25.5) approaches $f(x)$ as $n \to \infty$.

We will make this claim more precise and then prove it rigorously; but first we list some simple properties of the operator B^n.

THEOREM 25.1.

1) *B^n is a linear operator: for any functions f, g, and constants a, b, $B^n(af + bg) = aB^n f + bB^n g$.*
2) *If $f(x) \geq 0$ in $[0, 1]$, then $B^n f(x) \geq 0$ in $[0, 1]$.*
3) *If $f(x) \leq g(x)$ in $[0, 1]$, then $B^n f(x) \leq B^n g(x)$ in $[0, 1]$.*
4) *If $f(x) \equiv C$ is a constant function, then $B^n f \equiv f \equiv C$.*
5) *If $m \leq f(x) \leq M$ for x in $[0, 1]$, then $m \leq B^n f(x) \leq M$ for x in $[0, 1]$.*
6) *$B^n f(0) = f(0)$ and $B^n f(1) = f(1)$. We say that $B^n f$ interpolates f at $x = 0$ and at $x = 1$.*

PROOF.

1) Apply the distributive law to each term in the definition of $B^n(af + bg)$.
2) This holds because the basis polynomials are nonnegative in $[0, 1]$.
3) Use the fact that $B^n g - B^n f = B^n(g - f)$ and then apply 2).
4) This follows from $\sum_{k=0}^{n} B_k^n(x) \equiv 1$.
5) This is a consequence of 3) and 4).
6) $B^n f(0) = f(0)$ follows from the fact that $B_0^n(0) = 1$ and $B_k^n(0) = 0$ for $k > 0$. A similar situation exists at the other end.

Next we want to evaluate $B^n f$ for linear and quadratic functions. The following identity, which is an immediate consequence of the binomial coefficient identity

$$\binom{n}{k} = \frac{n}{k}\binom{n-1}{k-1},$$

will be useful:

$$\frac{k}{n} B_k^n(x) = x B_{k-1}^{n-1}(x). \tag{25.6}$$

(The value $k = 0$ is covered by our convention that for integers k outside the range $[0, n]$, $B_k^n(x) = 0$.)

THEOREM 25.2.

1) *The operator B^n leaves any linear function $a + bx$ unchanged.*

2) *If f is a quadratic function, $f(x) = a + bx + cx^2$, then*

$$B^n f(x) = f(x) + \frac{1}{n} cx(1-x). \tag{25.7}$$

PROOF. 1) By Theorem 25.1, 1) and 4), it suffices to show that the identity function is transformed into itself; i.e., that $B^n x = x$. The identity (25.6) gives

$$B^n f(x) = \sum_{k=0}^{n} \frac{k}{n} B_k^n(x) = x \sum_{k=0}^{n} B_{k-1}^{n-1}(x) = x$$

since by (25.2) the second sum is identically 1. Note that the definition $B_{-1}^n(x) = 0$ was useful here.

2) By the linearity of B^n and part 1), it suffices to show that $B^n x^2 = x^2 + \frac{1}{n} x(1-x)$. We have

$$B^n x^2 = \sum_{k=0}^{n} \frac{k^2}{n^2} B_k^n(x) = \sum_{k=0}^{n} \frac{k}{n} \frac{k-1}{n} B_k^n(x) + \sum_{k=0}^{n} \frac{k}{n^2} B_k^n(x).$$

We apply (25.6) twice to the terms in the first sum and once to the terms in the second sum. This gives

$$B^n x^2 = \sum_{k=0}^{n} \frac{n-1}{n} x^2 B_{k-2}^{n-2}(x) + \sum_{k=0}^{n} \frac{1}{n} x B_{k-1}^{n-1}(x) = \frac{n-1}{n} x^2 + \frac{1}{n} x,$$

as claimed.

Next we turn to deriving estimates for the difference between a function and its nth degree Bernstein approximant.

LEMMA 25.3. *Suppose f is twice differentiable in* $[0, 1]$ *and* $f''(x) \le M$ *there. Then*

$$B^n f(x) - f(x) \le \frac{1}{2n} Mx(1-x). \tag{25.8}$$

PROOF. For the purpose of this proof, let $\mathcal{D}$ denote the operator $B^n - I$, where I is the identity operator. Consider first the case $M = 0$. Let us denote by c the value of x for which we wish to prove (25.8); this will free the symbol x for other uses in the proof. With this change of notation the statement to be proved is

$$(\mathcal{D}f)(c) \le 0 \quad \text{if } c \in [0, 1] \quad \text{and} \quad f''(x) \le 0 \quad \text{for } x \in [0, 1]. \tag{25.9}$$

$\mathcal{D}$ is a linear operator and by Theorem 25.2 it maps linear functions into the identically 0 function. Thus neither $\mathcal{D}f$ nor f'' changes if we subtract a linear function from f. Hence it suffices to prove (25.9) for the function $f(x) - f(c) - f'(c)(x - c)$, which vanishes at $x = c$ together with its derivative.

Instead of introducing a new name for this function, we say that we assume without loss of generality that $f(c) = 0$ and $f'(c) = 0$. The condition $f''(x) \leq 0$ implies that $f(x)$ is concave downward on $[0, 1]$. Since the graph touches the x-axis at $x = c$, we have $f(x) \leq 0$ on $[0, 1]$. Hence $B^n f(x) \leq 0$ on $[0, 1]$ from which (25.9) follows.

Now consider the case when M is not 0. The function

$$g(x) = f(x) + \frac{1}{2}Mx(1 - x)$$

satisfies $g''(x) \leq 0$ on $[0, 1]$ and hence $(\mathcal{D}g)(x) \leq 0$. Since $\mathcal{D}$ is a linear operator,

$$\mathcal{D}f = \mathcal{D}g - \mathcal{D}\frac{1}{2}Mx(1 - x) = \mathcal{D}g + \frac{1}{2n}Mx(1 - x) \leq \frac{1}{2n}Mx(1 - x),$$

and the proof is completed.

If we apply Lemma 25.3 to $-f$, we get

$$\text{(25.10)} \quad \text{If } f''(x) \geq M \quad \text{in } [0, 1], \quad \text{then } B^n f(x) - f(x) \geq \frac{1}{2n}Mx(1 - x).$$

The formula (25.8) is useful to keep in mind, and it is worth writing it down in a different notation:

$$\text{(25.11)} \quad \text{If } f''(x) \begin{Bmatrix} \leq \\ = \\ \geq \end{Bmatrix} 0 \quad \text{in } [0, 1], \quad \text{then } B^n f(x) \begin{Bmatrix} \leq \\ = \\ \geq \end{Bmatrix} f(x) \quad \text{in } [0, 1].$$

The inequalities (25.8) and (25.10) imply the following theorem:

THEOREM 25.4. *If* $|f''(x)| \leq M$ *in* $[0, 1]$, *then*

$$\text{(25.12)} \qquad |B^n f(x) - f(x)| \leq \frac{1}{2n}Mx(1 - x).$$

Note that all these estimates are best possible in the sense that if f is a quadratic function, equality holds.

COROLLARY. *If a function f has a bounded second derivative in the interval $[0, 1]$, we can approximate it as closely as we wish by a polynomial.*

In fact, every continuous function can be so approximated. This is Weierstrass's approximation theorem. We can derive it by combining the above corollary with the fact that if $f(x)$ is continuous on $[0, 1]$ and ϵ is given, there is a function $g(x)$ such that $|f(x) - g(x)| \leq \epsilon$ and $g''(x) \leq M$ on $[0, 1]$, where M depends on f and ϵ. In more descriptive language, what we need is that we can approximate a continuous curve as closely as we wish by a smooth curve. This is rather obvious, but to prove it rigorously requires a more subtle argument than one would expect. In the chapter on moving averages, Chapter 29, we discuss one method of approximating a continuous function with functions having as many bounded derivatives as desired.

We wish to point out that our basic estimate, (25.8), was a consequence of the following four simple properties of the operator B^n:

1) B^n is linear;
2) If f is nonnegative, so is $B^n f$;
3) A linear function remains unchanged by B^n;
4) $B^n(1-x)x = \frac{n-1}{n}(1-x)x$.

We will not need Weierstrass's approximation theorem, but we will need an approximation estimate for piecewise linear, continuous functions. Functions with such polygonal graphs do not have derivatives at the corners. In order to be able to deal with such functions, we introduce the concept of Lipschitz continuity.

DEFINITION. A function $f(x)$ is *Lipschitz continuous* on an interval I if there is a constant L such that

$$|f(y) - f(x)| \le L|y - x| \quad \text{for all } x, y \in I, \tag{25.13}$$

i.e., the slopes of all the chords of the graph of f are bounded by some constant L. When we say L is a Lipschitz constant for f on I, we are not implying that L is the smallest constant which makes (25.13) true.

We leave it as an exercise to show that if the graph of f is a polygon with a finite number of segments then f is indeed Lipschitz continuous.

Examples: The function $f(x) = x^2$ is Lipschitz continuous on any finite interval. We have,

$$\frac{|f(y) - f(x)|}{|y - x|} = |x + y|.$$

Using similar identities, or derivatives, one can show that on any finite interval a polynomial is Lipschitz continuous, and on an infinite interval a polynomial of degree >1 is not Lipschitz continuous.

The function $\sqrt{1-x^2}$ is not Lipschitz continuous on the interval $[-1, 1]$ because near the ends of the interval there are chords of arbitrarily large slope.

We turn now to estimating the sum $W_{\text{far}}(x, \delta, n) = \sum_{|\frac{k}{n} - x| \ge \delta} B_k^n(x)$ introduced in Formula (25.3). We gave a reason to believe that for fixed x and δ, W goes to 0 as $n \to \infty$. Now we are ready to prove it. Again, the discussion will be easier to follow if we change the notation from x to c. We are going to obtain an upper bound for $W_{\text{far}}(c, \delta, n)$.

Consider

$$B^n\left((x-c)^2\right) = (x_0-c)^2 B_0^n(x) + (x_1-c)^2 B_1^n(x) + \cdots + (x_n-c)^2 B_n^n(x)$$
$$= (x-c)^2 + \frac{1}{n}x(1-x), \qquad \text{by Theorem 25.2.}$$

If we substitute $x = c$ in this formula, we get

$$(x_0-c)^2 B_0^n(c) + (x_1-c)^2 B_1^n(c) + \cdots + (x_n-c)^2 B_n^n(c) = \tfrac{1}{n}c(1-c). \tag{25.14}$$

In this sum, the coefficients of the B_k^n for which $|x_k - c| \ge \delta$ are $\ge \delta^2$. Using this, and replacing the symbol c which has served its role by x, we get

THEOREM 25.5. *Given any* $\delta > 0$,

$$W_{\text{far}}(x,\delta,n) = \sum_{|\frac{k}{n}-x|\ge\delta} B_k^n(x) \le \frac{1}{n\delta^2}x(1-x) \le \frac{1}{4n\delta^2} \tag{25.15}$$

for $0 \le x \le 1$.

Before we use this inequality to derive the desired estimate, let us state the interpretation of (25.15) in probability theory, where it is often used. It says that the probability that the proportion of successes in a sequence of n trials differs by at least δ from the probability x of success in one trial is $\le x(1-x)/\delta^2 n \le \frac{1}{4\delta^2 n}$. The first bound is smaller, but the second one can be used even if we do not know the true probability x.

As an example, let us use (25.15) to find an upper bound for the probability that in $n = 90$ throws of a die, the number of times we get 5 or 6 is outside the range 21–39. The probability of throwing a 5 or 6 in one throw is $1/3$, so the expected number of 5's and 6's in 90 throws is 30. What is the probability that the actual number differs from the expected number by at least 10—or that the proportion of successes differs from $1/3$ by at least $10/90 = 1/9 = \delta$? The estimate (25.15) tells us that this probability is

$$\le \frac{1}{90 \times \frac{1}{9^2}} \cdot \left(\frac{1}{3}\right) \cdot \left(\frac{2}{3}\right) = \frac{1}{5}.$$

THEOREM 25.6. *If f is Lipschitz continuous on the interval* $[0, 1]$ *with Lipschitz constant L then*

$$|f(x) - B^n f(x)| \le \frac{1.2L}{n^{1/3}}. \tag{25.16}$$

PROOF. Let $\delta > 0$ be given. For an arbitrary value of x, multiply (25.2) by $y = f(x)$ and subtract $B^n f(x)$. We get

$$
\begin{aligned}
y - B^n f(x) &= \sum_{k=0}^{n} (y - y_k) B_k^n(x) \\
&= \sum_{|x-x_k|<\delta} (y - y_k) B_k^n(x) + \sum_{|x-x_k|\geq\delta} (y - y_k) B_k^n(x) \qquad (25.17)\\
&= S_{\text{near}} + S_{\text{far}}.
\end{aligned}
$$

We find a bound for $|S_{\text{near}}|$ by using the inequality $|y - y_k| = |f(x) - f(x_k)| \leq L|x - x_k|$:

$$|S_{\text{near}}| \leq L\delta \sum_{0}^{n} B_k^n(x) \leq L\delta. \tag{25.18}$$

We get a bound for S_{far} using $|y - y_k| \leq L|x - x_k| \leq L$, and the bound (25.15) for the sum of the Bernstein polynomials:

$$S_{\text{far}} \leq \frac{L}{4n\delta^2}. \tag{25.19}$$

These inequalities show that for any $\delta > 0$,

$$|f(x) - B^n f(x)| \leq L\delta + \frac{L}{4n\delta^2}. \tag{25.20}$$

The parameter δ is arbitrary, and we want to choose it so as to make the right-hand side in (25.20) small. We get Theorem 25.6 by setting $\delta = (2n)^{-1/3}$ and noting that $2^{-1/3} + 16^{-1/3} < 1.2$.

Exercises

25.1 Prove that the maximum of $B_k^n(x)$ in $[0, 1]$ occurs at $x = x_k$.

25.2 Prove that $\max_i B_i^n(x_k) = B_k^n(x_k)$.

25.3 Show that for any constant c in $(0, 1)$, the ratio $B_k^n(x + c)/B_k^n(x)$ is a decreasing function of x for $0 < x < 1 - c$.

25.4 Show that $B_k^n(x)$ is monotone increasing in the interval $[0, x_k]$ and monotone decreasing in $[x_k, 1]$.

25.5 Prove that for $0 < K < n$,

$$\sum_{k=K}^{n} B_k^n(x)$$

is an increasing function on $[0, 1]$.

25.6 Show that if f is an increasing function on $[0, 1]$ then $B^n f$ is also an increasing function there.

25.7 Show that if the graph of the function f is a polygon with a finite number of sides then f is Lipschitz continuous.

25.8 Let $L_{\min} = \frac{1}{2}$[the difference between the least upper bound and the greatest lower bound of $(f(y) - f(x))/(y - x)$]. In most instances, $L_{\min} = \frac{1}{2}(\sup f'(x) - \inf f'(x))$. Show that Theorem 25.6 remains true if L is replaced by $L_{\min}$.

CHAPTER TWENTY-SIX

Properties of Bernstein Polynomials

The Bernstein polynomials studied in the previous chapter do not approximate $f(x)$ as closely as some other polynomials of the same degree but they have smoothing properties which are desirable in engineering applications. In this chapter we are going to present some of these properties. Some of the discussion will require that the reader know about derivatives but several important theorems will not require that.

We recall the abbreviations $f(i/n) = f(x_i) = y_i$. We are going to apply the identity operator I and the shift operator E, introduced in Chapter 8, to the sequence $y : \{y_0,\ y_1, \ldots, y_n\}$. The operator E maps y to the sequence Ey whose term with index i is y_{i+1}. We should write this as $(Ey)_i = y_{i+1}$ but we will just write $Ey_i = y_{i+1}$ because this is easier to read. However, strictly speaking, the shift operator E can not be applied to y_i, which is just a number or an algebraic expression; E operates on sequences, and so does the identity operator I. When we apply E to a finite sequence with last element y_n, Ey_n is not defined. We could say that the sequence Ey has one less term than the sequence y. In our work we can avoid special provisions for the first and last index by defining y_i to be 0 if i is outside the range $[0,\ n]$; we have already defined $B_i^n(x)$ to be 0 for those values of i.

In Chapter 8 we discussed sums and products of operators and also multiples of operators by scalar, that is, numerical, factors. We observed that for sums and products of the operators I and E, the associative, distributive, and commutative properties hold. Thus we can transform such expressions according to the familiar laws of algebra.

We have $y_k = E^k y_0$. Hence we can write the definition of $B^n f$ as

$$B^n f(x) = \left((1-x)^n I + \binom{n}{1} x(1-x)^{n-1} E + \binom{n}{2} x^2 (1-x)^{n-2} E^2 + \cdots \right) y_0.$$

Applying the rules of algebra to the operator in the above formula, we get

$$B^n f(x) = ((1-x)I + xE)^n\, y_0. \tag{26.1}$$

Keep in mind that the operator E increases the index in a sequence it operates on by 1; it does not change the factor x into $x+1$.

Formula (26.1) can be interpreted geometrically as follows. Draw the polygon Π with vertices

$$(x_0,\ y_0),\ (x_1,\ y_1), \dots, (x_n,\ y_n). \tag{26.2}$$

We call this the *control polygon* of the polynomial $B^n f$. Applying the operator $(1-x)I + xE$ to the sequence of ordinates, we get

$$\begin{aligned} y_0^1 &= (1-x)y_0 + xy_1, \\ y_1^1 &= (1-x)y_1 + xy_2, \\ \dots &= \dots, \\ y_{n-1}^1 &= (1-x)y_{n-1} + xy_n. \end{aligned} \tag{26.3}$$

These are the ordinates obtained by dividing each side of the control polygon Π in the ratio $x : 1-x$. Recall that $x_i = \frac{i}{n}$. Since we are dividing straight line segments, the abscissas

$$x_0^1 = x_0 + \frac{x}{n}, \quad x_1^1 = x_1 + \frac{x}{n}, \dots, \quad x_{n-1}^1 = x_{n-1} + \frac{x}{n} \tag{26.4}$$

of these points are obtained by applying the same operator $(1-x)I + xE)$ to the sequence $x_0,\ x_1, \dots, x_n$ of abscissas. Thus we obtain a polygon with n vertices, one fewer than Π, by applying the operator $(1-x)I + xE$ to the vertices of Π. To get the result of a second application of $(1-x)I+xE$, apply the operator

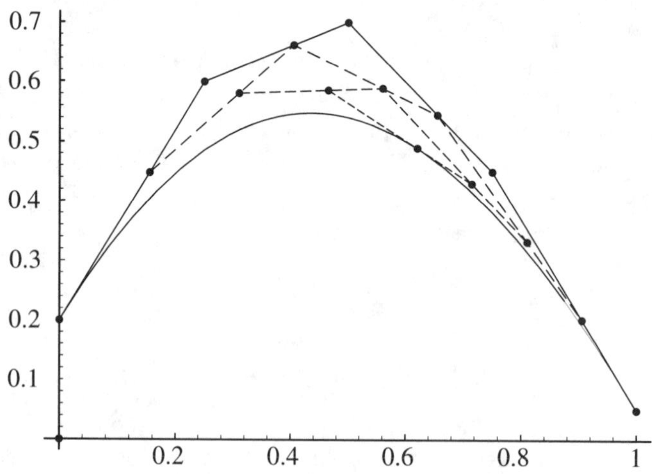

Figure 26.1. de Casteljau's construction with $n = 4$ and $x = 0.62$.

to the new polygon, etc. After n steps a single point (x_0^n, y_0^n) is obtained. The ordinate y_0^n is $B^n f(x)$, and we see from (26.4) that $x_0^n = x$, since $x_0 = 0$, so that the point (x_0^n, y_0^n) is on the curve $y = B^n f(x)$. The construction we just described is *de Casteljau's construction* or, in its algebraic form, *de Casteljau's algorithm.*

The figure indicates that the $(n-1)$st polygon, which is a single line segment, is the tangent to the curve $y = B^n f(x)$ at the point we are constructing. We will prove this when we discuss the subdivision construction.

We introduce the *forward difference operator* $\Delta = E - I$. To familiarize ourselves with Δ, let us apply it to some simple sequences.

$$\Delta\{1,\ 1,\ 1\ldots\} = \{0, 0, 0\ldots\}$$

$$\Delta\{1,\ 2,\ 3\ldots\} = \{1, 1, 1\ldots\}.$$

$$\Delta\{0,\ 1,\ 4,\ 9, \ldots, k^2, \ldots\} = \{1, 3, 5, 7, \ldots\}$$

Let us make a few observations about

$$\Delta\{\ldots, k^m, \ldots\} = \left\{\ldots, (k+1)^m - k^m, \ldots\right\}.$$

The general term of the sequence we get by applying Δ to the sequence $\{0, 1, 2^m, 3^m, \ldots\}$ is a polynomial in k of the form mk^{m-1} + lower degree terms. It follows that if the terms of a sequence are given by an mth degree polynomial in the index k, the result of applying Δ to it is a sequence given by an $(m-1)$st degree polynomial:

$$\Delta\{\ldots, c_m k^m + c_{m-1}k^{m-1} + \cdots\ \ldots\} = \{\ldots,\ mc_m k^{m-1} + \ldots,\ \ldots\}. \tag{26.5}$$

This is reminiscent of the rule for differentiating a polynomial of degree m. The similarity is to be expected because the differentiation operator is a limit of difference operators.

Every polynomial in Bernstein form

$$B^n f(x) = y_0(1-x)^n + y_1\binom{n}{1}x(1-x)^{n-1} + \cdots + y_i\binom{n}{i}x^i(1-x)^{n-i} + \cdots + y_n x^n$$

has an equivalent representation in the more familiar "power" form

$$p(x) = p_0 + p_1 x + p_2 x^2 + \cdots + p_m x^m.$$

Difference operators can be used to convert polynomials from Bernstein form to power form, that is, to find the coefficients p_i for which $p(x) \equiv B^n f(x)$. Formula (26.1) can be rewritten in terms of Δ as

$$B^n f(x) = (I + x\Delta)^n y_0. \tag{26.6}$$

Expanding the right side by using the binomial theorem we get

$$B^n f(x) = \sum_{k=0}^{n} \left(\binom{n}{k} \Delta^k y_0 \right) x^k$$

so the power-form coefficients are

$$(26.7) \qquad \begin{aligned} &p_0 = y_0; \qquad p_1 = n(y_1 - y_0); \\ &p_2 = \binom{n}{2}(y_2 - 2y_1 + y_0); \ \ldots\ ; \quad p_i = \binom{n}{k}\Delta^k y_0. \end{aligned}$$

If $f(x)$ is a polynomial of degree m, then $\{y_0, y_1, \ldots, y_n\}$ is a sequence generated by a degree m polynomial and by (26.5), $\Delta^k y_0 = 0$ for $k > m$. Hence, $p(x) \equiv B^n f(x)$ is of degree $\leq m$ for all n.

Another, at first sight surprising, consequence of (26.7) is that the coefficient of x^k in $B^n f(x)$ depends only on $y_0, y_1, \ldots, y_k$, since $\Delta^k y_0$ depends only on these quantities. (See, however, Exercise 26.3.)

Earlier we found by algebraic computation that a linear function is transformed into itself by B^n. Now we can prove this without any computation. We merely note that $B^n f$ is a polynomial of degree ≤ 1 and that it has the same values as $f(x)$ at $x = 0$ and at $x = 1$.

We can also apply (26.7) to compute $B^n(x^2)$ in a simpler way than we did in the last chapter. The sequence $\{y_0, y_1, y_2, \ldots\}$ is $\{0, 1/n^2, 4/n^2, \ldots\}$. Hence $\Delta y_0 = 1/n^2$, $\Delta y_1 = 3/n^2, \ldots$ and $\Delta^2 y_0 = 2/n^2$. Substituting in (26.7) we get

$$(26.8) \qquad B^n(x^2) = \frac{1}{n}x + \left(1 - \frac{1}{n}\right)x^2.$$

Next we present some nice formulas for the derivatives of $B^n f$. We start with an explanation of the formula

$$(26.9) \qquad \frac{d}{dx}(I + x\Delta)^n = n(I + x\Delta)^{n-1}\Delta.$$

We assume the reader knows that if u is a function of x, then

$$(26.10) \qquad \frac{d}{dx}(u^n) = nu^{n-1}\frac{du}{dx}.$$

In introductory calculus courses the function u is understood to be a function whose values are real numbers. The definition of derivative is also valid and of basic importance for functions with other kinds of values. In second or third semester calculus and in physics, derivatives of vector-valued functions are discussed. One has to go back to the definitions and proofs to see which properties of derivatives remain valid. (For instance, no geometrically meaningful definition of the nth power of a vector in 3 dimensions is known, so when u is a vector, the cherished Formula (26.10) falls by the wayside.)

For our operators I, E and Δ the concepts and computations used to derive (26.10) remain valid, and we encourage the reader to check this. This is how the Formula (26.9) is obtained.

Note that x enters (26.6) and the other forms of the formula for $B^n f$ only as a factor in the operator. The sequence $\{y_0, y_1, y_2, \dots\}$ is the same for all x. Thus we can take the x-derivative of (26.6) by taking the derivative of the operator.

To understand these things better it may help the reader to work out the formula for the case when both the operator and the sequence depend on x. What one gets is similar to the product rule $(uv)' = u'v + uv'$ in scalar calculus, and is derived by the same computation. There is one difference which one has to keep in mind. When we apply an operator to a sequence, what we write looks like a product and it has the distributive and associative properties of the ordinary product but one can not interchange the "factors". In our notation, if L is an operator and y is a sequence, then Ly is the result of L operating on y, but yL is undefined. Some textbook writers unnecessarily interchange the factors in the product rule, writing it as $(uv)' = uv' + vu'$. This form does not generalize.

Applying (26.9) to (26.6) we get

$$\frac{d}{dx} B^n f(x) = n(I + x\Delta)^{n-1} \Delta y_0 = n(I + x\Delta)^{n-1} \Delta y_0. \tag{26.11}$$

Compare this with (26.6). We see that $\frac{d}{dx} B^n f(x)$ is what we get by applying the operator B^{n-1} to a function $g(x)$ whose values at the n points

$$0, \quad \frac{1}{n-1}, \quad \frac{2}{n-1}, \dots, \quad \frac{n-2}{n-1}, \quad 1 \tag{26.12}$$

are

$$\begin{aligned} q_0 &= \frac{f\left(\frac{1}{n}\right) - f(0)}{\frac{1}{n}}, \qquad q_1 = \frac{f\left(\frac{2}{n}\right) - f\left(\frac{1}{n}\right)}{\frac{1}{n}}, \dots, \\ q_{n-1} &= \frac{f(1) - f\left(\frac{n-1}{n}\right)}{\frac{1}{n}}. \end{aligned} \tag{26.13}$$

By the mean value theorem, somewhere in the interval $[i/n, (i+1)/n]$ the difference quotient q_i is equal to $f'(x)$. The polynomial $\frac{d}{dx} B^n f(x)$ is the $(n-1)$st degree Bernstein polynomial of a function with the values q_0, q_1, $\dots, q_{n-1}$ assigned to the points $0,\ 1/(n-1),\ 2/(n-1), \dots,\ (n-2)/(n-1),\ 1$. The point $i/(n-1)$ lies in the interval $[i/n,\ (i+1)/n]$. It is at the left end for $i = 0$. The position of the point in its interval moves to the right at a uniform rate as i increases and reaches the right end for $i = n-1$. Therefore, not only do the Bernstein polynomials $B^n f$ approximate f, but their derivatives $\frac{d}{dx} B^n(x)$ approximate $f'(x)$ when it exists.

We should note that if $f''(x)$ does not change too rapidly, the point where $f'(x)$ is equal to q_i is approximately the midpoint of the interval $[i/n,\ (i+1)/n]$. Thus,

$\frac{d}{dx}B^n f(x)$ is slightly different from the Bernstein polynomial of degree $n-1$ for f' constructed from the data $\Delta y_0, \Delta y_1, \ldots, \Delta y_{n-1}$.

One can show in a similar way that as n increases the higher derivatives of $B^n f$ approach the higher derivatives of f when these exist.

We shall now discuss *shape preserving properties* of the Bernstein transformation. We have already established that $B^n(x)$ approximates the function $f(x)$ and its derivatives when n is large. In practice one uses values such as $n = 10$. This can be done because, as we will now show, $B^n(x)$ mimics the geometric behavior of $f(x)$ to a remarkable degree not just for large values of n, but for all n.

THEOREM 26.1. *If $f(x)$ is increasing (decreasing) in $[0, 1]$, then so is $B^n f(x)$.*

The reader may already have proved this theorem as part of a series of exercises in the last chapter. With derivatives at our disposal, we can give a shorter proof. Assume for definiteness that $f(x)$ increases in $[0, 1]$. Then from (26.11), since $y_{i+1} - y_i \geq 0$ for $i = 0, 1, \ldots, n-1$, and

$$\frac{d}{dx}B^n f(x) = n\sum_{i=0}^{n-1}(y_{i+1} - y_i)B_i^{n-1}(x),$$

we see that the first derivative of $B^n f(x)$ is nonnegative on $[0, 1]$. Hence $B^n f(x)$ increases on $[0, 1]$.

The geometric interpretation of the theorem is this. The Bernstein approximant of a function lies between the extreme values of the function itself, and the monotonicity of a function on $[0, 1]$ is maintained by its Bernstein approximants. To expand on this topic, we need the notion of convexity.

A function $f(x)$ is said to be *convex in an interval I* if for all $a \in I$, $b \in I$, every point of the chord connecting the points $(a, f(a))$ and $(b, f(b))$ is on or above the graph of $f(x)$. In algebraic language: convexity of f on I means that

$$(26.14) \quad f(x) \leq \frac{b-x}{b-a}f(a) + \frac{x-a}{b-a}f(b) \quad \text{for all } a, b \in I, \quad x \in [a, b].$$

A function g is *concave* if $-g$ is convex or equivalently, if (26.14) holds with the inequality reversed. Some people say *convex upwards* instead of *concave*, and to help people with poor memory, we can say *convex downwards* instead of *convex*.

We should mention a related but different meaning of "convex". A set S of points is said to be convex if, whenever $P \in S$ and $Q \in S$, all points of the line

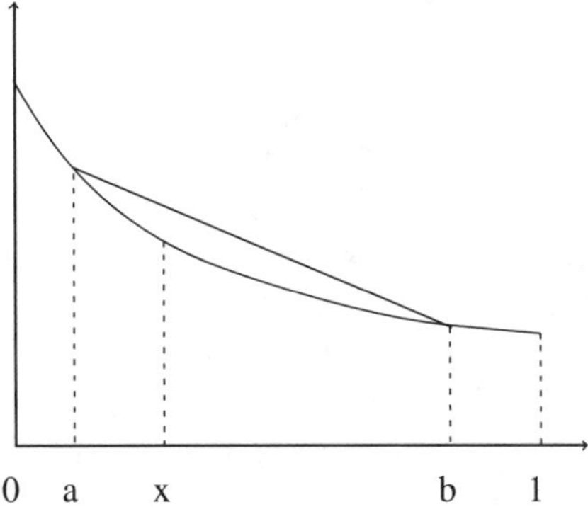

Figure 26.2. A convex function.

segment PQ are in S. See Exercise 26.10 and Appendix C for the relation between the two concepts.

THEOREM 26.2. *If $f(x)$ is convex in* $[0, 1]$ *then so is $B^n f(x)$.*

PROOF. We shall use the fact that a function having a nondecreasing derivative on an interval I is convex on I. This is rather clear intuitively. (For a proof, see the appendix on convexity.) In particular, if the second derivative of a function is nonnegative on I then the function is convex on I.

Let us differentiate equation (26.6) twice. We get

$$\frac{d^2}{dx^2} B^n f(x) = n(n-1)(I + x\Delta)^{n-2} \Delta^2 y_0. \tag{26.15}$$

The sequence $\Delta^2 y$ is $\{y_2 - 2y_1 + y_0,\ y_3 - 2y_2 + y_1, \dots,\ y_n - 2y_{n-1} + y_{n-2}\}$. Since f is convex in $[0, 1]$, all of these terms are nonnegative. The right side of (26.15) is $n(n-1)$ times the $(n-2)$nd degree Bernstein polynomial based on these values and is therefore nonnegative.

Next we present a smoothing property of the transformation B^n which is a generalization of Theorem 26.2. The proof does not utilize calculus.

THEOREM 26.3. *The number of times the graph of $B^n f$ crosses any straight line l does not exceed the number of times the graph of f crosses l.* We call this property of B^n the *variation-diminishing* property.

PROOF. Let the equation of l be $y = l(x)$. Since $B^n(f - l) = B^n f - l$, it suffices to prove the statement in the case where l is the x-axis.

One can derive our theorem from Descartes' Rule as follows. We write $B^n f(x)$ in the rather artificial form

$$(26.16)\quad B^n f(x) = (1-x)^n \sum_{k=0}^{n} \binom{n}{k} \left(\frac{x}{1-x}\right)^k y_k = (1-x)^n \sum_{k=0}^{n} \binom{n}{k} y_k u^k$$

where $u = x/(1-x)$. The factor $(1-x)^n$ does not produce sign changes in [0, 1]. The variable u goes from 0 to ∞ as x goes from 0 to 1.

If the polynomial $B^n f$ is identically 0 then it does not cross the x-axis, so the theorem holds. Otherwise, by Descartes' Rule, the number of roots of $B^n f(x)$ and hence the number of its sign changes in (0, 1) is at most equal to the number of sign changes of the sequence $\{y_0, y_1, \ldots, y_n\}$, which is $\leq$ the number of sign changes of f in (0, 1), q.e.d.

After we introduce degree elevation we will be able to give a more natural proof of Theorem 26.3.

We want to show that one can obtain Theorem 26.2 from Theorem 26.3, instead of by using the formula for the second derivative of the Bernstein polynomial. Suppose $f(x)$ is convex in [0, 1]. Then if $l(x)$ is a linear function, $f(x) - l(x)$ has at most two sign changes. Theorem 26.3 therefore tells us that $B^n f(x) - l(x)$ has at most two sign changes for any linear function l. This implies that $B^n f$ is either convex or concave. (This is intuitively clear but we included a detailed proof in the appendix on convexity, Proposition 7.) Let $l(x)$ be the linear function such that $l(0) = f(0)$ and $l(1) = f(1)$. Then $f(x) \leq l(x)$ since f is convex. Hence $B^n f(x) \leq B^n l(x) = l(x)$ and hence $B^n f$ is convex, not concave.

Degree elevation. We want to express the nth degree Bernstein basis polynomials $B_i^n(x)$ in terms of the $(n+1)$st degree polynomials $B_i^{n+1}(x)$. This is easy because of the identity

$$\begin{aligned} x^i(1-x)^{n-i} &= x^i(1-x)^{n-i}((1-x)+x) \\ &= x^i(1-x)^{n+1-i} + x^{i+1}(1-x)^{n+1-(i+1)}. \end{aligned}$$

In terms of Bernstein basis polynomials the identity is

$$(26.17)\qquad B_i^n(x) = \frac{1}{n+1}\left((n+1-i)B_i^{n+1}(x) + (i+1)B_{i+1}^{n+1}(x)\right).$$

If we substitute this in the definition of $B^n f$, we get the expression for $B^n f$ in terms of the basis polynomials $B_i^{n+1}(x)$ of degree $n+1$:

$$(26.18)\qquad B^n f(x) = \sum_{i=0}^{n+1}\left(\left(1-\frac{i}{n+1}\right) f\left(\frac{i}{n}\right) + \frac{i}{n+1} f\left(\frac{i-1}{n}\right)\right) B_i^{n+1}(x).$$

We call this the *degree elevation* formula. The occurrence of the function values $f\left(-\frac{1}{n}\right)$ and $f\left(1+\frac{1}{n}\right)$ is only apparent because both have coefficient 0.

Let us look at the degree elevation formula from a geometric point of view. Figure 26.3 shows the construction we are about to present performed 5 successive times. The polygon we start with has only 2 sides which are drawn as solid lines. The Bernstein polynomial with this control polygon is a parabola. The new sides of the polygons obtained by successive degree elevations are drawn with different kinds of dashing. Right now the reader needs to view only one of the polygons and the one just below it.

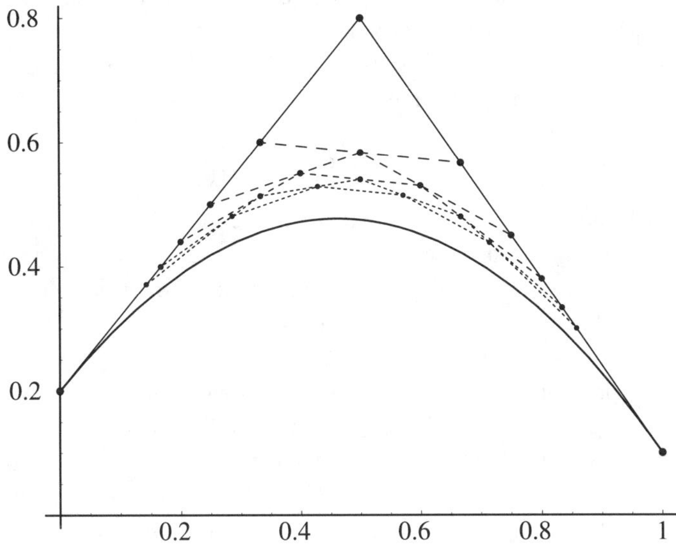

Figure 26.3. Iterated degree elevation.

Let Π be any control polygon with $n+1$ vertices, and let $p(x)$ be the nth degree Bernstein polynomial it defines. For $i = 1, 2, \ldots, n$, divide the ith side of Π in the ratio $1 - \frac{i}{n+1} : \frac{i}{n+1}$. If we take the endpoints of Π and these n interior points, which are the points of Π with abscissas $x = 0, \frac{1}{n+1}, \frac{2}{n+1}, \ldots, 1$ as the control polygon $\mathcal{E}\Pi$ of a nominally $(n+1)$st degree Bernstein polynomial, the latter is by (26.18) just the polynomial $p(x)$ defined by Π. Note that all the vertices of $\mathcal{E}\Pi$ lie on Π and have abscissas $x = 0, \frac{1}{n+1}, \frac{2}{n+1}, \ldots, 1$. We will denote by $\mathcal{E}^2\Pi$ the degree elevation of $\mathcal{E}\Pi$. In the rest of this discussion we shall use the notation Π, $\mathcal{E}\Pi, \ldots$ for both the polygons and the functions which have these polygonal graphs.

We might mention that although $B^{n+1}(\mathcal{E}\Pi)$ is the nth degree polynomial $B^n f$, the polynomial $B^n(\mathcal{E}\Pi)$ is in general different from $B^n f$.

The theoretically interesting aspect of degree elevation is what happens when we perform the construction repeatedly. Figure 26.3 gives support to a conjecture by A. Forrest for which G. Farin gave the proof below in his Master's thesis:

THEOREM 26.4. *The polygons* Π, $\mathcal{E}\Pi$, $\mathcal{E}^2\Pi$ *obtained by repeated degree elevation converge to the graph of* $B^n\Pi$.

PROOF. It is easy to show that the piecewise linear function $\Pi(x)$ is Lipschitz continuous, with the largest of the absolute values of the slopes of the segments of Π as Lipschitz constant L; see Exercise 26.8. The line segments which make up $\mathcal{E}\Pi$ are chords of Π; hence L is a Lipschitz constant for $\mathcal{E}\Pi(x)$ and for all the functions $\mathcal{E}^k\Pi(x)$. Since the polynomial $B^n\Pi(x)$ is the Bernstein approximant $B^{n+k}\mathcal{E}^k\Pi(x)$, we have by Theorem 25.6

$$|B^n\Pi(x) - \mathcal{E}^k\Pi(x)| \leq \frac{1.2L}{(n+k)^{1/3}}, \tag{26.19}$$

and the right side does indeed tend to 0 as $k \to \infty$.

Figure 26.1 suggests that the convergence is slow. However, it is not as slow as the right side of (26.19) might lead one to think. A more accurate estimate has exponent 1 instead of 1/3 in the denominator.

In the theorem just proved, we used degree elevation to get information about a sequence of polygons converging to a fixed Bernstein polynomial. Next, degree elevation will give us a result about the sequence of Bernstein approximants to a fixed function which other types of approximants do not have.

THEOREM 26.5. *Let* $f(x)$ *be convex on* $[0, 1]$. *Then, for* $n = 1, 2, 3, \ldots,$

$$B^n f(x) \geq B^{n+1} f(x) \geq f(x), \quad x \in [0, 1]. \tag{26.20}$$

PROOF. The degree elevation formula tells us that $B^n f$ can be obtained by applying B^{n+1} to the function $\Pi(x)$ whose graph is a polygon inscribed in the graph of f. If f is convex, then $\Pi(x) \geq f(x)$. Hence $B^n f = B^{n+1}\Pi \geq B^{n+1} f$ by part 3) of Theorem 25.1. The second inequality (26.20) holds because f is the limit of the decreasing sequence $B^n f(x)$.

All our theorems can of course be rephrased for concave functions. In particular, if f is concave in $[0, 1]$, then $f(x) \geq B^{n+1} f(x) \geq B^n f(x)$ in $[0, 1]$.

Alternative proof of the variation diminishing property stated in Theorem 26.3. Suppose we have a function $\phi(x)$ and a line l. Let us call a piecewise linear function $\Pi(x)$ whose vertices are on the graph of ϕ a *piecewise linear interpolant* of ϕ. The present proof is based on the obvious fact that the graph of a piecewise linear interpolant of ϕ crosses from one side of l to the other at most as many times as the graph of ϕ.

By Farin's Theorem 26.4 we can obtain $B^n f$ from f by constructing an infinite sequence of piecewise linear interpolants, as follows: First we construct the interpolant with vertices $(0, f(0))$, $(1/n, f(1/n)), \dots, (1, f(1))$, and then we perform an infinite sequence of degree elevations. In each of these steps, the number of times the resulting polygon crosses from one side of a given line l to the other does not increase. If $B^n f$ crosses from one side of l to the other c times, then any close enough approximant of $B^n f$ crosses at least c times. Hence c is not more than the number of crossings of f across l.

We end this chapter by giving another smoothing property of the operator B^n. We use the notation $(B^n)^k$ for the kth iterate of the transformation B^n.

THEOREM 26.6 [KELISKY AND RIVLIN '67]. *For any function $f(x)$ defined on $[0, 1]$, we have*

$$\lim_{k\to\infty} (B^n)^k f(x) = f(0) + [f(1) - f(0)]x. \tag{26.21}$$

PROOF. The proof given by Kelisky and Rivlin uses eigenvalues and eigenvectors of matrices. The following proof is much simpler.

Since one application of the operator B^n produces a bounded function, it suffices to prove the theorem for a bounded function f. Let $L(x) = f(0)(1 - x) + f(1)x$. We have $(B^n)^k(f(x) - L(x)) = (B^n)^k f(x) - L(x)$. So our theorem will be proved if we show that for any function $g(x)$ with $g(0) = g(1) = 0$, $(B^n)^k g(x) \to 0$ as $k \to \infty$.

Let M be such that $|g(x)| \le M$ in $[0, 1]$. Then

$$|B^n g(x)| \le \sum_{i=1}^{n-1} M B_i^n(x) = M\big(1 - x^n - (1 - x)^n\big) \le \left(1 - \left(\frac{1}{2}\right)^n\right) M,$$

since one of x and $1 - x$ is $\ge \frac{1}{2}$. Hence $|(B^n)^k g(x)| \le \left(1 - \left(\frac{1}{2}\right)^n\right)^k M$, which goes to 0 as $k \to \infty$.

COROLLARY. *The only functions invariant under the map B^n are the linear functions $f(x) = ax + b$.*

We have shown that the Bernstein polynomials enjoy attractive features, such as the shape-preserving and variation-diminishing properties. The price that must be paid for these beautiful approximation properties is that the convergence of Bernstein polynomials is very slow. For example, from (26.8) we conclude that $B^n(x^2)$ converges to x^2 no faster than $\frac{1}{4n}$. In his book *Interpolation and Approximation*, [Davis '75], Professor Philip Davis wrote:

"This fact (i.e., slow convergence) seems to have precluded any numerical application of Bernstein polynomials from having been made. Perhaps they will find application when the properties of the approximant in the large are of more importance than the closeness of the approximation."

This was a sound prediction. The first edition of his book appeared in 1963, at the same time as Professor Pierre Bézier, a French engineer at Renault, was developing his techniques for automobile design and manufacture. Bézier curves and surfaces, which are widely and successfully used in computer-aided geometric design, are vector-valued Bernstein polynomials. A detailed description of Bézier curves can be found in the next chapter.

Exercises

26.1 [CMO 1986] Given a sequence $a_0, a_1, a_2, \ldots$ which satisfies $a_{i-1} + a_{i+1} = 2a_i$ for $i = 1, 2, 3, \ldots$, show that the expression

$$\sum_{i=0}^{n} a_i \binom{n}{i} x^i (1-x)^{n-i}$$

is either a constant or a polynomial of degree 1 for any positive integer n.

26.2 Show that

$$\lim_{n\to\infty} n((B^n x^3)(x) - x^3) = 3(1-x)x^2.$$

26.3 Show directly from the definition $B^n f(x) = \sum y_k B_k^n(x)$ that the coefficient of x^i in $B^n f(x)$ depends only on $y_0, y_1, \ldots, y_i$.

26.4 Show that $B^n f \equiv B^n g$ if and only if $f(x_i) = g(x_i)$ for $i = 0, 1, \ldots, n$.

26.5 Show that every polynomial of degree $\le n$ is a sum of constant multiples of $B_0^n(x), B_1^n(x), \ldots$.

26.6 Let $f(x) = a(1-x)^2 + 2bx(1-x) + cx^2$. Find a necessary and sufficient condition for the positivity of $f(x)$ in $[0, 1]$.

26.7 Let $f(x) = ax^2 + bx + c$. Find necessary and sufficient conditions on a, b, c in order that $f(x)$ is nonnegative on $[-1, 1]$.

26.8 Prove, by replacing derivatives by differences in the proof of Theorem 3 of the last chapter, that if

$$\Delta^2 y_k \le M/n^2 \quad \text{for } k = 0, 1, \ldots, n-2, \tag{26.22}$$

then

$$B^n f(x_i) - f(x_i) \le \frac{M}{2n} x_i(1-x_i) \quad \text{for } i = 0, 1, \ldots, n. \tag{26.23}$$

26.9 Use Theorem 26.4 to prove that if f is convex in $[0, 1]$, then so is $B^n f$.

26.10 Show that if we define $B_k^m(x)$ to be identically 0 for all integer values of k outside the range $0 \ldots m$, then for all integers k

$$\frac{d}{dx} B_k^n(x) = n\left(B_{k-1}^{n-1}(x) - B_k^{n-1}(x)\right).$$

Use this to derive the descriptions (26.12) and (26.13) of the derivative of a Bernstein polynomial.

26.11 (H. Prautzsch) Show that the transformation B^n has the following variation diminishing property: *If f is continuous on* $[0, 1]$ *and $B^n f$ is not constant, then the number of relative maxima and minima of $B^n f$ on the interior of the interval is at most equal to the number of relative maxima and minima of f there.* (In some books the definition of a relative extremum includes the requirement that it be in the interior of the interval. Our statement is still true if we include extrema occurring at the endpoints.)

CHAPTER TWENTY-SEVEN

Bézier Curves

In the middle of the 1960's, a system for designing and manufacturing cars using free-form curves and surfaces was developed by P. Bézier at the Renault automobile company. This computerized system, named UNISURF, provides a general mathematical framework for defining arbitrarily shaped curves and surfaces.

Previously, designers of stamped parts such as car body panels had used manual tools such as French curves. (Of course, the Bézier curve could be referred to as a French curve also, but here we mean French curve in the traditional sense.) These designers defined the shape of a car body in terms of cross sections at most one hundred millimeters apart. In this process, the cross-sectional curves are carved into a three-dimensional model and interpolation is left to the experience of highly skilled pattern makers. The final standard, however, is the "master model", whose shape no longer coincides precisely with the curves originally traced on the drawing board. This inconsistency results in expenses and delays. No significant improvement could be expected in the absence of an accurate, complete mathematical definition of free-form shape.

A good designing system must allow for the interface between the underlying mathematical techniques and the designers, who have a good knowledge of descriptive geometry but may have little training in algebra or analysis. In order to be successful, a system must appeal to designers — it must be simple, intuitive and easy to use. It is crucial that such a design system make no mathematical demands on the users other than those to which they have been accustomed through the conventional design process.

UNISURF proved to be a highly successful system. The essence of its success was that it combined modern approximation theory and geometry in a way that provides the designers with computerized analogs of their conventional design and drafting tools. Due to his great contribution to the newly established branch of applied mathematics and computer science known as computer-aided geometric design, Professor Bézier has been recognized widely as one of the pioneers in this field. The curve and surface definitions which his UNISURF system popularized have come to be known as Bézier curves and surfaces. Bézier curves are available in most current drawing programs.

This chapter presents some practical examples of iterations involving Bézier curves. In particular, the de Casteljau algorithm for evaluating a point on a Bézier curve and the degree elevation algorithm are presented.

A Bézier curve is a vector-valued Bernstein polynomial. Much of what we present in this chapter is just a rephrasing in vector notation of what we did in the last chapter, but the point of view is different. Previously we started with a function $f(x)$ and viewed the Bernstein polynomials $B^n f$ as approximants of f. In this chapter we no longer have a function f. We consider only control polygons and the polynomial curves they define.

Before going farther, we want to tidy up a simple matter which may trouble some readers. In the following we shall have occasion to speak of weighted means of points, such as $P = uA + vB + wC$, where u, v, and w are scalars such that $u + v + w = 1$. The troubling feature of this definition of P is that the sum of points and the product of a point and a scalar depend on where we happen to put the origin of our coordinate system. We show now that, thanks to the condition $u + v + w = 1$, the point $P = uA + vB + wC$ does not depend on the choice of origin.

Suppose more generally that we are given $n + 1$ points $P_0, P_1, \ldots, P_n$ and real numbers $u_0, u_1, \ldots, u_n$. We then choose an origin O and define the linear combination

$$P = \sum_{i=0}^{n} u_i P_i = u_0 P_0 + u_1 P_1 + \cdots + u_n P_n$$

to be the endpoint of the vector $\overrightarrow{OP} = \sum_{i=0}^{n} u_i \overrightarrow{OP_i}$. We assert that P is independent of the choice of O if $\sum_{i=0}^{n} u_i = 1$. Indeed, suppose another point O' is selected as origin. With respect to this choice, we have $\sum_{i=0}^{n} u_i P_i = P'$, where $\overrightarrow{O'P'} = \sum_{i=0}^{n} u_i \overrightarrow{O'P_i}$. Now

$$\begin{aligned}\sum_{i=0}^{n} u_i \overrightarrow{O'P_i} = \sum_{i=0}^{n} u_i (\overrightarrow{O'O} + \overrightarrow{OP_i}) &= \left(\sum_{i=0}^{n} u_i\right) \overrightarrow{O'O} + \sum_{i=0}^{n} u_i \overrightarrow{OP_i} \\ &= \overrightarrow{O'O} + \overrightarrow{OP} = \overrightarrow{O'P},\end{aligned}$$

so that $P' = P$ as claimed.

We now define Bézier curves. We start with a sequence of $n + 1$ points $\mathbf{P}_i = (x_i, y_i, z_i)$, $i = 0, 1, 2, \ldots, n$. We associate with this sequence of points the *parametric space curve*:

$$\mathbf{P}(t) = \sum_{i=0}^{n} B_i^n(t) \mathbf{P}_i, \quad t \in [0, 1]. \tag{27.1}$$

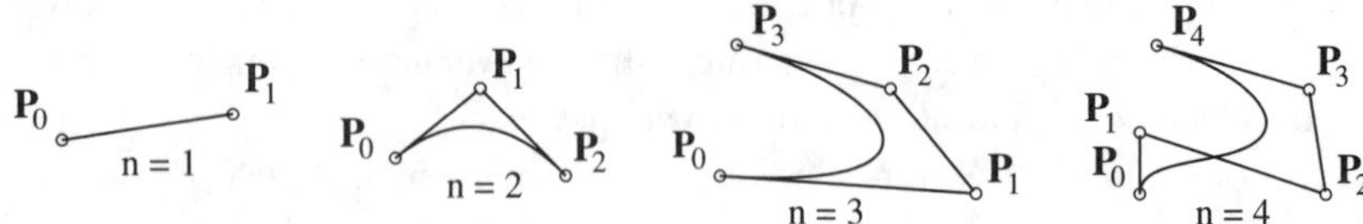

Figure 27.1. Bézier curves.

The curve (27.1) is called a *Bézier curve of degree n*, and the polygon $\mathbf{P}_0\mathbf{P}_1\ldots\mathbf{P}_n$ is called its *control polygon*. The $\mathbf{P}_i$ are the *control points* of the Bézier curve. We have thus defined a transformation which takes each n-gon to a parametric curve (Fig. 27.1).

In two dimensions, if the abscissas of the $\mathbf{P}_i$ are equidistant on the interval $[0, 1]$, i.e., if $x_i = \frac{i}{n}$, then the x-coordinate of the parametric plane curve $x(t) \equiv t$. For such a sequence of points, the Bézier curve is just the graph of a Bernstein polynomial.

Let us write down the vector versions of some of the basic properties of Bernstein polynomials, which we obtain by noting that the formulas of Chapter 26 apply to each component of a vector-valued function f.

The form of equation (27.1) in terms of the operators I and E is

$$\mathbf{P}(t) = ((1-t)I + tE)^n\,\mathbf{P}_0, \tag{27.2}$$

and we have

$$\mathbf{P}(0) = \mathbf{P}_0, \qquad \mathbf{P}(1) = \mathbf{P}_n. \tag{27.3}$$

This means that the Bézier curve joins the two endpoints of its control polygon. In particular, if the control polygon is closed, then so is its Bézier curve.

Special cases. For $n = 1$, the Bézier curve becomes

$$\mathbf{P}(t) = (1-t)\mathbf{P}_0 + t\mathbf{P}_1. \tag{27.4}$$

This is a line segment determined by two points $\mathbf{P}_0$ and $\mathbf{P}_1$. The equation (27.4) provides a formula for *linear interpolation*. The point $\mathbf{P}(t)$ divides the segment $\mathbf{P}_0\mathbf{P}_1$ in the ratio $t : 1-t$. For $n = 2$ we have

$$\mathbf{P}(t) = (1-t)^2\mathbf{P}_0 + 2(1-t)t\mathbf{P}_1 + t^2\mathbf{P}_2.$$

We show that if $\mathbf{P}_0, \mathbf{P}_1, \mathbf{P}_2$ are not collinear, this is an arc of a parabola. We have

$$\mathbf{P}(t) = ((\mathbf{P}_2 - \mathbf{P}_0) - 2(\mathbf{P}_1 - \mathbf{P}_0))t^2 + 2(\mathbf{P}_1 - \mathbf{P}_0)t + \mathbf{P}_0 = \mathbf{V}_2t^2 + \mathbf{V}_1t + \mathbf{P}_0.$$

If $\mathbf{P}_0, \mathbf{P}_1, \mathbf{P}_2$ are not collinear, then $\mathbf{V}_1 \neq 0$ and the first term in the expression for $\mathbf{V}_2$ is not a multiple of the second term, which is $\mathbf{V}_1$. Thus $\mathbf{V}_1$ and $\mathbf{V}_2$ are nonzero vectors pointing in different directions.

Now take a coordinate system with the origin at $\mathbf{P}_0$, the positive y-axis in the direction $\mathbf{V}_2$ and the x-axis in the plane of $\mathbf{P}_0, \mathbf{P}_1, \mathbf{P}_2$. Then $\mathbf{P}(t)$ is in the (x, y) plane for all t. The expressions for y and x components of $\mathbf{P}(t)$ are: $\mathbf{P}(t)_y = |\mathbf{V}_2|t^2 + (\mathbf{V}_1)_y t$; $\mathbf{P}(t)_x = (\mathbf{V}_1)_x t$. The last equation tells us that t is a constant multiple of $\mathbf{P}(t)_x$. Hence, by the previous equation, $\mathbf{P}(t)_y$ is a quadratic polynomial in $\mathbf{P}(t)_x$. Thus the locus of $\mathbf{P}(t)$ is a parabola.

We now discuss some properties of Bézier curves.

A region is called *convex* if it contains all line segments connecting any two of its points. The 3-dimensional regions enclosed by spheres or tetrahedrons are convex if we include all the boundary points, and also if we exclude all of them. The planar regions enclosed by triangles, circles, semicircles or ellipses are examples of convex planar figures. The interior of a simple polygon (i.e., one which does not intersect itself) is convex if and only if all the interior angles are $\leq 180°$.

The *convex hull* $\mathcal{H}$ of a nonempty set $\mathcal{S}$ of points is the intersection of all convex sets containing $\mathcal{S}$. In other words, $\mathcal{H}$ is the "smallest" convex set containing $\mathcal{S}$. It can be obtained by adjoining to $\mathcal{S}$ the line segments joining all pairs of points in $\mathcal{S}$, and repeating the operation indefinitely or until no new points are added. (In the plane and in three-dimensional space this will happen after no more than two steps but we shall not need this fact.)

For example, the convex hull of the set of three points A, B, and C, is the set of points inside or on the triangle ABC. Generally, the boundary of the convex hull of a set of finitely many points lying in a plane can be constructed physically by driving in nails at each point and wrapping a string around the resulting configuration.

We need the following simple fact:

The convex hull of the set of points $\mathbf{P}_0, \ldots, \mathbf{P}_n$ is the set of points of the form

$$u_0\mathbf{P}_0 + \cdots + u_n\mathbf{P}_n, \tag{27.5}$$

where

$$u_0 \geq 0, \ldots, u_n \geq 0, \quad \text{and} \quad u_0 + \cdots + u_n = 1.$$

The reader should be able to prove this, but for completeness we provide a proof in Appendix C.

An expression of the form (27.5) with $u_i \geq 0$, $\sum_{i=0}^{n} u_i = 1$ is called a *convex linear combination* of $\mathbf{P}_0, \ldots, \mathbf{P}_n$.

Since all the Bernstein polynomials of degree n are nonnegative in the interval $[0, 1]$ and sum up to unity, (27.1) says that for any $t \in [0, 1]$, $\mathbf{P}(t)$ is a convex linear combination of $\mathbf{P}_0, \mathbf{P}_1, \ldots, \mathbf{P}_n$. In other words, the Bézier curve lies entirely in the convex hull of its control points. This is called the *convex hull property* of Bézier curves. Simple consequences are that a control polygon with coplanar vertices always produces a planar Bézier curve, and a control polygon with collinear vertices always generates a line segment.

From the inspection of Fig. 27.1, one gets an idea in what sense the Bézier curve mimics its control polygon. Not only do the curve endpoints coincide with the polygon endpoints, but, as in the scalar case, the control polygon is also tangent to the curve at its endpoints. This follows by differentiating both sides of (27.2):

$$\mathbf{P}'(t) = n\,((1-t)I + tE)^{n-1}\,\Delta\mathbf{P}_0. \tag{27.6}$$

Putting $t = 0$ and $t = 1$ into (27.6), we obtain respectively

$$\mathbf{P}'(0) = n(\mathbf{P}_1 - \mathbf{P}_0), \qquad \mathbf{P}'(1) = n(\mathbf{P}_n - \mathbf{P}_{n-1}). \tag{27.7}$$

For the second derivatives at the endpoints, we have

$$\mathbf{P}''(0) = n(n-1)(\mathbf{P}_2 - 2\mathbf{P}_1 + \mathbf{P}_0), \tag{27.8}$$

$$\mathbf{P}''(1) = n(n-1)(\mathbf{P}_n - 2\mathbf{P}_{n-1} + \mathbf{P}_{n-2}). \tag{27.9}$$

To approximate a hand-drawn curve by a polynomial, we specify a control polygon that “exaggerates” the shape of the curve. One lets the computer draw the Bézier curve defined by the polygon. If the Bézier curve does not match the hand-drawn curve well enough, one adjusts the location of the polygon’s vertices, possibly deleting some old vertices and inserting new ones. An experienced person will usually get close enough to the given curve by doing this two or three times.

We may put two or more Bézier curves together to form a *composite curve*. Let $\mathbf{P}_0, \mathbf{P}_1, \ldots, \mathbf{P}_n$ and $\mathbf{Q}_0, \mathbf{Q}_1, \ldots, \mathbf{Q}_n$ be the vertices of two control polygons which generate the Bézier curves $\mathbf{P}(t)$ and $\mathbf{Q}(t)$. If $\mathbf{P}_n = \mathbf{Q}_0$ then by (27.3) we have $\mathbf{P}(1) = \mathbf{Q}(0)$. If moreover the three points $\mathbf{P}_{n-1}, \mathbf{P}_n, \mathbf{Q}_1$ are collinear, then by (27.7) we see that the composite curve is smooth at the junction, i.e., it does not have a corner there. For a pleasing appearance it is also necessary that the curvature of the composite curve should not jump from one value to another at the junction. This is called *second order continuity* of the composite curve and it can be achieved by taking into account the condition in (27.9).

Perhaps the most widespread use of Bézier curves is in the definition of font outlines. The characters on most printed pages that are typeset using a computer are defined using Bézier curves. In fact, every letter in this very book contains dozens of Bézier curves! Fig. 27.2 shows the outline of a lower-case “g”, with the control points of three of the Bézier curves that comprise it. This composite Bézier curve is formed from 23 cubic Bézier curves and four linear Bézier curves

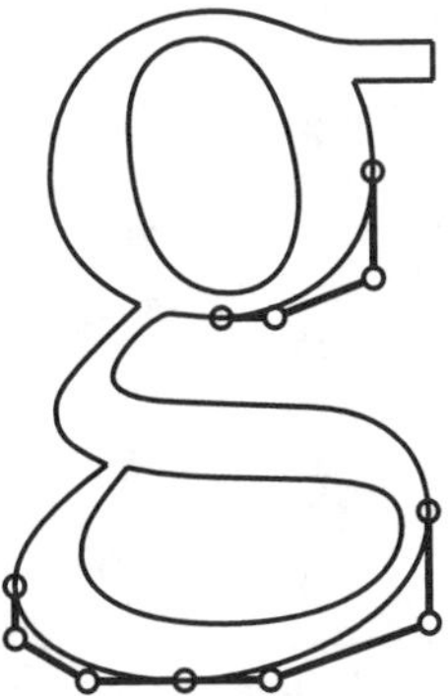

Figure 27.2. Outline of a character "g".

(line segments). Notice that most of the cubic Bézier curves meet with first order continuity.

To illustrate the approximation of a curve by a Bézier curve, we will discuss the approximation of the first quadrant of the unit circle by a cubic Bézier curve. (An arc of a circle can not be parametrized exactly by polynomials, see Ex. 4.) To make the curve start at the point $(1, 0)$ and end at $(0, 1)$, set $\mathbf{P}_0 = (1, 0)$ and $\mathbf{P}_3 = (0, 1)$. Since the circular arc has a vertical tangent at $(1, 0)$ and a horizontal tangent at $(0, 1)$, to match the tangent lines at these two points we must have $\mathbf{P}_2 = (1, s_1)$ and $\mathbf{P}_3 = (s_2, 1)$. Considering the symmetry of the circular arc, it is reasonable to put $s_1 = s_2 = s$. Hence we consider cubic curves with control points

$$\mathbf{P}_0 = (1, 0), \qquad \mathbf{P}_1 = (1, s), \qquad \mathbf{P}_2 = (s, 1), \qquad \mathbf{P}_3 = (0, 1).$$

Let $\mathbf{P}(t) = (x(t),\ y(t))$. The coefficients of the Bernstein basis polynomials in $x(t)$ and in $y(t)$ are $(1, 1, s, 0)$ and $(0, s, 1, 1)$, i.e.

$$x = (1-t)^3 + 3t(1-t)^2 + 3s\,t^2(1-t), \qquad y = 3s\,t(1-t)^2 + 3t^2(1-t) + t^3.$$

We see that $x'(0) = y'(1) = 0$, as it should be. Symmetry with respect to the line $y = x$ is expressed by the fact that $x(t) = y(1-t)$ for all $t \in [0, 1]$ which is also easy to read off from the formulas.

Our Bézier curve depends on a single parameter $s \geq 0$. As s varies, different curves with various shapes are obtained. We look at these briefly before returning to the problem of approximating a quarter-circle.

First of all let $s = 0$. The second vertex of the control polygon now coincides with the first vertex, $(1, 0)$, and the third vertex coincides with the fourth, $(0, 1)$. The cubic curve reduces to a line segment joining $(1, 0)$ and $(0, 1)$. (This is not the usual parametrization of a line segment, but one can easily verify that $x + y = 1$.)

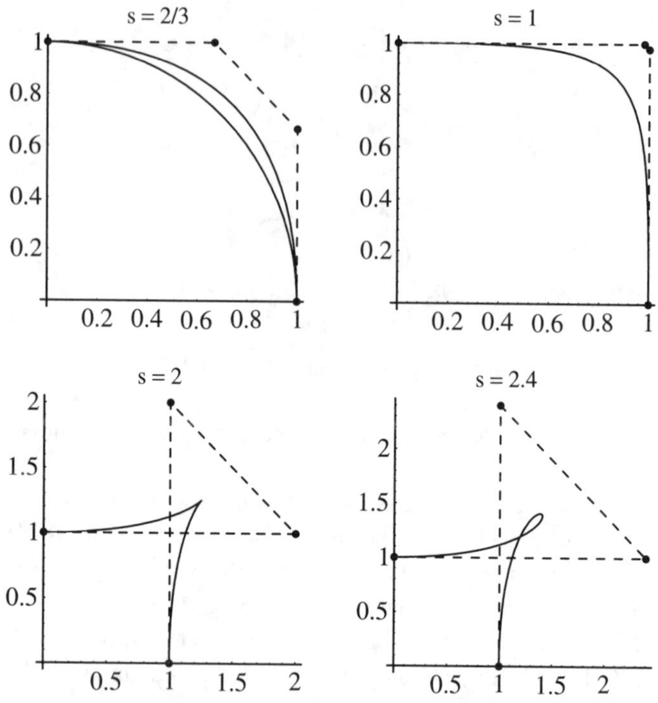

Figure 27.3. Four different values of s.

The next value we consider is $s = 2/3$. In Fig. 27.3, the first of the four graphs shows this curve and also the quarter-circle with the same endpoints. This value of s has some special interest: the cubic curve with these four control points is the same as the quadratic curve with the control points $(1, 0)$, $(1, 1)$, $(0, 1)$. This follows from the degree elevation formula (27.11) below. Hence the cubic curve reduces to a quadratic Bézier curve which we know to be a parabola. Of course, we do not need a general theorem to see that for $s = 2/3$ the cubic terms cancel out; we can verify that by computation.

If $s = 2$, the Bézier curve has a cusp at the point $(1.25, 1.25)$, as in this case we have $\mathbf{P}'(1/2) = 0$, $\mathbf{P}''(1/2) \neq 0$. If $s > 2$, the Bézier curve follows the control polygon in having a loop.

We return to the problem of approximating a quarter-circle. Following S. A. Coons, another pioneer of computer-aided geometric design who shared honor and reputation with Bézier, we choose s so that the cubic Bézier curve passes through the midpoint of the quarter circle. The condition for this is $\mathbf{P}(1/2) = (1/\sqrt{2},\ 1/\sqrt{2})$ [Coons '64]. This means that $s = 4(\sqrt{2} - 1)/3 \approx 0.55228475$.

Coons knew that this approximation was good, but did not present an error estimate.

It is convenient to measure the deviation of $\mathbf{P}(t)$ from the quarter circle by the quantity $\epsilon(t) = x^2(t) + y^2(t) - 1$. The error function $\epsilon(t)$ is a polynomial of degree six. It has double zeros at $t = 0$ and at $t = 1$ because the Bézier curve is tangent to the quarter-circle at the ends. The symmetry of the construction with respect to the line $y = x$ implies that $\epsilon(t)$ is an even function of $t - \frac{1}{2}$, hence the root at $\frac{1}{2}$ must also be a double root. Since the leading coefficient of ϵ is $K = 2(3s - 2)^2 = 8(3 - 2\sqrt{2})^2$, we have

$$\epsilon(t) = Kt^2(1-t)^2\left(t - \tfrac{1}{2}\right)^2.$$

We see that $\epsilon(t) \geq 0$ for all $t \in [0, 1]$, which means that all points of the Bézier curve always lie on or outside the circle. Note that the graph of $u = \epsilon(t)$ is symmetric with respect to the straight line $t = 1/2$.

To find the maximum of $\epsilon(t)$, we find the roots of $\epsilon'(t) = 0$. The polynomial $\epsilon'(t) = 0$ of degree five has three obvious zeros at 0, $1/2$, and 1; and two additional roots of $\epsilon'(t) = 0$ satisfy $t^2 - t + 1/6 = 0$. Taking the least nonzero root $t^* = (1 - \sqrt{3}/3)/2$, we get

$$\epsilon(t) \leq \epsilon(t^*) = 0.000545134\ldots.$$

Since $\epsilon(t)$ is the amount by which the square of the distance of $\mathbf{P}(t)$ from the origin exceeds 1, the Bézier curve is everywhere within a distance $[\epsilon(t*) + 1]^{1/2} - 1 = 0.00028\ldots$ from the circle. This is a really good approximation. Fig. 27.4 shows

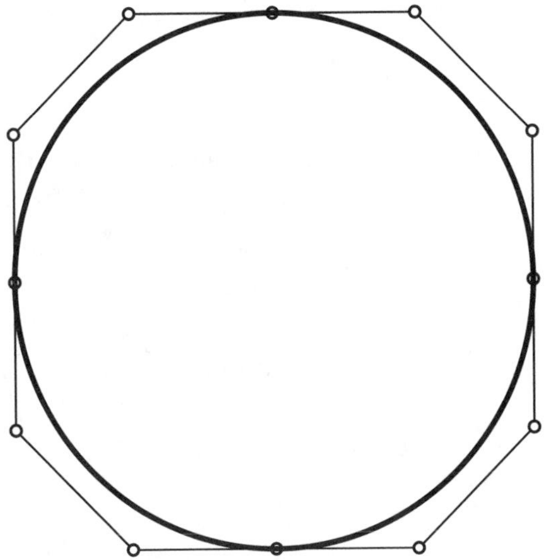

Figure 27.4. Circle approximated by four Bézier curves.

an approximation to the whole unit circle by four such cubic Bézier curves. We cannot tell the difference between the whole circle and the approximation with the naked eye. At each junction the two Bezier arcs have a common tangent, and their curvatures are also equal there. We say that the closed approximation consisting of four pieces of cubic curves is *geometrically continuous of second order*.

We now look at the quantities implicit in the Formula (27.2) in more detail. We arrange them in a triangular array. The top row of the array consists of the sequence of control points $\mathbf{P}_0$, $\mathbf{P}_1, \ldots, \mathbf{P}_n$. If we work graphically, this means we draw the control polygon. For a fixed value of t in [0, 1] we apply the operator $(1-t)I + tE$ to the first row to produce the second row. (Our notation does not show that the rows after the first depend on the parameter t.) Graphically, the points of each second row are obtained by dividing the sides of the control polygon in the ratio $t : 1-t$. We get a polygon with one fewer vertices than the first row. Each succeeding row of the triangular array is obtained from the previous row in the same way. If we define $P_i^0 = P_i$, then for $j = 1, 2, \ldots, n$ the individual array elements are given by the formulas $P_i^j = (1-t)P_i^{j-1} + tP_{i+1}^{j-1}$. By (27.2) the entry at the bottom vertex of the triangle is $\mathbf{P}(t) = \mathbf{P}_0^n = (1-t)\mathbf{P}_0^{n-1} + t\mathbf{P}_1^{n-1}$. This is *de Casteljau's algorithm,* contained in a 1959 internal memorandum of the Citroen car company and thus preceding the work of Bézier at Renault.

$$
\begin{array}{ccccccccccc}
\mathbf{P}_0 & & \mathbf{P}_1 & & \mathbf{P}_2 & \cdots\cdots & \mathbf{P}_{n-2} & & \mathbf{P}_{n-1} & & \mathbf{P}_n \\
& \mathbf{P}_0^1 & & \mathbf{P}_1^1 & & \cdots\cdots\cdots & & \mathbf{P}_{n-2}^1 & & \mathbf{P}_{n-1}^1 & \\
& & \mathbf{P}_0^2 & & \mathbf{P}_1^2 & \cdots\cdots & \mathbf{P}_{n-3}^2 & & \mathbf{P}_{n-2}^2 & & \\
& & & & \cdots\cdots & \cdots\cdots & \cdots\cdots & & & & \\
& & & & \mathbf{P}_0^{n-1} & & \mathbf{P}_1^{n-1} & & & & \\
& & & & & \mathbf{P}_0^n & & & & &
\end{array}
$$

The geometrical de Casteljau algorithm is illustrated in Fig. 27.5.

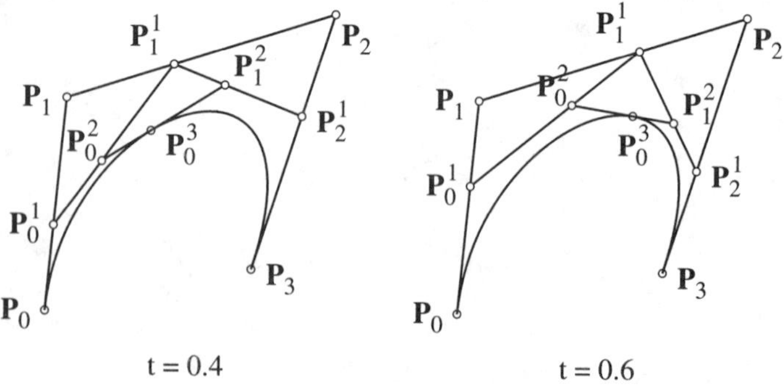

Figure 27.5. The de Casteljau Algorithm.

The algorithm of de Casteljau could be used as the definition of Bézier curves. The convex hull property would be an immediate consequence. We are now going to use it as a point of departure for splitting a given Bézier curve into two Bézier curves — a process which has many applications. In Fig. 27.5, the Bézier curve has been divided into two pieces by the point $\mathbf{P}_0^n$. One of them, denoted by $\mathbf{P}_l(t)$, has $\mathbf{P}_0$ and $\mathbf{P}_0^n$ as its endpoints; the other, denoted by $\mathbf{P}_r(t)$, has $\mathbf{P}_0^n$ and $\mathbf{P}_n$ as endpoints. It looks like these two curves, treated independently as Bézier curves, have control points taken from the two non-horizontal sides of the above triangular array. In fact, we can prove the following:

SUBDIVISION THEOREM. The curves $\mathbf{P}_l(t)$ and $\mathbf{P}_r(t)$ are the Bézier curves with control points

$$\mathbf{P}_0,\ \mathbf{P}_0^1, \dots, \mathbf{P}_0^n; \quad \text{and} \quad \mathbf{P}_0^n, \mathbf{P}_1^{n-1}, \dots, \mathbf{P}_{n-1}^1,\ \mathbf{P}_n,$$

respectively.

PROOF. Since the parameter t has a fixed value in [0, 1], we introduce a new parameter s to represent the Bézier curve determined by the first set of control points. These points are

$$\text{(27.10)} \qquad \mathbf{P}_0, \qquad ((1-t)I + tE)\,\mathbf{P}_0, \qquad ((1-t)I + tE)^2\,\mathbf{P}_0, \dots.$$

Thus, applying the shift operator E to the above sequence of points gives the same result as applying the operator $(1-t)I + tE$ to the sequence $\mathbf{P}_0,\ \mathbf{P}_1, \dots$. Use (27.2) with the control polygon (27.10) to define $\mathbf{P}_l(s)$. Replacing the sequence (27.10) by $\mathbf{P}_0,\ \mathbf{P}_1, \dots$ and the operator E by $(1-t)I + tE$, we get

$$\mathbf{P}_l(s) = ((1-s)I + s((1-t)I + tE))^n\,\mathbf{P}_0 = ((1-st)I + stE)^n\,\mathbf{P}_0 = \mathbf{P}(st).$$

As s goes from 0 to 1, $\mathbf{P}(st)$ describes the portion of the original Bézier curve between $\mathbf{P}(0)$ and $\mathbf{P}(t)$. Thus the statement about the left half is proved, and the one about the right half follows by symmetry.

In Fig. 27.5, which shows the de Casteljau construction, we can also see the curves $\mathbf{P}_l(t)$ and $\mathbf{P}_r(t)$ and their control polygons.

The subdivision theorem is useful in designing with Bézier curves. It enables designers to isolate a part of a Bézier curve with which they are already satisfied, and continue modifying the rest of it.

As a theoretical application we note the following.

THEOREM 27.1. *The tangent to the Bézier curve at the point $\mathbf{P}(t)$ is the last line segment $\mathbf{P}_0^{n-1}\mathbf{P}_1^{n-1}$ constructed in the de Casteljau algorithm.*

PROOF. We know that the last segment of the control polygon of a Bézier curve is the tangent at the endpoint. The last segment of the control polygon of the left half of our curve is $\mathbf{P}_0^{n-1}\mathbf{P}(t)$, and this is part of the segment $\mathbf{P}_0^{n-1}\mathbf{P}_1^{n-1}$.

We can apply degree elevation to a Bézier curve in the same way as we applied it to a scalar Bernstein polynomial. Suppose that a given Bézier curve of degree n is determined by the control points $\mathbf{P}_0, \mathbf{P}_1, \ldots, \mathbf{P}_n$. We recall that degree elevation increases the number of vertices of the control polygon in such a way that the Bézier curve remains the same. The $n+2$ control points of the degree elevated polygon are

$$(27.11)\quad \mathbf{Q}_i = \frac{i}{n+1}\mathbf{P}_{i-1} + \left(1 - \frac{i}{n+1}\right)\mathbf{P}_i, \qquad i = 0, 1, \ldots, n+1.$$

Again, $\mathbf{P}_{-1}$ and $\mathbf{P}_{n+1}$ can be defined arbitrarily, since they have coefficient 0 in the formula. The operator B^{n+1} applied to the sequence (27.11) gives the same Bézier curve as B^n applied to $\mathbf{P}_0, \mathbf{P}_1, \ldots, \mathbf{P}_n$. The vertices $\mathbf{Q}_i$ of the elevated control polygon lie on the original control polygon, see Fig. 27.6. Formula (27.11) represents linear interpolations between consecutive terms $\mathbf{P}_{i-1}$ and $\mathbf{P}_i$ of the sequence $\mathbf{P}_0, \mathbf{P}_1, \ldots, \mathbf{P}_n$; the ratio $1 - \frac{i}{n+1}$ decreases at a uniform rate as we go from side to side of the polygon.

Degree elevation defines a transformation taking an n-gon to an $(n+1)$-gon. Obviously the new control polygon is contained in the convex hull of the old one, and therefore the convex hull of the new polygon is also contained in that of the old one. The process of degree elevation can be repeated over and over again, and a sequence of polygons is obtained, see Fig. 27.6. One can prove the following theorem in the same way as in the scalar case:

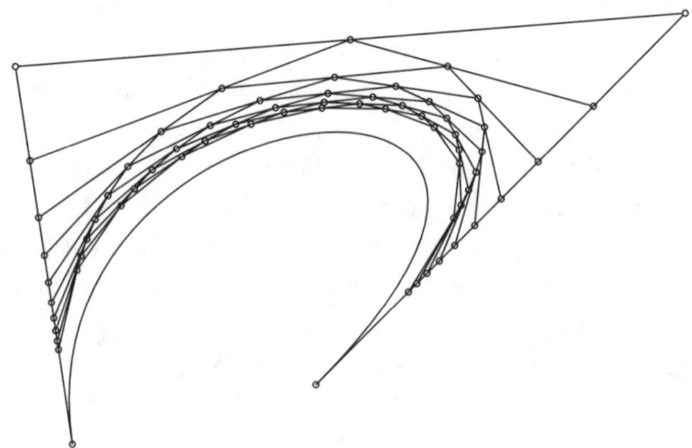

Figure 27.6. Repeated degree elevation: all polygons define the same curve.

THEOREM 27.2. *The control polygons generated by repeated degree elevation converge to the Bézier curve that each of them defines.*

Figure 27.6 gives an indication that the convergence is quite slow. Our previous work indicates that the magnitude of the gap between the curve and the polygon with n sides is of order $1/n$.

We mention several applications of degree elevation.

1. Since the elevated control polygon is closer to the Bézier curve than its original control polygon, it gives us better control of the curve if we want to modify it.
2. Since the elevated control polygon has one more vertex than the original one, the designer has one more degree of freedom in manipulating the curve.
3. Several algorithms that produce surfaces from control polygons of curves require that the polygons have the same number of vertices. We can use degree elevation to raise the number of vertices in some of the control polygons to achieve equality.

To close this chapter, we show that the planar Bézier curve possesses the convexity preserving property. If a continuous curve is a portion of the boundary of a planar convex region, the curve is called *convex*. We have to admit that this use of the word is different from the definition we used in the last chapter. There a curve was called convex if it was the graph of a convex function. That definition is not appropriate when our primary interest is in the curve as a geometric object, rather than the function it represents because it involves not only the shape of the curve but also its orientation with respect to the coordinate system. A curve which is convex in the sense of the last chapter is also convex according to the definition we are using here, but not conversely. In particular, a control polygon in a plane is convex if it is part of the boundary of a convex region. We have the following theorem:

THEOREM 27.3. *A planar convex control polygon produces a convex Bézier curve.*

PROOF. Joining the endpoints of the given control polygon, we obtain a convex closed polygon. By Theorem 27.2 a Bézier curve can be obtained from its control polygon as the limit of an infinite sequence of degree elevations. In this case degree elevation is geometrically equivalent to a process of cutting some corners from the closed convex polygon, i.e. we replace a part of the polygon by the chord connecting the two endpoints of the part. The result of such an operation is again a convex polygon. It is easy to see that even the limit of an infinite sequence of corner-cuttings is convex, and this completes the proof.

This theorem is due to D. Y. Liu (see [Liu '82]).

Exercises

27.1 A point on a curve $\mathbf{P}(t)$ at which $\mathbf{P}'(t) = 0$ and $\mathbf{P}''(t) \neq 0$ is a cusp pointing in the direction $-\mathbf{P}''(t)$. Verify that the curve in Fig. 27.3 with $s = 2$ has a cusp pointing in a direction bisecting the angle between the negative x and y axes.

27.2 Verify that the curve in Fig. 27.3 with $s = 2/3$ is a parabola.

27.3 Find $\mathbf{P}'(1)$ for the cubic Bézier curve for which

$$\mathbf{P}(0) = (0, 0), \qquad \mathbf{P}(1) = (3, 0), \qquad \mathbf{P}'(0) = (0, 3), \qquad \mathbf{P}''(0) = (12, -24).$$

27.4 Show that a pair of polynomials $x(t)$, $y(t)$ can not represent an arc of a circle exactly. Find a pair of rational functions which does represent an arc of a circle.

27.5 A degree two Bézier curve is elevated twice to produce a degree four Bézier curve with the control points

$$\mathbf{P}_0 = (0, 0), \qquad \mathbf{P}_1 = (3, 3), \qquad \mathbf{P}_2 = (5, 4), \qquad \mathbf{P}_3 = (6, 3), \qquad \mathbf{P}_4 = (6, 0).$$

What are the control points $\mathbf{Q}_0$, $\mathbf{Q}_1$, $\mathbf{Q}_2$ of the original curve?

CHAPTER TWENTY-EIGHT

Cubic Interpolatory Splines

The following interpolation problem often arises in both engineering and mathematics: draw a smooth curve which passes through $n + 1$ points

$$(x_i, y_i), \quad i = 0, 1, \ldots, n \tag{28.1}$$

in the plane. For instance, ship designers in the 19th century had to do this all the time. Draftsmen used, and still use, long thin elastic strips of wood or some other material to form a smooth curve passing through specified points. These strips, or *splines*, are held in place by lead weights called "ducks" at the specified points, see Fig. 28.1. For theoretical purposes we asume there is no friction between the spline and the ducks. The resulting curves usually have a pleasing appearance in addition to being smooth.

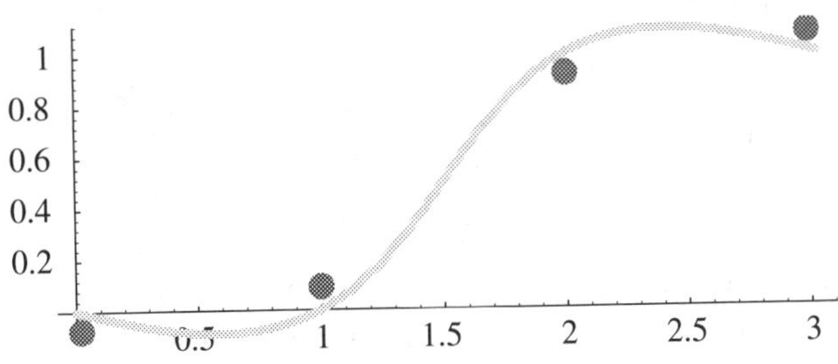

Figure 28.1. Ducks on a spline.

One way to find a smooth curve through the given points is to compute the shape of an elastic spline through them. However, this is quite complicated. Moreover, if the data points are scattered too much, frictionless ducks can not force a spline to go through them at all; the equations have no solution. This is a reflection of the physical situation. If we somehow force a spline to go through such a sequence of data points and then leave only frictionless ducks to hold it, the spline will balloon out between some ducks and no matter how long it is, it will pop out and get away from the ducks.

The Lagrange interpolation polynomials of Chapter 17 are simple to compute for any given sequence of points (28.1). However, if we have more than a handful

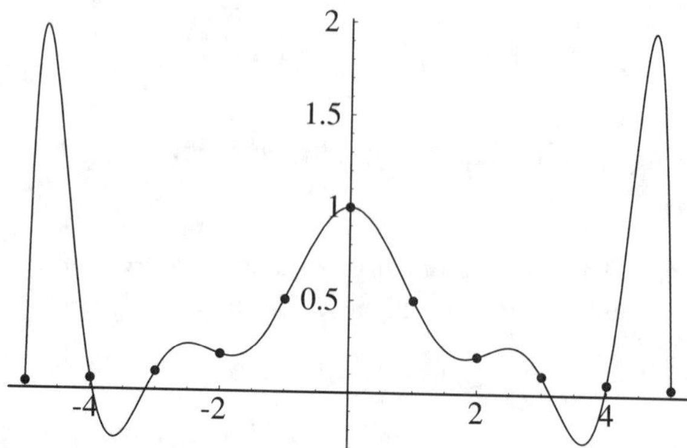

Figure 28.2. Oscillating interpolation polynomial.

of points the Lagrange polynomials tend to *oscillate*. For quite reasonable data points, the polynomial interpolant exhibits wild wiggles that are not inherent in the data. An example of this is shown in Fig. 28.2 which shows the graph of the polynomial of degree 10 through the 11 marked points. The points lie on the curve $y = 1/(1 + x^2)$. One may say that polynomial interpolation is not shape preserving.

Cubic splines are smooth curves through the points (28.1). They are mathematically simpler than the shapes of elastic splines. We are going to discuss them in detail. They could be discussed without reference to elastic splines but the relation between the two is instructive. We are going to present briefly the equations of elastic splines. This discussion involves the moment of a force. The reader who is not familiar with moments can skip this part and still follow the rest of the chapter.

If the deformation is not too great, the curvature of the elastic spline at a point P is proportional to the *bending moment* at that point; this is the sum of the moments with respect to P of all the forces acting on one of the two parts into which P divides the spline. (This must be equal and opposite to the sum of the moments acting on the other part of the spline, since otherwise the spline would move.) If the deformed curve can be written in the form $y = y(x)$, we can write:

$$\kappa(x) = \frac{y''}{(1 + y'^2)^{3/2}} = \frac{M(x, y)}{E}, \tag{28.2}$$

where $\kappa(x)$ is the curvature of the spline with a sign that is required in the present discussion, E is a measure of the rigidity of the spline and $M(x, y)$ is the sum of the moments of the forces acting on the part of the beam to the right of the point (x, y). The bending moment $M(x, y)$ is a linear function of x and y as long as we do not cross a point where a force acts on the spline, i.e., a duck. In other words, $M(x, y) = a_k x + b_k y + c_k$ on the segment of the spline between the kth and the $k + 1$st ducks, where a_k, b_k and c_k are constants. $M(x, y)$ is not a linear function of x alone even between consecutive ducks because y is not a linear function of x.

The definition (28.2) of κ gives a signed quantity. The sign is positive if, moving along the curve in the direction of increasing x coordinate, the curve bends to the left. The sign of $\kappa(s)$ thus depends not just on the shape of the curve but also on the way it is placed in the coordinate system. For this reason the customary definition of the curvature κ is the absolute value of the quantity we defined. However, in the present context the sign is useful.

Our intuition tells us that an elastic spline is curved as little as possible, given that it has to go through the points (28.1). One can make this mathematically precise as follows. Let V be the work required to bend a straight elastic rod into a given shape. One can show that

$$V = \frac{1}{2} E \int \kappa(s)^2 \, ds.$$

A shape for which V is smaller than for all other curves close to it through the given points is an equilibrium shape for a spline. One can show that if there is no friction between the ducks and the spline, the spline will assume such shape, see Fig. 28.1. Our discussion of the elastic spline does not amount to a rigorous mathematical derivation, so we do not call the following summary of our result a theorem

The shape of an elastic spline is the "straightest" curve through a given set of points in the sense that it minimizes the integral of the square of the curvature, $\int \kappa(s)^2 \, ds$.

The exact solution of the nonlinear differential equation (28.2) cannot be expressed in terms of elementary functions, and it is quite laborious to satisfy the condition that the curve should go through the data points. The mathematics becomes simpler in the case of *small deflections*, when all the data points are close to the x-axis and $|y'|$ and $|y|$ are small compared to 1. Then y'^2 is even smaller, and we neglect it. The equation (28.2) can be approximately replaced by the much simpler equation

$$y'' = M(x)/E \tag{28.3}$$

where $M(x)$ is linear between ducks and continuous. Thus *if the deflections are small, the curve $y = y(x)$ is approximately a cubic curve between consecutive ducks and it has a continuous second derivative everywhere, even at ducks.*

We relegate the last statement and the physics to the backs of our minds and begin a discussion of the mathematics of cubic splines. We want to interpolate the data points (28.1) with n different cubic polynomials which form a *piecewise* polynomial function $S(x)$ satisfying:

1. $S(x_i) = y_i$ for $i = 0, 1, \ldots, n$.
2. $S(x)$ is a cubic polynomial in each interval $[x_{i-1}, x_i]$ $i = 1, 2, \ldots, n$.
3. $S(x)$, $S'(x)$, and $S''(x)$ are continuous functions on $[x_0, x_n]$.

Let $m_i = S'(x_i)$ for $i = 0, 1, \ldots, n$. On the interval $[x_{i-1}, x_i]$, $S(x)$ is a cubic polynomial with specified function values y_{i-1}, y_i and first derivatives m_{i-1}, m_i at the endpoints. Consider first the case where $x_0 = 0$ and $x_1 = 1$. In this case it is particularly simple to express $S(x)$ in terms of these four given values y_{i-1}, y_i, m_{i-1}, m_i by using the *cubic Hermite basis*. This basis consists of four cubic polynomials $F_0(x)$, $F_1(x)$, $G_0(x)$ and $G_1(x)$ such that one of the four given data is 1 and the others are 0 for each of them:

$$\begin{array}{llll} F_0(0) = 1, & F_0'(0) = 0, & F_0(1) = 0, & F_0'(1) = 0; \\ G_0(0) = 0, & G_0'(0) = 1, & G_0(1) = 0, & G_0'(1) = 0; \\ F_1(0) = 0, & F_1'(0) = 0, & F_1(1) = 1, & F_1'(1) = 0; \\ G_1(0) = 0, & G_1'(0) = 0, & G_1(1) = 0, & G_1'(1) = 1. \end{array}$$

We can construct a cubic polynomial with any given values and derivatives at $x = 0$ and $x = 1$ by multiplying the appropriate Hermite basis polynomials by the given data and adding.

The expressions for the cubic Hermite interpolation polynomials could be found by elementary algebra, but we prefer to use a geometric argument using control polygons of Bernstein polynomials. Remember that at the endpoints $x = 0$ and $x = 1$ the value and the slope of a Bernstein polynomial is the same as the value and the slope of the control polygon. Hence the control polygon of $F_0(x)$ goes through the point $(0, 1)$ and the segment ending there is horizontal.

The other endpoint of the polygon is $(1, 0)$ and the segment ending there is also horizontal. The two interior vertices of the control polygon of $F_0(x)$ have abscissas $1/3$ and $2/3$, so they are $(1/3,\ 1)$ and $(2/3,\ 0)$. Hence $F_0(x) = B_0^3(x) + B_1^3(x)$.

The control polygon of $G_0(x)$ can be determined in a similar way. The curve goes through $(0, 0)$ and has slope 1 there, so the leftmost segment of the control polygon starts at the origin and has slope 1. Hence the second vertex of the polygon is $(1/3,\ 1/3)$ and the other two vertices are again $(2/3,\ 0)$ and $(1, 0)$, so that $G_0(x) = \frac{1}{3}B_1^3(x)$. We show these two configurations in Fig. 28.3. The

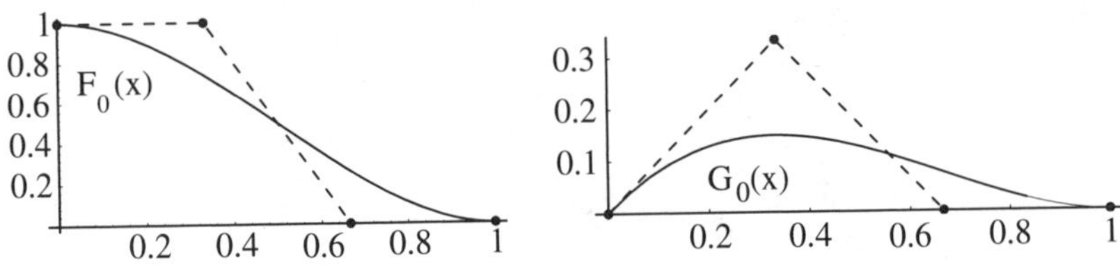

Figure 28.3. $F_0(x)$ and $G_0(x)$.

polynomials $F_1(x)$ and $G_1(x)$ can be determined similarly. We get

$$\text{(28.4)}\qquad \begin{aligned} F_0(x) &= 1 - 3x^2 + 2x^3, & F_1(x) &= 3x^2 - 2x^3, \\ G_0(x) &= x - 2x^2 + x^3, & G_1(x) &= -x^2 + x^3. \end{aligned}$$

We introduce the notation $h_i = x_i - x_{i-1},\ i = 1, 2, \ldots, n,$ for the lengths of the x-intervals. Then it is easy to see that the function $S(x)$ given by the following expression is a cubic polynomial with the required values and derivatives at the endpoints of the interval $[x_{i-1}, x_i]$:

$$\text{(28.5)}\qquad \begin{aligned} S(x) = {} & y_{i-1} F_0\left(\frac{x - x_{i-1}}{h_i}\right) + y_i F_1\left(\frac{x - x_{i-1}}{h_i}\right) \\ & + m_{i-1} h_i G_0\left(\frac{x - x_{i-1}}{h_i}\right) + m_i h_i G_1\left(\frac{x - x_{i-1}}{h_i}\right) \end{aligned}$$

for $x \in [x_{i-1}, x_i]$. At each point $x_i,\ i = 1, \ldots, n-1,$ the two polynomial segments joined at x_i have the same value y_i and the same derivative m_i. Hence the function $S(x)$ defined above interpolates the $n+1$ data points and is continuously differentiable in $[a, b]$. Next we derive the conditions which the $m_i,\ i = 0, 1, \ldots, n,$ have to satisfy in order for $S(x)$ to have a continuous second derivative.

The function $S(x)$ is twice continuously differentiable in $[a, b]$ if and only if for each $i,\ i = 1, \ldots, n-1$, the two polynomials joined at x_i have the same second derivative there. The second derivative at x_i of the polynomial representing $S(x)$ on $[x_{i-1},\ x_i]$ is

$$\frac{y_{i-1}}{h_i^2} F_0''(1) + \frac{y_i}{h_i^2} F_1''(1) + \frac{m_{i-1}}{h_i} G_0''(1) + \frac{m_i}{h_i} G_1''(1).$$

Since

$$\begin{aligned} F_0''(x) &= -6 + 12x; & F_0''(x) &= 6 + 12x; \\ G_0''(x) &= -4 + 6x, & G_0''(x) &= -2 + 6x, \end{aligned}$$

the last expression becomes

$$-6\frac{y_i - y_{i-1}}{h_i^2} + 2\frac{m_{i-1}}{h_i} + 4\frac{m_i}{h_i}. \tag{28.6}$$

If the second derivative of the polynomial representing $S(x)$ on $[x_i, x_{i+1}]$ is evaluated at x_i, we get

$$\frac{y_i}{h_{i+1}^2}F_0''(0) + \frac{y_{i+1}}{h_{i+1}^2}F_1''(0) + \frac{m_i}{h_{i+1}}G_0''(0) + \frac{m_{i+1}}{h_{i+1}}G_1''(0),$$

i.e.,

$$6\,\frac{y_{i+1} - y_i}{h_{i+1}^2} - 4\,\frac{m_i}{h_{i+1}} - 2\,\frac{m_{i+1}}{h_{i+1}}. \tag{28.7}$$

For the second derivative of S to exist and be continuous at $x = x_i$ the quantities in (28.6) and (28.7) must be equal:

$$-6\,\frac{y_i - y_{i-1}}{h_i^2} + 2\,\frac{m_{i-1}}{h_i} + 4\,\frac{m_i}{h_i} = 6\,\frac{y_{i+1} - y_i}{h_{i+1}^2} - 4\,\frac{m_i}{h_{i+1}} - 2\,\frac{m_{i+1}}{h_{i+1}}.$$

After rearrangement and simplification, we obtain the condition

$$\begin{aligned} &\frac{1}{h_i}m_{i-1} + 2\left(\frac{1}{h_i} + \frac{1}{h_{i+1}}\right)m_i + \frac{1}{h_{i+1}}m_{i+1} \\ &\qquad = 3\left(\frac{y_{i+1} - y_i}{h_{i+1}^2} + \frac{y_i - y_{i-1}}{h_i^2}\right). \end{aligned} \tag{28.8}$$

For given values $y_0, \ldots, y_n$, (28.8) is a system of $n - 1$ linear algebraic equations with $n+1$ unknowns $m_0, m_1, \ldots, m_n$. To determine the unknowns uniquely, two additional equations are needed. In general these equations, called *endpoint conditions*, specify two of the following four quantities: the first and second derivatives at x_0 and x_n. For example, if the first derivatives are specified at both endpoints, i.e., m_0 and m_n are given, then (28.8) reduces to a system of $n - 1$ linear equations in $n - 1$ unknowns $m_1, m_2, \ldots, m_{n-1}$.

In order to get the endpoint conditions which give us in some sense the smoothest cubic spline through the given points, we look again at the equations of an elastic spline. The curvature of an elastic spline at a point P is proportional to the sum of the moments of the forces acting on the spline on either side of P. As P approaches an endpoint A, the bending moment approaches 0, since the only force acting on one side of P is the force acting at A, and its arm goes to 0. Hence the curvature of an elastic spline tends to 0 as we approach an endpoint. The formula (28.2) for the curvature shows that $y'' \to 0$ as x approaches an endpoint of the spline. This suggests that if we want to make our cubic spline as straight as possible, we should impose the endpoint conditions that the second derivative is 0 at the endpoints. By (28.7) this gives

$$2m_0 + m_1 = 3\,\frac{y_1 - y_0}{h_1} \quad \text{and} \quad m_{n-1} + 2m_n = 3\,\frac{y_n - y_{n-1}}{h_n}. \tag{28.9}$$

These two equations, together with the $n-1$ equations (28.8), form a system of $n+1$ linear equations for the unkowns $(m_0, m_1, \ldots, m_n)$. The coefficent matrix of the system is

$$A = \begin{bmatrix} \frac{2}{h_1} & \frac{1}{h_1} & 0 & \ldots & 0 & 0 & 0 \\ \frac{1}{h_1} & 2\left(\frac{1}{h_1} + \frac{1}{h_2}\right) & \frac{1}{h_2} & \ldots & 0 & 0 & 0 \\ 0 & \frac{1}{h_2} & 2\left(\frac{1}{h_2} + \frac{1}{h_3}\right) & \ldots & 0 & 0 & 0 \\ \vdots & \vdots & \vdots & & \vdots & \vdots & \vdots \\ 0 & 0 & 0 & \ldots & \frac{1}{h_{n-1}} & 2\left(\frac{1}{h_{n-1}} + \frac{1}{h_n}\right) & \frac{1}{h_n} \\ 0 & 0 & 0 & \ldots & 0 & \frac{1}{h_n} & \frac{2}{h_n} \end{bmatrix}.$$

This matrix has two properties which make our system much easier to deal with than a general linear system.

1) A is *tridiagonal*, i.e., the only nonzero entries are on the main diagonal or next to it.

2) A has a *dominant diagonal*, i.e., in each row the absolute value of the diagonal entry is greater than the sum of the absolute values of the other entries. (In the present case it is equal to twice their sum.)

If the coefficient matrix of a system of linear equations has a dominant diagonal, it has a unique solution for any given right sides. There are many ways to prove this very useful fact. Conceptually, the simplest way is to relate it to the contracting mapping property which we encountered in Chapter 20, Theorem 1, and used in Chapter 21. We divide each equation by the entry on the main diagonal and take all the other terms to the right side. The system of equations, (28.8) and (28.9) in our case, assumes the form $\mathbf{m} = C\mathbf{m} + \mathbf{r} = \phi(\mathbf{m})$. Here $\mathbf{m}$ is the column vector with coordinates $m_0, m_1, \ldots, m_n$. In the present case the sum of the absolute values of the entries in each row of C is $\frac{1}{2}$. (For a general system of linear equations with a dominant diagonal, the sum would be $\leq k < 1$.) The column vector $\mathbf{r}$ is independent of the unknowns m_i. We want to show that the function ϕ has a contracting property.

This property becomes visible only if we measure the lengths of vectors by means of the *maximum norm,* which differs from the one customarily used in geometry. Let $\mathbf{x}$ be a column vector with coordinates $x_0, x_1, \ldots, x_n$. The maximum norm of $\mathbf{x}$ is defined by

$$|\mathbf{x}|_{\max} = \max_i\{|x_i|\}. \tag{28.10}$$

If the sum of the absolute values of the entries in each row of the matrix C is $\leq k$, we have for the ith coordinate of $C\mathbf{x}$

$$\begin{aligned}|c_{i0}x_0 + c_{i1}x_1 + \cdots + c_{in}x_n| &\leq (|c_{i0} + |c_{i1}| + \cdots + |c_{in}|)| \max_j(|x_j|)\\ &\leq k \max_j(|x_j|).\end{aligned}$$

Consequently,

$$|C\mathbf{x}|_{\max} \leq k|\mathbf{x}|_{\max}. \tag{28.11}$$

Now we can formulate the contracting property of the ϕ function of our system of equations: If $\mathbf{x}$, $\mathbf{y}$ are any two vectors, then

$$|\phi(\mathbf{x}) - \phi(\mathbf{y})|_{\max} = |C(\mathbf{x} - \mathbf{y})|_{\max} \leq \tfrac{1}{2}|\mathbf{x} - \mathbf{y}|_{\max}. \tag{28.12}$$

It follows that the system of equations consisting of (28.8) and (28.9) has a unique solution which can be approximated by starting with any vector and repeatedly applying ϕ to it.

The above is a simple and instructive way of seeing that the system always has a unique solution but it is not the best way to solve it in practice. For numerical computation, Gauss elimination is a much better method. Leave all the unknowns on the left side. Eliminate each entry just under the main diagonal of the augmented matrix by subtracting a multiple of the row above it, going from top to bottom. Then eliminate the entries just above the main diagonal by subtracting a multiple of the row below, going from bottom to top. We give as an example a computation of the cubic spline through the points $(0, 0)$, $(1, 0)$, $(2, 1)$, $(3, 1)$. The graph is shown in Fig. 28.4.

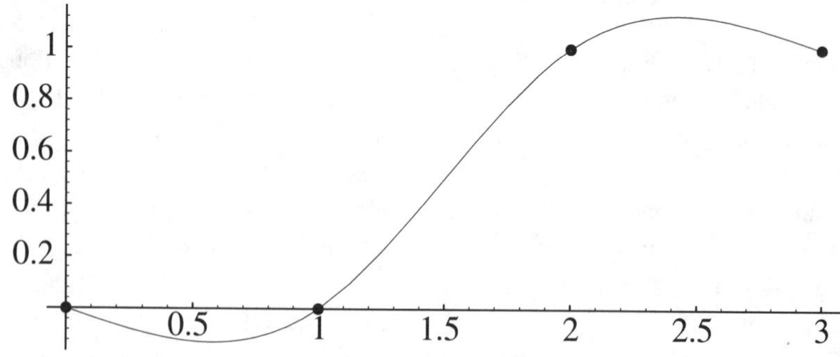

Figure 28.4. A cubic spline.

In this example, all the differences h_i are equal to 1. The system of equations (28.8) and (28.9) is, in matrix notation,

$$\begin{pmatrix} 2 & 1 & 0 & 0 \\ 1 & 4 & 1 & 0 \\ 0 & 1 & 4 & 1 \\ 0 & 0 & 1 & 2 \end{pmatrix} \begin{pmatrix} m_0 \\ m_1 \\ m_2 \\ m_3 \end{pmatrix} = \begin{pmatrix} 0 \\ 3 \\ 3 \\ 0 \end{pmatrix}.$$

The result of the downward eliminations, after first dividing each row by its diagonal entry, is

$$\begin{pmatrix} 1 & 1/2 & 0 & 0 \\ 0 & 1 & 2/7 & 0 \\ 0 & 0 & 1 & 7/26 \\ 0 & 0 & 0 & 1 \end{pmatrix} \begin{pmatrix} m_0 \\ m_1 \\ m_2 \\ m_3 \end{pmatrix} = \begin{pmatrix} 0 \\ 6/7 \\ 15/26 \\ -1/3 \end{pmatrix}.$$

The upward elimination, or back substitution in the triangular system, gives the values

$$m_0 = -1/3, \qquad m_1 = 2/3, \qquad m_2 = 2/3, \qquad m_3 = -1/3.$$

We can use (28.4) and (28.5) to write the polynomials representing $S(x)$:

$$S(x) = \begin{cases} -\frac{1}{3}x + \frac{1}{3}x^3 & \text{in } [0, 1]; \\ \frac{2}{3}(x-1) + (x-1)^2 - \frac{2}{3}(x-1)^3 & \text{in } [1, 2]; \\ 1 + \frac{2}{3}(x-2) - (x-2)^2 + \frac{1}{3}(x-2)^3 & \text{in } [2, 3]. \end{cases}$$

From now on, when we say "cubic spline", we mean the cubic spline with the end conditions (28.9). We imposed those end conditions guided by the physics of elastic splines, in the hope that these conditions would give the straightest curve through the data points. We can now state and prove a smoothness property of the cubic spline function without utilizing concepts of physics.

THEOREM 28.1. *The cubic spline function $S(x)$ passing through the points $(x_0, y_0), (x_1, y_1), \ldots, (x_n, y_n)$, $a = x_0 < x_1 < \cdots < x_n = b$, is the unique smoothest function interpolating these points, in the sense that if $f(x)$ is any other twice continuously differentiable function through these points, then*

$$\int_a^b (f''(x))^2 dx > \int_a^b (S''(x))^2 dx. \tag{28.13}$$

PROOF. Let $f(x) = g(x) + S(x)$. In terms of g the interpolatory properties of f are

$$g(x_i) = 0 \quad \text{for } i = 0, 1, \ldots, n. \tag{28.14}$$

We have

$$\int_a^b (f''(x))^2 dx = \int_a^b (g''(x) + S''(x))^2 dx$$

(28.15)
$$= \int_a^b (g''(x))^2 dx + \int_a^b (S''(x))^2 dx + 2\int_a^b g''(x)S''(x)dx.$$

We show that the last integral vanishes. We have

$$\int_a^b g''(x)S''(x)dx = \sum_{i=1}^{n} \int_{x_{i-1}}^{x_i} g''(x)S''(x)dx.$$

Integrating twice by parts, we get

$$\int_{x_{i-1}}^{x_i} g''(x)S''(x)dx = g'(x)S''(x)|_{x_{i-1}}^{x_i} - \int_{x_{i-1}}^{x_i} g'(x)S^{(3)}(x)dx$$

(28.16)
$$= g'(x)S''(x)|_{x_{i-1}}^{x_i} - g(x)S^{(3)}(x)|_{x_{i-1}}^{x_i} + \int_{x_{i-1}}^{x_i} g(x)S^{(4)}(x)dx.$$

Since $g(x_i) = 0$ and $S^{(4)} = 0$, this simplifies to

$$\int_{x_{i-1}}^{x_i} g''(x)S''(x)dx = g'(x)S''(x)|_{x_{i-1}}^{x_i}.$$

Hence

$$\int_a^b g''(x)S''(x)dx = \sum_{i=1}^{n} g'(x)S''(x)|_{x_{i-1}}^{x_i} = g'(b)S''(b) - g'(a)S''(a) = 0$$

by virtue of the endpoint conditions $S''(a) = S''(b) = 0$. Thus, by (28.16), $\int_a^b f''(x)^2 dx \geq \int_a^b S''(x)^2 dx$ with equality only if $g''(x) \equiv 0$. The last equation implies that $g(x)$ is a linear function. Since $g(a) = g(b) = 0$, it follows that $g(x) \equiv 0$, and our proof is complete.

In Fig. 28.5 the elastic spline of Fig. 28.1 and the cubic spline of Fig. 28.4 are superimposed. (Because of the symmetry, the left half of each of these curves is the same as the spline through the three points $(0, 0)$, $(1, 0)$ and $(1.5, 0.5)$.) The cubic spline is the curve which dips lower in $(0, 1)$. We compare the two curves in the light of the minimum principles presented above. We used *Mathematica* to compute a numerical solution of the differential equations of the elastic spline, draw the figures and obtain the values in the following table:

	cubic spline	elastic spline
Arclength	3.504	3.4736
$\int_0^3 f''^2(x)\,dx$	4	4.985
$\int \kappa^2(s)ds$	2.375	2.129

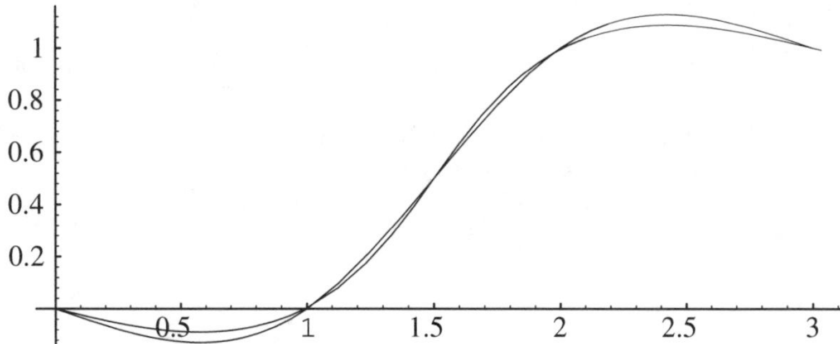

Figure 28.5. The cubic and the elastic spline through the same points.

All values except the 4 are rounded. The angle between the positive x-axis and the elastic spline at $x = 0$ is -0.235 radians. With the stiffness of the spline taken to be 1, the force acting on it at $(0,\ 0)$ has a magnitude of approximately 1.363 units and is perpendicular to the spline. The force acting at $(1,\ 0)$ has a magnitude of 2.72576 units and acts at an angle of 0.4655 radians from the y-axis. The forces acting at $(3,\ 2)$ and at $(2,\ 1)$ are equal and opposite to the above. Given these data, the differential equation for the elastic spline can be solved fairly easily with one of the simple mathematics packages, or by programming it from scratch.

At any point, we have

$$(28.17)\quad f''(x) = \kappa(s)\left(\frac{ds}{dx}\right)^3 \quad \text{and} \quad f''(x)^2\,dx = \kappa(s)^2\left(\frac{ds}{dx}\right)^5 ds.$$

Here s denotes arc length. The fifth power on the right side explains the large difference between the integral of $\kappa(s)^2$ and $f''(x)^2$, despite the fact that the two are nearly equal where the curve is close to horizontal; in addition, $\kappa(s)^2$ is integrated over the total arc length, which is a longer range than the x-interval of definition.

The formula (28.17) also explains why the cubic spline is more curved than the elastic spline where both curves are close to horizontal and less curved where they are steep. The second derivative must be fairly large somewhere in order to get from the two lower points to the two higher ones. Look at the elastic spline which minimizes the integral of $\kappa^2(s)$, and consider how one would have to change it to minimize the integral of $(f''(x))^2$ instead. Since $(f''(x))^2$ is disproportionately larger than $\kappa^2(s)$ where the curve is steep, when we minimize the integral of $(f''(x))^2$, rather than that of $\kappa^2(s)$, it is advantageous to make $|f''(x)|$ larger where the slope is smaller, so that it can be smaller where the slope and hence the factor in (28.17) is larger.

Exercises

28.1 Show that if $f(x)$ is a linear function, the cubic spline through its values at $n+1$ points is $f(x)$.

28.2 Show that if $f(x)$ is a nonlinear quadratic or cubic function, the cubic spline through any $n+1$ points of its graph is not equal to $f(x)$.

CHAPTER TWENTY-NINE

Moving Averages

In Chapter 28 we discussed cubic spline functions. We interpolated given data points with functions which were piecewise cubic polynomials and had continuous second derivatives everywhere. In this chapter we discuss one way of approximating a function $f(x)$ by a function with continuous kth derivatives and a related method of finding interpolating functions which are piecewise polynomials of degree $k+1$ and have continuous derivatives of order k.[1]

Our work is based on taking moving averages k times. Moving averages are used in attempts to discern long term trends in data which are also subject to fluctuating short term influences, such as stock prices and weather data. Alas, we are not able to predict stock prices but we will present a variety of mathematical properties of moving averages and their iterations.

For any positive number h we define the operator S_h which replaces a function f by its average over an interval with length h and center x:

$$(S_h f)(x) = \frac{1}{h}\int_{x-h/2}^{x+h/2} f(t)dt \tag{29.1}$$

If f is continuous at a point x, then for any fixed x, $\lim_{h\to 0} S_h f(x) = f(x)$, since we are taking averages over shorter and shorter intervals around x. The operator S_h is a linear operator. This is obvious but we display it to make sure the reader has it in mind when we make use of it:

$$S_h(af + bg) = aS_h f + bS_h g \tag{29.2}$$

for any functions f, g and constants a, b. Another easily verified property of the operator S_h is that it leaves linear functions unchanged:

$$S_h(ax+b) = ax+b. \tag{29.3}$$

In the rest of this chapter we assume that the functions under consideration all vanish for sufficiently large negative values of x. The object of our discussion

[1] A good source of more information on this subject is [Schoenberg '73]. We should alert readers who consult that book to a difference in the notations we are using: Schoenberg's $M_i(x)$ is our $M_{i-1}(x)$. Hoschek and Lasser '93 is an excellent reference work.

is the smoothing of a function over a finite interval, and we may set the function equal to 0 outside the interval of interest even if the function in the interval of interest is given by a formula which gives nonzero values for large negative values of x. The advantage of this provision is this. We will want to consider the integral of a function. The indefinite integral of a function is not a function but a set of functions, because it has an arbitrary constant. For this reason it is simpler to use a definite integral with some fixed lower limit. It turns out that using $-\infty$ as the lower limit makes the formulas simpler than a finite limit would.

With the provision we have just made, we can rewrite (29.1) as

$$(29.1a) \qquad (S_h f)(x) = \frac{1}{h}\left(\int_{-\infty}^{x+h/2} f(t)dt - \int_{-\infty}^{x-h/2} f(t)dt\right).$$

If a function is continuous the derivative of its integral is the function itself. Hence, if f is continuous,

$$(29.4) \qquad \frac{d}{dx}(S_h f)(x) = \frac{f\left(x+\frac{1}{2}h\right) - f\left(x-\frac{1}{2}h\right)}{h}.$$

This formula tells us:

THEOREM 29.1. *If f is continuous, $S_h f$ has one more derivative than f.*

For example, if f is a piecewise linear function, i.e., the graph of f is a polygon, then f does not have a continuous derivative but $S_h f$ has a continuous derivative and $S_h^2 f$ has a continuous first and second derivative.

If f is not continuous but $|f(x)| \leq M$ then $S_h f$ is Lipschitz continuous with Lipschitz constant $2M/h$.

It is worth noting that although the Formula (29.4) for the derivative of $S_h f(x)$ involves only f, not f', $\frac{d}{dx}S_h f(x)$ approaches $f'(x)$ if h goes to 0 and f is differentiable at x.

For simplicity, we take $h = 1$ and write $S_1 = S$ for most of the rest of this chapter, unless varying the parameter h is an essential part of the topic under discussion. We can write Sf in an alternative form by means of the function

$$M_0(x) = \begin{cases} 1 & \text{if } |x| < \frac{1}{2} \\ \frac{1}{2} & \text{if } |x| = \frac{1}{2} \\ 0 & \text{if } |x| > \frac{1}{2} \end{cases}.$$

The graph of $M_0(x)$ is shown in Fig. 29.1.

For a general value of h we would use the function $M_0(x/h)/h$.

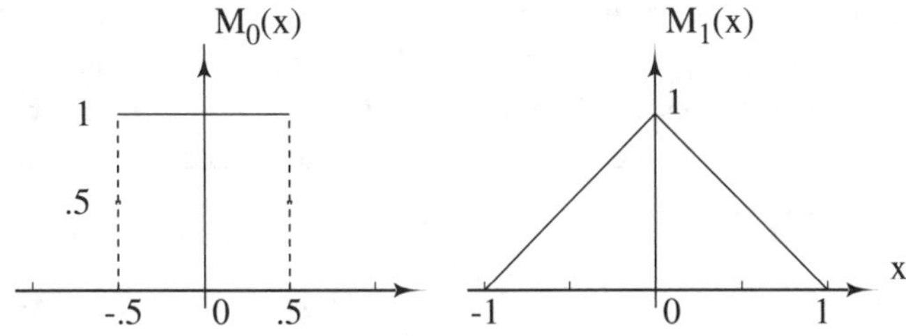

Figure 29.1. $M_0(x)$ and $M_1(x)$.

The operator S can be expressed as

$$(Sf)(x) = \int_{-\infty}^{\infty} M_0(x-t)f(t)dt. \tag{29.5}$$

The factor $M_0(x-t)$ singles out the interval over which we want to average, so we do not need to have the variable x in the limits of the integral. The infinities in the limits of integration can cause no problems in this formula even if f does not vanish for large negative t because M_0 cuts off the integrand.

Integrals of the type (29.5) are often used in mathematics. We call the right-hand side of 29.5 the *convolution* of f with the *kernel* M_0. The only occurrence of the variable x in the convolution is in the argument of the kernel M_0. A definite integral is a limit of sums, but one can understand its simpler properties most easily by thinking of it as a sum. Think of the integral (29.5) as a sum of functions of x, each of which is a copy of M_0 shifted to the right by t and multiplied by a factor $f(t)dt$. With this in mind, we see that we can apply a linear operator to a convolution by applying the operator to the kernel.

The scheme of representing a function as a convolution with a simple kernel and evaluating a linear operation by applying it to the kernel has been used with great success in many problems. As a first application of this idea, integrate (29.9) over the entire real axis and use that $\int_{-\infty}^{\infty} M_0(x-t)dx \equiv 1$. We get

$$\int_{-\infty}^{\infty} (Sf)(x)dx = \int_{-\infty}^{\infty} f(x)dx. \tag{29.6}$$

Intuitively, one would expect this because Sf is obtained by "smearing out" the values of f, but the total is not changed. One can argue more rigorously as follows.

Let $f(x)$ have compact support, i.e., suppose it vanishes outside a finite interval. We have

$$\int_{-\infty}^{\infty} (Sf)(x)dx = \int_{-\infty}^{\infty} \left(\int_{-\infty}^{\infty} M_0(x-t)f(t)dt \right) dx. \tag{29.7}$$

The function $M_0(x-t)f(t)$ is bounded and vanishes outside a finite region in the (x,t) plane. A theorem in 2-variable calculus says that the repeated integral in (29.7) is equal to the integral of the function $M_0(x-t)f(t)$ over the (x,t) plane, and this integral can also be evaluated as a repeated integral with the integral over x inside. Hence

$$\int_{-\infty}^{\infty}(Sf)(x)dx = \int_{-\infty}^{\infty}\left(\int_{-\infty}^{\infty} M_0(x-t)f(t)\,dx\right)dt = \int_{-\infty}^{\infty} f(t)dt,$$

wherein the last step we used $\int_{-\infty}^{\infty} M_0(x-t)\,dx = 1$. This completes the proof of (29.6).

In a similar way we can obtain from (29.5) the integral formula for the iterated averages:

$$(S^2 f)(x) = \int_{-\infty}^{\infty}(SM_0)(x-t)f(t)dt \tag{29.8}$$

and

$$(S^k f)(x) = \int_{-\infty}^{\infty}(S^k M_0)(x-t)f(t)dt. \tag{29.9}$$

We introduce the notation $M_k(x) = (S^k M_0)(x)$ for the kernel function in (29.9).

The functions $M_k(x)$ have the following meaning in probability theory. $M_0(x)$ is the probability density function of a random variable X which is uniformly distributed in the interval $(-\frac{1}{2}, \frac{1}{2})$. $M_k(x)$ is the density of the sum of $k+1$ independent evaluations of X.

We are going to derive a formula for $M_k(x)$. Before we get into the general case, we compute $M_1(x)$ from the definition and use it to approximate polygonal functions by smooth ones. We have

$$M_1(x) = \int_{-\infty}^{\infty} M_0(x-t)M_0(t)dt = \begin{cases} 1-|x| & \text{if } |x| \le 1 \\ 0 & \text{if } |x| \ge 1. \end{cases}$$

A simple way to arrive at the value of the above integral is to note that the area under the graph of the integrand is the overlap of the unit squares with centers $(x, \frac{1}{2})$ and $(0, \frac{1}{2})$. The graph of $M_1(x)$ is shown in Fig. 29.1.

Approximation of $f(x)$ by the smooth functions $S_h^2 f(x)$

We are going to use the theory developed so far to approximate a Lipschitz continuous function $f(x)$, with Lipschitz constant L by functions $g_h(x) = S_h^2 f(x)$ with continuous second derivatives. Here h is a parameter. We assume that $f(x)$ is defined for all real values of x.

There are two quantities we are trying to make small in such an approximation: the closeness of the approximation, which we shall measure by $\max(|f(x) - g_h(x)|)$, and the smoothness of $g_h(x)$ which we shall measure by $\max(|g_h''(x)|)$. An easily visualized case is when the graph of $f(x)$ is a polygon. The simplest way to approximate such a function by a function with continuous second derivatives is to replace the corners by smooth curves. The operator S_h^2 relieves us of specifying such a construction, and it works equally well for more general types of Lipschitz continuous functions.

We have, using the convolution form of the operator S_h^2 and the fact that the integral of the kernel is 1,

$$
\begin{aligned}
|g_h(x) - f(x)| &= \left| \int_{-\infty}^{\infty} \frac{1}{h} M_1\left(\frac{x-t}{h}\right) (f(t) - f(x))\, dt \right| \\
&\leq \int_{-\infty}^{\infty} \frac{1}{h} M_1\left(\frac{x-t}{h}\right) L|t-x|\, dt \qquad (29.10) \\
&= \frac{L}{h} \int_{-h}^{h} M_1(u)|u|\, du = \frac{L}{3} h.
\end{aligned}
$$

Applying (29.4) twice, we get

$$
(29.11) \qquad |g_h''(x)| = \left| \frac{1}{h} \left(\frac{f(x+h) - f(x)}{h} - \frac{f(x) - f(x-h)}{h} \right) \right| \leq \frac{2L}{h}.
$$

Thus we get

THEOREM 29.2. *Let $f(x)$ be a Lipschitz continuous function with Lipschitz constant L. Let $h > 0$ and $g_h = S_h^2 f$. Then*

$$
(29.12) \qquad |f(x) - g_h(x)| \leq \frac{Lh}{3} \quad \textit{and} \quad |g_h''(x)| \leq \frac{2L}{h}.
$$

This theorem can be used to improve the estimate we had for the approximation of Lipschitz continuous functions by Bernstein polynomials, see the exercises.

Basic properties of $M_k(x)$

THEOREM 29.3.

1. $M_k(x) \begin{cases} > 0 & \textit{if } |x| < \frac{k+1}{2} \\ = 0 & \textit{if } |x| > \frac{k+1}{2} \end{cases}$
2. *$M_k(x)$ is an even function.*
3. *For $k \geq 2$, $M_k(x)$ has continuous derivatives of orders up to and including $k-1$.*

4. *$M_k(x)$ is increasing in the interval* $(-\frac{k+1}{2}, 0)$ *and decreasing in* $(0, \frac{k+1}{2})$.
5.

$$\int_{-\infty}^{\infty} M_k(x)dx = 1.$$

These properties clearly hold for $k=0$ and $k=1$. From the definition $M_{k+1}=SM_k$ of M_{k+1} as a moving average of M_k the set of properties can easily be extended from k to $k+1$. We leave 1–3 to the reader and show the derivation for 4 and 5.

We have from the definition of S or from (29.1a): $M'_{k+1}(x) = M_k(x+\frac{1}{2}) - M_k(x-\frac{1}{2})$. Since $M_k(x)$ is even we get $M'_{k+1}(x) = M_k(|x+\frac{1}{2}|) - M_k(|x-\frac{1}{2}|)$. The function $M_k(x)$ is decreasing in $x>0$. Since $|x+\frac{1}{2}| > |x-\frac{1}{2}|$ for $x>0$, we get that $M'_{k+1}(x) \leq 0$ there, so $M_{k+1}(x)$ is decreasing in $x>0$. We can prove the other half of assertion 4 similarly.

Property 5 can be derived by applying (29.6) repeatedly.

We introduce the *central difference* operator

$$(\delta_h f)(x) = f\left(x+\frac{h}{2}\right) - f\left(x-\frac{h}{2}\right) \tag{29.13}$$

and the shift operators for functions:

$$(E^h f)(x) = f(x+h) \tag{29.14}$$

We put the h in the position of an exponent because all the laws of exponents are satisfied, as the reader can easily verify. The central difference operator is a difference of two shift operators:

$$\delta_h = E^{h/2} - E^{-h/2} \tag{29.15}$$

The fact that the operators E^h and $\int_{-\infty}^{x}$ commute is pretty obvious, but we verify it in detail:

$$E^h \int_{-\infty}^{x} f(t)dt = \int_{-\infty}^{x+h} f(t)dt = \int_{-\infty}^{x} f(t+h)dt = \int_{-\infty}^{x} (E^h f)(t)dt$$

If we had chosen a finite lower limit for the integral, say 0, then the operators would not commute.

Since the central difference operator δ_h is a difference of two shifts, it too commutes with $\int_{-\infty}^{x}$:

$$S_h = \delta_h \int_{-\infty}^{x} = \int_{-\infty}^{x} \delta_h. \tag{29.16}$$

By (29.1), it is easy to show that if $f(x)$ is nonnegative in $(-\infty, \infty)$ then so is $S_h f(x)$; and if $f(x)$ is an even function (i.e., $f(-x)=f(x)$) then so is $S_h f(x)$.

We write $\delta_1 = \delta$ and $E^1 = E$, to simplify discussions in which the parameter is kept constant.

We are going to obtain a formula for $M_k(x)$ by applying operators to *Heaviside's*[2] unit function

$$\mathbf{1}(x) = \begin{cases} 0 & \text{if } x < 0 \\ \frac{1}{2} & \text{if } x = 0 \\ 1 & \text{if } x \geq 0. \end{cases}$$

Clearly

$$M_0(x) = (E^{1/2} - E^{-1/2})\mathbf{1}(x) = \delta\mathbf{1}(x). \tag{29.17}$$

Hence

$$M_k(x) = S^k M_0(x) = \left(\delta \int_{-\infty}\right)^k \delta\, \mathbf{1}(x) = \delta^{k+1}\left(\int_{-\infty}\right)^k \mathbf{1}(x) \tag{29.18}$$

where in the last step we made multiple use of the fact that the difference operator commutes with integration from $-\infty$.

The iterated integrals of the unit function are the *truncated power functions*

$$\left(\int_{-\infty}\right)^k \mathbf{1}(x) = \frac{1}{k!}x_+^k \quad \text{where} \quad x_+^k = \begin{cases} 0 & \text{if } x < 0; \\ x^k & \text{if } x \geq 0\,. \end{cases}$$

Substituting this in (29.18) we get

THEOREM 29.4.

$$M_k(x) = \delta^{k+1}\frac{1}{k!}x_+^k = \frac{1}{k!}\left(E^{1/2} - E^{-1/2}\right)^{k+1} x_+^k \tag{29.19}$$

$$= \frac{1}{k!}\sum_{j=0}^{k+1}(-1)^j \binom{k+1}{j}\left(x + \frac{k+1}{2} - j\right)_+^k, \tag{29.20}$$

$M_k(x)$ has up to $k-1$st continuous derivatives, but its kth derivative has jump discontinuities at

$$-\frac{k+1}{2},\ -\frac{k-1}{2}, \ldots, \frac{k-1}{2},\ \frac{k+1}{2}.$$

Let us compute $M_2(x)$. From (29.20) we have

$$M_2(x) = \tfrac{1}{2}\left((x+3/2)_+^2 - 3(x+1/2)_+^2 + 3(x-1/2)_+^2 - (x-3/2)_+^2\right).$$

---

[2]Oliver Heaviside (1850–1925) pioneered the use of operators to solve differential equations. He was content to leave a rigorous justification of his calculus of operators to others.

If we interpret the definition of the truncated power functions for the appropriate intervals, we get:

$$M_2(x) = 0 \qquad \text{for } x \le -\tfrac{3}{2};$$

$$M_2(x) = \tfrac{1}{2}\left(x+\tfrac{3}{2}\right)^2 = \tfrac{1}{2}x^2 + \tfrac{3}{2}x + \tfrac{9}{8} \qquad \text{for } -\tfrac{3}{2} \le x \le -\tfrac{1}{2};$$

$$M_2(x) = \tfrac{1}{2}\left(\left(x+\tfrac{3}{2}\right)^2 - 3\left(x+\tfrac{1}{2}\right)^2\right) = -x^2 + \tfrac{3}{4} \qquad \text{for } -\tfrac{1}{2} \le x \le \tfrac{1}{2};$$

$$M_2(x) = \tfrac{1}{2}\left(\left(x+\tfrac{3}{2}\right)^2 - 3\left(x+\tfrac{1}{2}\right)^2 + 3\left(x-\tfrac{1}{2}\right)^2\right)$$

$$= \tfrac{1}{2}x^2 - \tfrac{3}{2}x + \tfrac{9}{8} \qquad \text{for } \tfrac{1}{2} \le x \le \tfrac{3}{2};$$

$$M_2(x) \equiv 0 \qquad \text{for } \tfrac{3}{2} \le x.$$

Graphs of $M_2(x)$ and $M_3(x)$ are illustrated in Fig. 29.2.

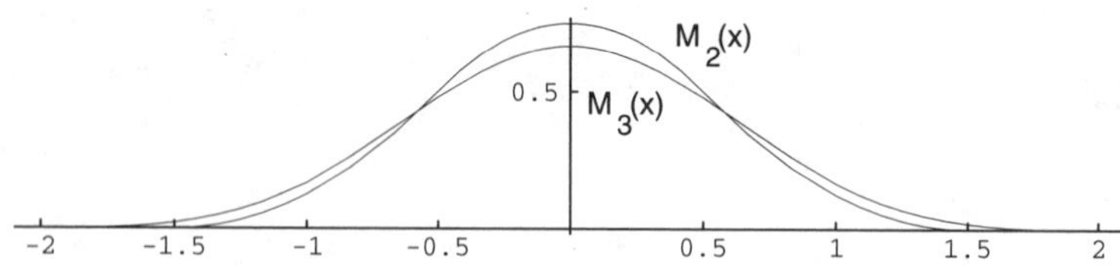

Figure 29.2. $M_2(x)$ and $M_3(x)$.

The graphs of $M_2(x)$ and $M_3(x)$ are reminiscent of the bell curve of probability theory, given by $Z(x) = \frac{1}{\sqrt{2\pi}}e^{-x^2/2}$. There is in fact a connection between the $M_k(x)$ and $Z(x)$, but we have to scale the curves $M_k(x)$ to obtain it. The relation is

$$\lim_{k\to\infty} \sqrt{\frac{k+1}{12}}\, M_k\left(\sqrt{\frac{k+1}{12}}\,x\right) = Z(x).$$

This is a consequence of a general theorem in probability theory which says that the density functions of sums of many terms, suitably scaled, approach the normal density. k does not have to be large to get reasonably close to the limit. For example, for $k = 9$ and $x = 0.3$, the quantity inside the limit is $0.376\ldots$ and $Z(0.3) = 0.381\ldots$.

We defined the function $M_k(x)$ as an integral of $M_{k-1}(x)$. Our formula 29.19 enables us to obtain an algebraic recursion formula for M_k in terms of M_{k-1} which is of some interest in itself and will be needed later on. We need the following commutation formula. For the purpose of this derivation only, let μ

denote the mean operator

$$\mu f(x) = \frac{1}{2}\left(f\left(x - \frac{1}{2}\right) + f\left(x + \frac{1}{2}\right)\right).$$

Then for any function $f(x)$,

$$\delta\,(xf(x)) = x\,\delta f(x) + \mu f(x). \tag{29.21}$$

The identity results from rearranging terms:

$$\begin{aligned}\delta(xf(x)) &= \left(x + \frac{1}{2}\right) f\left(x + \frac{1}{2}\right) - \left(x - \frac{1}{2}\right) f\left(x - \frac{1}{2}\right)\\ &= x\left(f\left(x + \frac{1}{2}\right) - f\left(x - \frac{1}{2}\right)\right) + \frac{1}{2}\left(f\left(x + \frac{1}{2}\right) + f\left(x - \frac{1}{2}\right)\right).\end{aligned}$$

We can also express (29.21) in a more concise form as an identity

$$\delta x = x\delta + \mu \tag{29.22}$$

between operators. Unfortunately the notation in (29.22) is ambiguous because here x denotes the operator of multiplication by the function x, rather than the function x itself. If we apply the operator δ to the function x we get the constant function which is identically 1. That is different from the operator δx above. We hope that, this having been pointed out, the reader will not be confused if we apply the operator equation (29.22) repeatedly:

$$\delta^{k+1} x = \delta^k x \delta + \delta^k \mu = \delta^{k-1} x \delta^2 + \delta^{k-1} \mu \delta + \delta^k \mu \quad \text{etc.} \tag{29.23}$$

We use now that the operators δ and μ are both expressions in powers of E and hence they commute. Thus the last two terms in (29.23) are equal to $\mu\delta^k$. If we move x all the way to the front, we get the operator equation

$$\delta^{k+1} x = x\delta^{k+1} + (k+1)\mu\delta^k. \tag{29.24}$$

We apply this formula to (29.19), using that $x_+^k = x\, x_+^{k-1}$. We get

$$M_k(x) = \frac{1}{k}(x\delta M_{k-1}(x) + (k+1)\mu M_{k-1}(x)),$$

i.e.,

$$\begin{aligned}M_k(x) &= \frac{\frac{1}{2}(k+1) + x}{k}\, M_{k-1}\left(x + \frac{1}{2}\right)\\ &\quad + \frac{\frac{1}{2}(k+1) - x}{k}\, M_{k-1}\left(x - \frac{1}{2}\right).\end{aligned} \tag{29.25}$$

We wish to discuss this formula briefly. The first function on the right-hand side is obtained by shifting $M_{k-1}(x)$ one half unit to the left and multiplying it with a linear factor which vanishes at the left endpoint of the support of $M_{k-1}(x)$, i.e., the left endpoint of the set where the function is different from 0. The same can be said about the second term except that left and right have to be interchanged. At each end of the support $(-k - \frac{1}{2},\ k + \frac{1}{2})$ there is an interval of length 1 where one of the two terms is 0. We know from our previous work that $M_k(x)$ is represented by a polynomial of degree k in intervals of unit length between $-(k+1)/2$ and $(k+1)/2$. This can be read off the recursion formula (29.25). We also know that $M_k(x)$ has continuous derivatives up to order $k - 1$ everywhere on the x-axis. This is not at all obvious from the formula (29.25) because the $k-1$st derivatives of each summand on the right side have jumps at $k+1$ points. On the other hand, it can be read off the closely related formula (29.19) because the right-hand side is a linear combination of translates of x_+^k, and this function does have a continuous $k - 1$st derivative.

Interpolation with the functions $M_k(x)$

The following identity is of some interest in itself, and it shows that translates of $M_k(x)$ can be used to represent the constant function, a fact we shall need in the following.

THEOREM 29.5.

$$\sum_{i=-\infty}^{\infty} M_k(x - i) \equiv 1. \tag{29.26}$$

The sum is not infinite, since for any x only the terms with $|i - x| \leq (k + 1)/2$ are different from 0*. On the other hand, we do need infinitely many terms if we want the formula to hold for all values of x.*

PROOF. For $k = 0$, (29.26) is an obvious consequence of the definition of $M_0(x)$. Then apply the operator S to the equation k times and note that a constant function is invariant under S.

A function $p(x)$ which is represented by a polynomial of degree $\leq k$ in each of the intervals $[0,\ 1],\ [1,\ 2], \ldots, [n - 1,\ n]$ and has continuous derivatives of orders $\leq k - 1$ in $(0,\ n)$ is called a *spline* function of degree k with *nodes* $0,\ 1 \ldots, n$.

THEOREM 29.6. *Let $p(x)$ be a spline function of degree k with nodes $0,\ 1 \ldots, n$. Let κ be the distance from the origin to the endpoints of the support* *of* $M_k(x)$, *i.e.*, $\kappa = \frac{1}{2}(k+1)$. *Then* $p(x)$ *can be represented in* $[0,\ n]$ *uniquely as*

$$p(x) = \sum_{i=1}^{n+k} c_i M_k(x + \kappa - i). \tag{29.27}$$

The sum is over all indices i such that the support of $M_k(x+\kappa-i)$ *overlaps the interval* $(0,\ n)$.

PROOF. We prove this by induction on the degree k.

For $k = 1$, $p(x)$ is a continuous piecewise linear function in $[0, n]$. Then

$$p(x) = p(0)M_1(x) + p(1)M_1(x-1) + \cdots + p(n)M_1(x-n) \tag{29.28}$$

because the right-hand side is a piecewise linear function which assumes the prescribed values at $x = 0, 1, \ldots, n$, and clearly the set of coefficients is uniquely determined.

$k > 1$. The function $p'(x)$ is a spline function of degree $k-1$ with nodes $0, 1, \ldots, n$. By the induction hypothesis, in the interval $[0,\ n]$, $p'(x)$ has a unique representation

$$p'(x) = a_1 M_{k-1}\big(x + \kappa - \tfrac{1}{2} - 1\big) + a_2 M_{k-1}\big(x + \kappa - \tfrac{1}{2} - 2\big) + \cdots \tag{29.29}$$

The equations the c_i in (29.27) have to satisfy for $p'(x)$ to be equal to (29.29) are obtained from

$$M_k'(x+\kappa-i) = M_{k-1}\big(x+\kappa-\tfrac{1}{2}-(i-1)\big) - M_{k-1}\big(x+\kappa-\tfrac{1}{2}-i\big).$$

We get

$$c_i - c_{i-1} = a_{i-1}. \tag{29.30}$$

The derivative of (29.27) also contains the terms $c_1 M_{k-1}(x + \kappa - \frac{1}{2})$ and $-c_{k+n} M_{k-1}(x + \kappa - \frac{1}{2} - k - n)$ but the supports of these terms are outside the interval $[0,\ n]$. Hence, (29.29) will be satisfied in $[0,\ n]$ if and only if (29.30) is satisfied. Thus the requirement that the sum (29.27) have the derivative $p'(x)$ in $[0,\ n]$ determines the coefficients $c_1, c_2, \ldots, c_{n+k}$ up to an additive constant. By (29.26), adding a constant C to all coefficients c_i adds the constant function C to the sum (29.27), and that is exactly what we need to do once we have an expression whose derivative is $p'(x)$. This completes the proof of Theorem 29.6.

Theorem 29.6 tells us that the functions $M_k(x+\kappa-i)$ whose support overlaps the interval $[0,\ n]$ form a basis of the linear space of splines of degree k with nodes $0,\ 1, \ldots,\ n$. For this reason the functions $M_k(x+\kappa-i)$ are sometimes called *B-splines*. This is short for basis-splines.

Theorem 29.6 provides a unique representation for any spline of degree k with nodes $0, 1, \dots, n$. A problem closer to what is needed in practice is to find *some* spline of degree k which will have given values at these nodes. These data give us only $n+1$ equations for the $n+k$ coefficients. As in the case of cubic splines, we need additional equations to tie up the loose ends, so to speak. In the cubic case minimizing the integral of the square of the second derivative of the approximant gave us a good way to pick a particularly smooth interpolant. We have more continuous derivatives available now. Minimizing the integral of the square of one of them would be a reasonable way of obtaining enough additional linear equations to pick one particular set of coefficients c_i from among those which produce curves going through the given points.

As an example, we revisit interpolation with a twice continuously differentiable piecewise cubic polynomial function. In the chapter on cubic splines we expressed the result in terms of Hermite polynomials. Here we shall use translates of $M_3(x)$. We have to assume that the interpolation points are equally spaced, which was not necessary in the previous chapter. (B-splines can be defined for arbitrary sets of nodes, but for that we have to refer the reader to [Hoschek and Lasser '93].) We have

$$M_3(\pm 2) = 0, \qquad M_3(\pm 1) = \tfrac{1}{6}, \qquad M_3(0) = \tfrac{2}{3}.$$

We write down Formula 29.27; we changed the numbering of the coefficients to make the indices equal to the abscissas of the peaks of the curves they multiply:

$$\begin{aligned} p(x) &= c_{-1}M_3(x+1) + c_0 M_3(x) \\ &\quad + c_1 M_3(x-1) + \cdots + c_{n+1}M_k(x-n-1). \end{aligned} \tag{29.31}$$

Substituting the prescribed values into this formula gives

$$\tfrac{1}{6}c_{i-1} + \tfrac{4}{6}c_i + \tfrac{1}{6}c_{i+1} = y_i, \quad i = 0, 1, 2, \dots, n. \tag{29.32}$$

This is a system of $n+1$ linear equations in $n+3$ unknowns. To get a particular solution we may impose two additional equations as long as they are not inconsistent with the ones we already have. The simplest condition from the algebraic point of view would be $c_{-1} = c_{n+1} = 0$. We leave it as an exercise to show that this leads to a consistent system. In the chapter on cubic splines, we saw that the condition which gives the solution which is smoothest in a certain sense is that the second derivative of the interpolant should vanish at the ends. To be able to write this down, we need $M_3''(x)$ for integer values of x. Applying (29.4) twice we get $M_3''(x) = M_1(x+1) - 2M_1(x) + M_1(x-1)$ and hence

$$M_3''(\pm 2) = 0, \qquad M_3''(\pm 1) = 1, \qquad M_3''(0) = -2.$$

Thus the remaining pair of equations for the representation of the "smoothest" cubic spline through the given points in terms of M_3 is

$$c_{-1} - 2c_0 + c_1 = 0, \qquad c_{n-1} - 2c_n + c_{n+1} = 0. \tag{29.33}$$

Functions invariant under the transformation S

Linear functions are invariant under the transformation S. There are many other invariant functions, see the exercises, but the only bounded functions which are invariant are constants. To prove this, we need the following theorem which says that if we smear out $M_0(x)$ more and more times it gets very flat.

THEOREM 29.7.

$$\max_x (M_k(x)) \to 0 \quad as \quad k \to \infty. \tag{29.34}$$

PROOF. We obtain $M_{k+1}(x)$ from $M_k(x)$ by "smearing" all its values over an interval of length 1. It is very plausible that the graph will get ever flatter and wider as the process is continued. A proof along these lines is straightforward, but the facts we have already assembled enable us to shorten it. We shall use the following simple inequality, which is far from sharp.

LEMMA. *Every function $M_k(x)$ is bounded by the inequality*

$$M_k(x) \leq \frac{1}{2|x|} \tag{29.35}$$

PROOF. We have $\int_0^x M_k(t)dt \leq \frac{1}{2}$ because the integral to infinity is $\frac{1}{2}$. To get (29.35), replace the integrand by the constant $M_k(x)$, which is smaller than $M_k(t)$ because $M_k(t)$ is decreasing on $t > 0$. Clearly any function which is nonnegative and nonincreasing on $(0, \infty)$ and has integral $\frac{1}{2}$ will satisfy this inequality, so we cannot expect it to be particularly sharp.

By Theorem 29.3, (4.), $M_k(x)$ attains its maximum at $x = 0$, so we have to prove $\lim_{k\to\infty} M_k(0) = 0$. The average of a function over an interval is $\leq$ its maximum, hence $M_{k+1}(0) \leq M_k(0)$. The sequence $M_0(0),\ M_1(0), \ldots$ is a nonincreasing sequence of positive numbers, hence it has a limit μ. We have to show $\mu = 0$.

We use (29.24). Putting $x = 0$ into the formula and using that $M_{k-1}(x)$ is an even function we get

$$M_k(0) = \frac{k+1}{k} M_{k-1}\left(\frac{1}{2}\right).$$

If we solve for $M_{k-1}(\frac{1}{2})$ and go to the limit we get $\lim_{k\to\infty} M_k(\frac{1}{2}) = \mu$. Now we put $x = \frac{1}{2}$ into (29.24) and we obtain $\lim_{k\to\infty} M_k(1) = \mu$. We can move forward by steps of $\frac{1}{2}$ to prove that the limit of $M_k(x)$ for all integer and half-integer values of x is μ. This is compatible with (29.35) only if $\mu = 0$, and the proof of Theorem 29.7 is complete.

Intuition tells us that smearing a function makes it smoother and flatter. Thus the following theorem is very plausible:

THEOREM 29.8. *A bounded function $f(x)$ such that $Sf = f$ is a constant function.*

PROOF. Let C be such that $f(x) \le C$ for all x. We have to show that for any value of a, $f(a) - f(0) = 0$. Since f is invariant under the operator S, we have

$$f(x) = \int_{-\infty}^{\infty} M_n(x-t) f(t) dt \quad \text{for all } n. \tag{29.36}$$

Hence

$$\begin{aligned} |f(a) - f(0)| &= \left| \int_{-\infty}^{\infty} (M_n(a-t) - M_n(-t)) f(t)\, dt \right| \\ &\le C \int_{-\infty}^{\infty} |M_n(a-t) - M_n(-t)|\, dt \\ &= C \int_{-\infty}^{\infty} |M_n(t) - M_n(t-a)|\, dt. \end{aligned} \tag{29.37}$$

The last quantity in (29.38) is an even function of a. Assume for definiteness that $a > 0$. Since $M_n(x)$ is an even function and is increasing in the interval $x < 0$, we have

$$|M_n(x) - M_n(x-a)| = \begin{cases} M_n(x) - M_n(x-a) & \text{in} \quad x \le a/2 \\ M_n(x-a) - M_n(x) & \text{in} \quad x \ge a/2, \end{cases} \tag{29.38}$$

Consequently

$$\int_{-\infty}^{\infty} |M_n(t) - M_n(t-a)|\, dt = 2 \int_{-a/2}^{a/2} M_n(t)\, dt \le 2a M_n(0).$$

Substituting this in (29.38) we get $|f(a) - f(0)| \le 2aCM_n(0)$. Since $M_n(0)$ can be made as small as we want by choosing n large enough, it follows that $f(a) = f(0)$.

Exercises

29.1 Give an example of a nonconstant bounded function $f(x)$ such that $Sf(x) \equiv 0$.

29.2 Show that there are nonlinear functions which are invariant under the operator S. You may need a nonzero function which has infinitely many derivatives everywhere and vanishes on an interval. The function

$$f(x) = \begin{cases} e^{-1/x} & \text{for } x > 0 \\ 0 & \text{for } x \le 0 \end{cases}$$

is such a function.

29.3 Show that a polynomial $p(x)$ of degree $\le k$ can be represented in exactly one way in the form

$$p(x) = a_0 x^k + a_1(x+1)^k \ldots + a_k(x+k)^k. \tag{29.39}$$

29.4 Show that if the function $\phi(x)$ has the properties

1. $\phi(x)$ is piecewise polynomial of degree $\le k$ on each interval $(-\frac{1}{2}(k+1)+i, -\frac{1}{2}(k+1)+i+1, i = 0, 1, \ldots, k;$
2. $\phi(x) \equiv 0$ for $|x| \ge \frac{1}{2}(k+1)$,
3. $\phi(x)$ has continuous derivatives of order $\le k-1$ everywhere,

then $\phi(x)$ is a constant multiple of $M_k(x)$.

29.5 Show that the system (29.32) together with the equations $c_1 = c_{n+3} = 0$ form a consistent system with a unique solution.

29.6 We know from our previous work on the "smoothest" cubic spline that the system of equations consisting of (29.32) and (29.33) has a unique solution. Show that this is so by using only that a linear system with a dominant diagonal has a unique solution.

29.7 Use Theorem 29.2 and Theorem 4 of Chapter 25 to prove the following improvement of Theorem 6 of Chapter 25:

If L is a Lipschitz constant for the function $f(x)$ in $0 \le x \le 1$, then

$$|B^n f(x) - f(x)| \le \sqrt{\frac{2}{3}} \frac{L}{n^{1/2}}. \tag{29.40}$$

One can check that, apart from the constant factor, this estimate is best possible in the sense that for a Lipschitz continuous function with corners, such as $|x - \frac{1}{2}|$, the difference at the corners goes to 0, as a function of n, no faster than $n^{-1/2}$.

CHAPTER THIRTY

Approximation of Surfaces

Computer-aided geometric design focuses on the representation and design of surfaces in a computer graphics environment. A popular way to represent surfaces in such an environment is with Bézier surface patches. In this chapter we extend most of the results about Bernstein polynomials and Bézier curves to surfaces.

We have seen in Fig. 27.2 how complicated curved shapes can be defined by joining together several Bézier curves. Each is really a segment of a curve, restricted to parameter values $0 \le t \le 1$. Similarly, Bézier surface patches are portions of surfaces that can be pieced together to form complicated shapes. The two most common Bézier patches are called "three-sided" (or, *triangular*) and "four-sided" (or, *rectangular*). Of course, these are not triangles and rectangles in the traditional sense, but regions of surfaces with three- or four-sided boundaries.

The conceptually simplest way of extending the univariate Bernstein polynomials to several variables is to use rectangular patches. (See [Lorentz '53].) We want to approximate a function $f(x, y)$ in the square $0 \le x \le 1,\ 0 \le y \le 1$, using values of f on the grid points $x = i/m,\ y = j/n$ where m, n are positive integers and $i = 0, 1, \ldots, m,\ j = 0, 1, \ldots, n$. We replace, for each fixed y, $f(x, y)$ as a function of x by its mth degree Bernstein approximant. We get an mth degree polynomial in x with coefficients which are functions of y. Next we replace the coefficients of the powers of x by their nth degree Bernstein approximants. The result is a polynomial of degree m in x and degree n in y.

Let us do this computation for $m = n = 1$. We want the Bernstein approximant which is a linear function of x for fixed y, with coefficients which are linear functions of y. For any fixed value of y, replacing $f(x, y)$ by its Bernstein approximant of degree 1 in x gives $f(0, y)(1 - x) + f(1, y)x$. Next we replace the coefficients $f(0, y)$ and $f(1, y)$ of the linear Bernstein basis polynomials in x by their linear Bernstein approximants $f(0, 0)(1 - y) + f(0, 1)y$ and $f(1, 0)$ $(1 - y) + f(1, 1)y$. We get

$$f(0,0)\,(1-x)(1-y) + f(0,1)\,(1-x)y + f(1,0)\,x(1-y) + f(1,1)\,xy.$$

Let us see what we get for the example $z = f(x, y) = \sqrt{x^2 + y^2}$. The graph of this function is an inverted cone with vertex at the origin and the z-axis as its axis. The Bernstein approximant of the restriction of this function to the square $0 \le x \le 1,\ 0 \le y \le 1$ is

$$(1-x)y + x(1-y) + \sqrt{2}xy = x + y - (2-\sqrt{2})xy. \tag{30.1}$$

If we apply the procedure with polynomials of degree m in x and degree n in y, the result is

$$\sum_{i=0}^{m}\sum_{j=0}^{n} f\left(\frac{i}{m}, \frac{j}{n}\right) B_i^m(x)\, B_j^n(y)\,.$$

We can extend the one-dimensional approximation estimate and the formulas for degree elevation, subdivision and the de Casteljau algorithm to the two-dimensional case without difficulty by working on one variable at a time. The formulas for derivatives as Bernstein polynomials of difference quotients also generalize to the 2-variable case.

We shall not provide more details about those properties of Bernstein polynomials which *can* easily be extended to rectangular patches. Instead, we note the rather surprising fact that two important properties—the variation-diminishing and convexity preserving properties—do *not* apply to rectangular patches.

In the one variable case we called the property that $B^n f$ has at most as many sign changes as f the variation diminishing property of the operator B^n. It is not clear what the 2-variable analogue of this might be. H. Prautzsch considered some possibilities, for instance, that the operator does not increase the number of local minima and maxima, but he did not find any that the 2-dimensional operator B^n possesses. (1985 SIAM Conference on Geometric Modeling and Robotics, Albany, N.Y.)

Convexity of a function of two variables $f(x, y)$ is defined as in the case of one variable, namely f is convex in a region R if every point of every chord is on or above the graph. (The region R should be convex, since otherwise parts of some chords would be above points in the (x, y)-plane where there is no graph.) The function $f(x, y) = \sqrt{x^2 + y^2}$, whose graph is an inverted cone, is an example of a convex function; it is intuitively obvious that a line segment connecting two points on the mantle is entirely inside the cone, although a rigorous proof requires some thought.

In contrast to the one-variable case, the Bernstein approximants of a convex function of two variables need not be convex. An example is the bilinear approximant (30.1) to the function $f(x, y) = \sqrt{x^2 + y^2}$. The bilinear approximant is a convex function of x for each fixed value of y, and also a convex function of y for each fixed value of x, since it is linear in each variable when the other variable is kept constant, and a linear function is convex. However, being a convex function of each of its variables separately does not ensure that the function is convex. Along the line $x = y$, the function (30.1) has the value $2x - (2 - \sqrt{2})x^2$, whose graph is a parabola which is concave downwards.

In the rest of this chapter we develop the 2-variable theory based on triangles. This was actually how it was first done in practice. One advantage of a triangle over a rectangle in approximating functions is that if values at the three vertices of a triangle are prescribed, there is a unique linear function which assumes those values. In contrast, there is in general no linear function which assumes prescribed values at the corners of a rectangle. Triangular grids have a similar advantage over rectangular grids if we want to interpolate or approximate with polynomials of total degree $\leq n$ for some integer n. If the triangle is divided into subtriangles with linear dimensions $1/n$ times the dimensions of the original triangle, there is exactly one polynomial of total degree n which is equal to the given function at the grid points. The number of data and the number of available terms for a Bernstein approximation match in the same way. (In the rectangular approximation the number of data and the number of available terms also match but the polynomials we construct have degree m in x and degree n in y.)

A *domain triangle* ABC is arbitrarily given in a plane. It is best to express triangular Bernstein polynomials in terms of barycentric coordinates (u, v, w), $u+v+w = 1$, with respect to the triangle ABC. We introduced barycentric cordinates in Chapter 15. We will often drop the distinction between the barycentric coordinates of a point and the point itself and write $P = (u, v, w)$.

Let $f(P)$ be a function defined on the triangle ABC. In terms of barycentric coordinates, f can be expressed as $f(u, v, w)$. Let n be any positive integer. Let

$$P_{i,j,k} = (i/n,\ j/n,\ k/n)$$

where the nonnegative integers i, j, and k satisfy $i + j + k = n$. The number of these points is $(n+1)(n+2)/2$. They lie on $n+1$ equally spaced lines, parallel to any one of the three sides. They divide the triangle ABC into congruent subtriangles. We call this the *n-triangulation* of the triangle ABC and denote it by $\mathcal{T}^n$. The points $P_{i,j,k}$, $i+j+k = n$, are called the *nodes* of the triangulation. The triangulations for $n = 1, 2, 3, 4$ are shown in Fig. 30.1. The subtriangles in $\mathcal{T}^n$ are congruent to each other and similar to the domain triangle ABC. More than half of them have the same orientation as the domain triangle. We shall call these triangles *upright.* They have the form $T^n_{i,j,k} = P_{i+1,j,k},\ P_{i,j+1,k},\ P_{i,j,k+1},\ 0 \leq i,\ j,\ k \leq n,\ i+j+k = n-1$. In words, for each index, two of the vertices lie on a line where the index has a certain value and the third lies on the line where that index is greater by 1. The number of such triangles is $n(n+1)/2$. The triangles between the upright triangles are turned by 180° with respect to the others. We call these triangles *inverted.* They have the form $T^n_{i,j,k} = P_{i-1,j,k},\ P_{i,j-1,k},\ P_{i,j,k-1},$

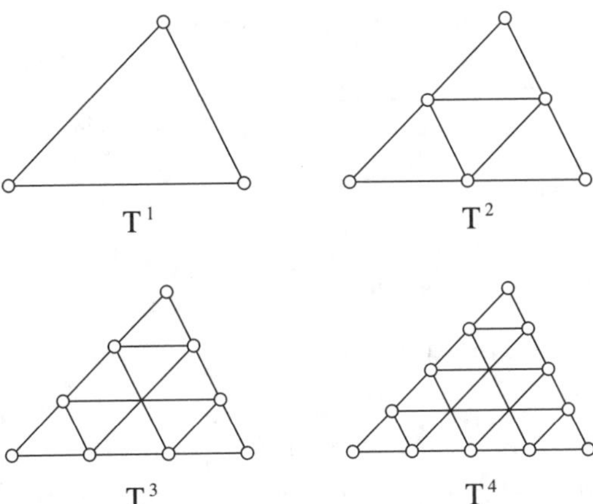

Figure 30.1. $\mathcal{T}^n$ for $n = 1, \ldots, 4$.

$1 \leq i,\ j,\ k \leq n,\ \ i + j + k = n + 1$, i.e., in the inverted triangles each index occurs twice with one value and once with a value one less. There are $n(n-1)/2$ of these. The total number of small triangles is n^2, since the sides of each are $1/n$ times those of $\triangle ABC$, and hence their areas are $1/n^2$ times the area of $\triangle ABC$. The upright triangle $T^n_{i,j,k}$ $(i + j + k = n - 1)$ is related to the node $P^{n-1}_{i,j,k}$ of $\mathcal{T}^{n-1}$ and the inverted triangle $T^n_{i,j,k}$ $(i + j + k = n + 1)$ is related to the node $P^{n+1}_{i,j,k}$ of $\mathcal{T}^{n-1}$, see Exercise 31.4.

We return to polynomial approximations of a function defined in some domain triangle. We shall use barycentric coordinates $(u, v, w),\ u + v + w = 1$. The *Bernstein basis polynomial* $B^n_{i,j,k}(P)$ is the term with $u^i v^j w^k$ in the expansion of $(u + v + w)^n$. The definition immediately gives us the identity

$$\sum_{i+j+k=n} B^n_{i,j,k}(P) = 1. \tag{30.2}$$

From the trinomial theorem, or two applications of the binomial theorem, we get

$$B^n_{i,j,k}(P) = \frac{n!}{i!j!k!} u^i v^j w^k, \quad i + j + k = n. \tag{30.3}$$

Note that we have defined $B^n_{i,j,k}(P)$ only when $i+j+k = n$. There are altogether $(n+1)(n+2)/2$ polynomials in the basis of degree n. Each of these is nonnegative over the domain triangle.

Since $u + v + w = 1$, only two of the variables u, v, w are independent. Let us consider what type of polynomials we can obtain as combinations of the $B^n_{i,j,k}(P)$ for a fixed value of n if we express everything in terms of two independent variables, say u and v. Substituting $1 - u - v$ for w, we get a combination of terms $u^i v^j$ with $i + j \leq n$.

The number of $B^n_{i,j,k}(P)$ is equal to the number of pairs i, j with $i + j \leq n$ so we would expect that all polynomials of total degree $\leq n$ in u and v can be expressed in terms of the $B^n_{i,j,k}(P)$ or equivalently, in terms of $u^i v^j w^k$ with $i + j + k = n$, provided that $u + v + w = 1$. To verify this we need to show that every function $u^i v^j$ with $i + j \leq n$ can be represented. This can indeed be done, by expanding the right-hand side of the identity $u^i v^j = u^i v^j (u + v + w)^{n-i-j}$.

The nth degree Bernstein polynomial, or Bernstein approximant, of the function $f(P)$ is defined by the formula

$$(B^n f)(P) = \sum_{i+j+k=n} f(P_{i,j,k}) B^n_{i,j,k}(P). \tag{30.4}$$

We can regard B^n as a mapping of the surface patch $f(P)$ to a patch which is the graph of a polynomial of total degree n. Both functions are over the domain triangle. Just as in the case of univariate Bernstein polynomials, we can see that B^n is a linear transformation which is also positive in the sense that if $f(P) \geq 0$ for all P in the domain triangle, then so is $(B^n f)(P)$.

We shall show a little later that if f is continuous, $B^n f$ approaches $f(P)$ as $n \to \infty$. If f has bounded second derivatives, an estimate holds:

$$\frac{|B^n f(P) - f(P)| \approx c}{n}$$

where c is some constant (see [Lorentz '53]) and the explicit formula for the Bernstein polynomial of $f(u, v, w) = u^2$ shows that $|B^n f(P) - f(P)|$ can in fact go to 0 as slowly as that. The advantage of the Bernstein polynomial patches is not that they are especially close to f but that they are very smooth and that their tangent planes also approach the tangent planes of f.

The boundary curve of the polynomial patch on the side BC of the domain triangle is obtained by substituting $u = 0$ into (30.4). Since in this case the terms associated with positive i become zero, we have

$$\begin{aligned} B^n f(0, v, w) &= \sum_{j+k=n} f\left(0, \frac{j}{n}, \frac{k}{n}\right) \frac{n!}{j!k!} v^j w^k \\ &= \sum_{j=0}^{n} f\left(0, \frac{j}{n}, \frac{n-j}{n}\right) \binom{n}{j} v^j (1-v)^{n-j}, \end{aligned} \tag{30.5}$$

a univariate Bernstein polynomial. This implies that the boundary curves of the polynomial patch are the univariate Bernstein polynomial curves determined by the boundary curves of the surface patch $f(P)$. Observe also that the triangular Bernstein polynomial interpolates the surface $f(P)$ at the three corners.

The function which is linear on each subtriangle of the n-triangulation of the domain triangle and assumes the value $f(P_{i,j,k})$ at the point $P_{i,j,k}$ is of special

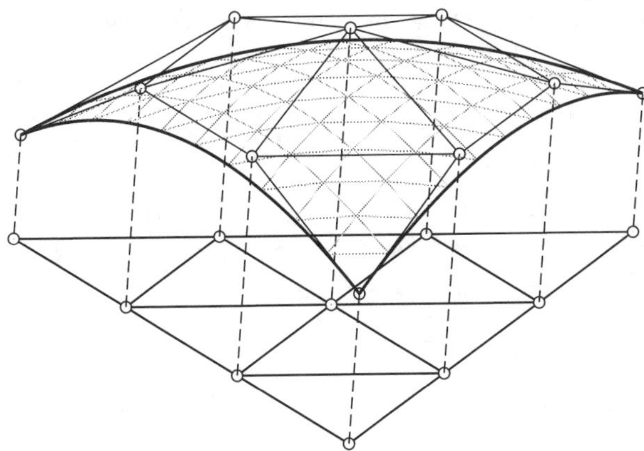

Figure 30.2. Triangular patch and its control net.

importance. This piecewise linear function is said to be the *control net* of the Bernstein triangular polynomial $B^n f(P)$. In Fig. 30.2, a control net and the corresponding triangular patch are illustrated. The figure shows how the surface patch $B^n f(P)$ is "drawn toward" its control net.

Let us direct our attention now to estimating the difference $|f(P) - B^n f(P)|$. The inequality we shall obtain, formula (30.11), is similar to what we had in the one variable case, and it will not take too much work to derive it. A rough explanation of why $B^n f(P) \to f(P)$ as $n \to \infty$ is the following. Formulas (30.4) and (30.2) tell us that $B^n f(P)$ is a weighted mean of the function values at the points $P_{i,j,k}$, with weights depending on P. As in the 1-dimensional case, most of the weight is concentrated at points $P_{i,j,k}$ close to P. Thus, if f is continuous, $B^n f(P)$ is for the most part a mean of function values close to $f(P)$ and hence it differs from $f(P)$ by a small amount. We present the details for readers who wish to see them.

We introduce the notation $u_i = \frac{i}{n}$, $v_j = \frac{j}{n}$, $w_k = \frac{k}{n}$, in analogy with the x_i notation of Chapter 25. The maximum of $B^n_{i,j,k}(u, v, w)$ over the domain triangle occurs at the point $P_{i,j,k} = (u_i, v_j, w_k)$. This can be deduced from the fact that the geometric mean is never greater than the arithmetic mean, in the same way as in the one-variable case. Moreover, we shall show that for any given u, v, w, most of the sum (30.2) comes from the terms peaking at nodes $P_{i,j,k}$ close to (u, v, w) because the sum of the terms whose maxima are far from $P_{i,j,k}$ is very small. We can transfer the estimates obtained in Chapter 25 to the triangular basis polynomials as follows.

The triangular Bernstein basis polynomials $B^n_{i,j,k}(u, v, w)$ are the terms in the expansion of $(u+v+w)^n$, where $u+v+w = 1$. The 1-variable Bernstein basis polynomials $B_i^n(u)$ are the terms of the binomial expansion of $(u + (1 - u))^n$. Hence if we collect all the terms $B^n_{i,j,k}(u, v, w)$ with a given value of i, we get $B^n(u)$:

$$\sum_{\substack{j,k \\ j+k=n-i}} B^n_{i,j,k}(u, v, w) = B_i^n(u). \tag{30.6}$$

Using this and the inequality (25.15), we get

$$\sum_{\substack{i,j,k \\ |u_i-u|\geq\delta}} B^n_{i,j,k}(u, v, w) = \sum_{\substack{i \\ |u_i-u|\geq\delta}} B_i^n(u) \leq \frac{1}{4n\delta^2}\,. \tag{30.7}$$

We have similar estimates for the sums over the terms with $|v_j - v| \geq \delta$ and the terms with $|w_k - w| \geq \delta$. Consequently,

$$\sum_{\substack{i,j,k \\ \max(|u_i-u|,|v_j-v|,|w_k-w|)\geq\delta}} B^n_{i,j,k}(u, v, w) \leq \frac{3}{4n\delta^2}\,. \tag{30.8}$$

Next we look at Lipschitz continuity and how to express it conveniently in triangular coordinates. For the purpose of obtaining estimates involving triangular patches we define the *triangular distance* between two points $P(u, v, w)$ and $P'(u', v', w')$ in a given triangular coordinate system by

$$|PP'|_t = \tfrac{1}{2}(|u - u'| + |v - v'| + |w - w'|). \tag{30.9}$$

The identity $(u - u') + (v - v') + (w - w') = 0$, which follows from the fact that the sum of both sets of coordinates is 1, can be used to express $|PP'|_t$ in a different form. If the sum of three real numbers is 0, the sum of their absolute values is twice the absolute value of the largest. Hence

$$|PP'|_t = \max(|u - u'|,\ |v - v'|,\ |w - w'|). \tag{30.9a}$$

In geometry the distance between two points is the length of the line segment connecting the points, but a city with a rectangular layout is an example when another measure of distance is more useful. In mathematics itself various kinds of distance turned out to be useful. While triangular distance does have a property (Exercise 30.8) which may seem inappropriate for a distance function, it does share with ordinary distance the following three properties

$$|PP|_t = 0; \qquad |PQ|_t = |QP|_t > 0 \quad \text{if } Q \neq P; \qquad |PQ|_t + |QR|_t \geq |PR|_t.$$

It would not help our thinking to call something a distance unless it had these properties.

A geometric interpretation of triangular distance is given in Exercise 30.2

We say the real number L_t is a *triangular Lipschitz constant* for the function $f(P)$ in a domain D if for all pairs of points P, P' in D,

$$|f(P) - f(P')| \leq L_t\, |PP'|_t. \tag{30.10}$$

THEOREM 30.1. *If L_t is a triangular Lipschitz constant for the function f in a given domain triangle, then at every point P in the triangle*

$$|f(P) - B^n f(P)| \le \frac{2L_t}{n^{1/3}}. \tag{30.11}$$

PROOF. Let $\delta > 0$ be given. For an arbitrary value of P, multiply (30.2) by $y = f(P)$ and subtract $B^n f(P)$. Using the abbreviation $y_{i,j,k} = f(P_{i,j,k})$, we get

$$\begin{aligned} y - B^n f(P) &= \sum_{i+j+k=n} (y - y_{i,j,k}) B^n_{i,j,k}(P) \\ &= \sum_{|P-P_{i,j,k}|<\delta} (y - y_{i,j,k}) B^n_{i,j,k}(P) + \sum_{|P-P_{i,j,k}|\ge\delta} (y - y_{i,j,k}) B^n_{i,j,k}(P) \qquad (30.12) \\ &= S_{\text{near}} + S_{\text{far}}. \end{aligned}$$

We find a bound for $|S_{\text{near}}|$ by the Lipschitz continuity of f : $|y - y_{i,j,k}| \le L_t |P - P_{i,j,k}|_t$:

$$|S_{\text{near}}| \le L_t \delta \sum_{i+j+k=n} B^n_{i,j,k}(P) \le L_t \delta. \tag{30.13}$$

We estimate S_{far} by using (30.8) to bound the sum of the Bernstein polynomials. For the factors $y - y_{i,j,k}$ we use the crude estimate L_t, which is true because $|P - P_{i,j,k}|_t \le 1$ since both points are in the domain triangle.

$$S_{\text{far}} \le \frac{3L_t}{4n\delta^2}. \tag{30.14}$$

These inequalities show that for any $\delta > 0$,

$$|f(P) - B^n f(P)| \le L_t \delta + \frac{3L_t}{4n\delta^2}. \tag{30.15}$$

This holds for arbitrary positive values of δ. We complete the proof of (30.11) by choosing the value $\delta = 1/n^{1/3}$, which, for our purposes, comes close enough to minimizing the sum on the right side of (30.15).

As in the 1-variable case, it is possible to get better bounds than (30.11), especially when f has bounded second derivatives, but (30.11) will suffice for our purposes.

Exercises

30.1 Let ABC be an arbitrarily given triangle. A line passing through its centroid divides the triangle into two parts. Show that the absolute value of the difference between the areas of these two parts is less than or equal to one ninth of the area of triangle ABC (a problem from the Mathematical Olympiad in Anhui Province, China, 1978).

30.2 Show that if all three sides of the domain triangle have length 1, the triangular distance $|PP'|_t$ is the least length a polygon from P to P' with sides parallel to the sides of the domain triangle can have.

30.3 Prove that if the function f depends on only one of the variables, e.g., if $f(u, v, w) = \phi(u)$, then $B^n f(u, v, w) = B^n \phi(u)$. The operator B^n on the right-hand side is the one-variable operator we defined in Chapter 25. Use this to deduce that the transformation B^n leaves constant and linear functions of u, v, w unchanged. Show also that $B^n uv = (1 - \frac{1}{n})uv$.

30.4 Show that if $B^n f(u, v, w) = B^n g(u, v, w)$ then $f_{i,j,k} = g_{i,j,k}$ for all triples i, j, k with $i + j + k = n$. Equivalently, show that if $B^n f(P)$ is identically 0 in the domain triangle then all the coefficients $f_{i,j,k}$ with $i + j + k = n$ are 0.

30.5 Prove that the function $f(x, y) = \sqrt{x^2 + y^2}$, whose graph is an inverted cone, is convex.

30.6 Take the function $f(P) = |OP| = \sqrt{x^2 + y^2}$ in the domain triangle (0, 0), (1, 0), (0, 1). Express $B^2 f$ as a function of x and y and show that, although f is convex in the domain triangle, $B^2 f$ is not convex.

30.7 Let the domain triangle have side lengths a, b, c, so that the equations of these sides in triangular coordinates are $u = 0,\ v = 0,\ w = 0$. Show that

$$\overrightarrow{CP} = u\,\overrightarrow{CA} + v\,\overrightarrow{CB} \tag{30.16}$$

with similar formulas in terms of v and w, and w and u. Deduce that if $P' = (u', v', w')$,

$$\begin{aligned} |PP'|^2 &= (u' - u)^2 b^2 + (v' - v)^2 a^2 + (u' - u)(v' - v)(a^2 + b^2 - c^2) \\ &= -a^2(v' - v)(w' - w) - b^2(u' - u)(w' - w) - c^2(u' - u)(v' - v). \end{aligned}$$

30.8 Show that if the points $P,\ Q,\ R$ lie on a line in the given order then

$$|PQ|_t + |QR|_t = |PR|_t. \tag{30.17}$$

Show also that, unlike the similar equation for ordinary distances, (30.17) does not imply that the three points are collinear.

30.9 Let $h_{\min}$ be the smallest height of the domain triangle. Prove the following inequalities between the ordinary and the "triangular" distance between two points:

$$h_{\min}|PP'|_t \ \le\ |PP'|\ \le\ \max(a, b, c)|PP'|_t. \tag{30.18}$$

Hence show that if L is a Lipschitz constant for f in the domain triangle ABC (i.e., for any two points $P,\ P'$ in the triangle, $|f(P') - f(P)| \le L|P' - P|$), then $\max(a, b, c)L$ is a triangular Lipschitz constant for f. Show also that if L_t is a triangular Lipschitz constant for f then $L_t/h_{\min}$ is an ordinary Lipschitz constant.

CHAPTER THIRTY-ONE

Properties of Triangular Patches

In this chapter we extend many of the properties of one variable Bernstein polynomials to triangular Bernstein polynomials, or Bernstein triangular patches as we tend to call them when we are thinking of the surfaces they represent.

We introduce an operator representation of triangular Bernstein polynomials. To see the similarity between the formulas in this chapter and our formulas for one-variable Bernstein polynomials, it helps to keep in mind that one can rewrite the one-variable polynomials with u and v in place of x and $1-x$, with the added condition $u+v=1$.

Let

$$f_{i,j,k} = f\left(\frac{i}{n}, \frac{j}{n}, \frac{k}{n}\right), \qquad i+j+k=n.$$

We introduce three *partial shifting operators* E_1, E_2 and E_3 as follows:

$$E_1 f_{i,j,k} = f_{i+1,j,k}, \qquad E_2 f_{i,j,k} = f_{i,j+1,k}, \qquad E_3 f_{i,j,k} = f_{i,j,k+1}.$$

We mentioned when we introduced operators acting on sequences that it is simpler and clearer to use a notation that makes it look as if the operator were acting on an element of a sequence although, strictly speaking, the shifting operator E acts on the whole sequence $\{y_i\}$ and Ey_i is shorthand for the entry, with index i, of the shifted sequence. The same applies to the operators we have just introduced. They act on three-index arrays, but our notation makes it seem that an operator acts on a single element of an array. We have to mention an additional point in the present context. Our array $f_{i,j,k}$ is defined only for triples $(i,\ j,\ k)$ for which $i+j+k=n$, where n is some fixed integer. Yet in our formulas there will be expressions of the form $f_{i,j,k}$ with $i+j+k \neq n$; indeed the definition of the operators we just gave has different sums of the indices on the two sides of the equation. However, in all our results the sum of the indices is the same, n, in all terms. Since the elements with other index sums do not enter the final result, we may assign arbitrary values, e.g., 0, to them.

In the one-variable situation we defined addition and multiplication by scalars and multiplication by other operators for the shifting operator E and the identity operator I. These operations had the associative, distributive and commutative properties, and so we were able to substitute operators into algebraic identities such as the binomial theorem. The partial shifting operators $E_1,\ E_2,\ E_3$ also have these algebraic properties. We note in particular that our operators commute: $E_iE_j = E_jE_i$ for all i and j.

We are ready to write down the definition of the triangular Bernstein polynomials in terms of operators, in analogy with the useful formula (26.1) for the one-variable case. The formula is:

$$z = B^n f(P) = (uE_1 + vE_2 + wE_3)^n f_{0,0,0}. \tag{31.1}$$

The undefined quantity $f_{0,0,0}$ which appears in this formula is an example of what we have said above: if we apply all the operators, the sum of the indices of f in the resulting terms is n, and those are the quantities which really occur in (31.1). To see that the right-hand side of (31.1) is the same as the definition (30.4) of $B^n f(u, v, w)$ in the previous chapter, expand the operator by means of the trinomial theorem.

We observed in the last chapter that on the three line segments bounding the domain triangle, $B^n f(P)$ is equal to the nth degree Bernstein polynomial formed by using the $n + 1$ values of f at equally spaced points on the boundary segment. We are now going to relate the triangular Bernstein polynomial for general values of the argument to one-variable polynomials and draw some geometric conclusions.

At the beginning of the previous chapter we introduced rectangular Bernstein approximants to a function $f(x, y)$ as follows. Think of y as fixed and replace $f(x, y)$ as a function of x by its Bernstein polynomial of degree m. We get a polynomial whose coefficients are functions of y. The rectangular Bernstein polynomials are obtained by replacing these coefficients by their Bernstein polynomials of degree n. We want to discuss how this process can be adapted to the triangular case.

We first discuss representations of points in the domain triangle by means of two parameters which can vary independently from 0 to 1. As usual, we denote the vertices (1, 0, 0), (0, 1, 0), and (0, 0, 1) by A, B, and C. Along a line parallel to BC the coordinate u is constant. At the point where the line intersects AC, $v = 0$ and where it intersects AB, $v = 1 - u$. Hence the quantity $v/(1 - u)$ goes from 0 to 1 as we move along a line $u = const$ from one side of the triangle to the other. We have

$$\frac{v}{1-u} = \frac{v}{v+w} = \frac{1}{1+w/v}$$

which shows that the parameter $\frac{v}{1-u}$ depends only on w/v. Hence the lines $\frac{v}{1-u} = const$ are the lines through A.

The two parameters $u,\ v/(1 - u)$ represent points in the domain triangle in a manner which does not retain the symmetry of the triangular representation; with only two parameters the symmetry can not be retained. Where $u = 1$, the second parameter is meaningless. Fortunately it is not needed as long as we are interested only in points in the domain triangle, since only one point, A, has $u = 1$ there.

The representation of $B^n f(u, v, w)$ in terms of 1-variable Bernstein polynomials can now be stated as follows.

THEOREM 31.1. *Let $P_1^{i,n}(x)$ be the ith degree Bernstein polynomial with $f_{n-i,0,i}, f_{n-i,1,i-1}, f_{n-i,2,i-2}, \ldots$ as the coefficients of the basis polynomials. The triangular Bernstein polynomial $B^n f(u, v, w)$ can be written in terms of the one-variable Bernstein polynomials $P_1^{i,n}(x)$ as*

$$B^n f(u, v, w) = \sum_{i=0}^{n} B_{n-i}^n(u) P_1^{i,n}\left(\frac{v}{1-u}\right). \tag{31.2}$$

We have of course similar formulas based on the subarrays of $f_{i,j,k}$ in which j or k is kept constant, with polynomials $P_2^{j,n}(\frac{w}{1-v})$ and $P_3^{k,n}(\frac{u}{1-w})$ replacing $P_1^{i,n}(\frac{v}{1-u})$ in (31.2).

PROOF. We want to collect terms with like powers of u in $B^n f(u, v, w)$. This could be done directly from the original definition, but it is even more transparent to use the operator formula (31.1). We expand $(uE_1+vE_2+wE_3)^n = (uE_1 + (vE_2 + wE_3))^n$ using the binomial theorem:

$$\begin{aligned} B^n f(u, v, w) &= \sum_{i=0}^{n} u^{n-i} E_1^{n-i}(vE_2 + wE_3)^i f_{0,0,0} \\ &= \sum_{i=0}^{n} \binom{n}{n-i} u^{n-i}(vE_2 + wE_3)^i f_{n-i,0,0}. \end{aligned} \tag{31.3}$$

We have $v + w = 1 - u$. Hence $\frac{v}{1-u} + \frac{w}{1-u} = 1$. Thus, recalling the operator formula for the one-variable Bernstein polynomial, we see that

$$\left(\frac{v}{1-u}E_2 + \left(1 - \frac{v}{1-u}\right)E_3\right)^i f_{n-i,0,0} = P_1^{i,n}\left(\frac{v}{1-u}\right).$$

Multiply this by $(1-u)^i$ and substitute in (31.3). Then the binomial coefficient and the powers of u and $1-u$ combine to form a Bernstein basis polynomial in u and Theorem 31.1 is proved.

We are in a position to investigate the geometric relationship between the triangular patch and its control net.

THEOREM 31.2 [SEDERBERG '84].

1. *If for all the values $i = 0, 1, \ldots, n-1$ the sequences*

$$f_{n-i,0,i}, \quad f_{n-i,1,i-1}, \quad f_{n-i,2,i-2}, \ldots, \quad f_{n-i,i,0} \tag{31.4}$$

are nondecreasing, then, in the domain triangle, on any line where u is constant, $B^n f(u, v, w)$ is a nondecreasing function of v.

2. *If we replace the assumption that all the sequences* (31.4) *are nondecreasing by the assumption that they are all convex, then, in the domain triangle, on all the lines where u is constant, $B^n f(u, v, w)$ is a convex function of v.*

Of course, analogous statements hold for any pair of distinct variables in place of u and v.

PROOF. We have proved in the chapters on one-variable Bernstein polynomials that if the sequence $\{f_i\}$ is increasing then $B^n f$ is increasing in $[0, 1]$, and if $\{f_i\}$ is convex, then $B^n f$ is convex in $[0, 1]$. Thus, under the first assumption, for a fixed value of u the right side of (31.2) is a sum of nondecreasing functions of v, and under the second assumption, it is the sum of convex functions of v. A sum of nondecreasing functions is nondecreasing, and a sum of convex functions is convex.

Next we discuss degree elevation in the triangular case. The triangulations $\mathcal{T}^n$ and $\mathcal{T}^{n+1}$ will both occur in our discussion. We denote the node $(\frac{i}{n}, \frac{j}{n}, \frac{k}{n})$ of $\mathcal{T}^n$ by $P^n_{i,j,k}$.

A triangular Bernstein basis polynomial of degree n can be expressed in terms of basis polynomials of degree $n + 1$ by use of the identity $u^i v^j w^k = u^i v^j w^k (u + v + w)$. In terms of Bernstein basis polynomials the identity is

$$\begin{aligned} B^n_{i,j,k}(u, v, w) &= \frac{n!}{i!j!k!} u^i v^j w^k \\ &= \frac{i+1}{n+1} B^{n+1}_{i+1,j,k} + \frac{j+1}{n+1} B^{n+1}_{i,j+1,k} + \frac{k+1}{n+1} B^{n+1}_{i,j,k+1}. \end{aligned}$$

If we substitute this into the definition (30.4) of $B^n f$ and collect the terms in which $B^{n+1}_{i,j,k}(P)$ occurs, we get (note that on the right side in the formula above, $i + j + k = n$ but in Formula (31.5) below, $i + j + k = n + 1$)

$$\begin{aligned} B^n f(P) = \sum_{i+j+k=n+1} \frac{1}{n+1} \big(i\, f\big(P^n_{i-1,j,k}\big) \\ + j\, f\big(P^n_{i,j-1,k}\big) + k\, f\big(P^n_{i,j,k-1}\big)\big) B^{n+1}_{i,j,k}(P). \end{aligned} \tag{31.5}$$

The expression in the sum (31.5) contains function values which are undefined when an index is 0, because they have a $-1/n$ as an argument. We could have avoided this by writing a more complicated formula, but the terms that are undefined and should not be there have coefficient 0, so they cause no error.

The coefficients of the Bernstein basis polynomials of degree $n + 1$ in (31.5) have a simple and useful interpretation. One can obtain the interpretation by

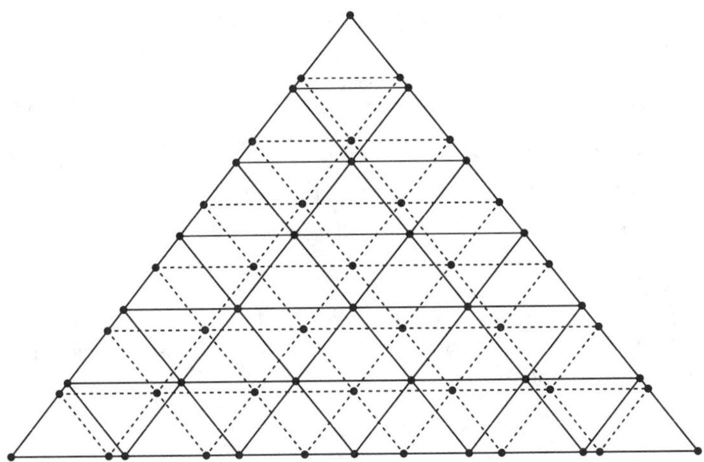

Figure 31.1. Triangulations of orders 6 and 7.

straightforward algebraic computation or by utilizing some simple results in the exercises of the last chapter. We will utilize those results. Part of the argument concerns the way $\mathcal{T}^n$ and $\mathcal{T}^{n+1}$ are related. For this it will help to view Fig. 31.1.

Consider the case when f is a linear function. Then $B^n f = B^{n+1} f = f$, see Exercise 30.3. We also ascertained (Exercise 30.4) that there is only one set of coefficients of the Bernstein basis polynomials of a given degree which produces a given Bernstein triangular patch. It follows that *the coefficient of* $B^{n+1}_{i,j,k}(P)$ *in* (31.5) *is the value of* f *at the point* $P^{n+1}_{i,j,k}$. If we apply this to the coordinate functions we get

$$(31.6) \qquad P^{n+1}_{i,j,k} = \frac{1}{n+1}\left(i\, P^n_{i-1,j,k} + j\, P^n_{i,j-1,k} + k\, P^n_{i,j,k-1}\right),$$

$i + j + k = n + 1$.

The implication for the case when f is not linear is as follows. The linear combination of the three values of f which is the coefficient of $B^{n+1}_{i,j,k}(P)$ in (31.5) is the value at the node $P^{n+1}_{i,j,k}$ of the linear function interpolating $f(u, v, w)$ at the three corners of the triangle

$$T^n_{ijk} = \triangle P^n_{i-1,j,k} P^n_{i,j-1,k} P^n_{i,j,k-1},$$

i.e., the triangle of $\mathcal{T}^n$ with sides $u = \frac{i}{n}$, $v = \frac{j}{n}$, $w = \frac{k}{n}$. (When one of i, j, k is 0, the interior of T^n_{ijk} is outside the domain triangle but in such a case the linear interpolation is only along a segment of the boundary of the domain triangle.)

We note that the coefficients of the f-values in (31.5) are nonnegative. Hence the point $P^{n+1}_{i,j,k}$ is inside or on the boundary of the triangle T^n_{ijk} over which we interpolate. We can summarize our results as follows.

THEOREM 31.3. DEGREE ELEVATION FOR TRIANGULAR PATCHES. *Let $Cn(P)$ be the control net of f corresponding to $\mathcal{T}^n$, i.e., $Cn(P) = f(P)$ at each node of $\mathcal{T}^n$ and $Cn(P)$ is a linear function on each triangle. Then the polynomial $B^{n+1}Cn$ is the same as $B^n Cn = B^n f$.*

We digress here to point out a geometric relation between successive triangulations, such as those in Fig. 31.1. Formula (31.6) says that the barycentric coordinates of $P^{n+1}_{i,j,k}$ with respect to the inverted triangle T^n_{ijk} of $\mathcal{T}^n$ are

$$\frac{i}{n+1}, \qquad \frac{j}{n+1}, \qquad \frac{k}{n+1}.$$

These are also the barycentric coordinates of $P^{n+1}_{i,j,k}$ with respect to the domain triangle. We conclude that the position of $P^{n+1}_{i,j,k}$ relative to the triangle T^n_{ijk} is the same as its position relative to the domain triangle. Consequently, T^n_{ijk} is inversely similar to the domain triangle, with center of similarity $P^{n+1}_{i,j,k}$. In particular, $P^{n+1}_{i,j,k}$ is the intersection of the lines connecting the vertices of T^n_{ijk} with the corresponding vertices of the domain triangle. We leave it as an exercise to prove theorem 31.3 and the last statement by algebraic computation, instead of using the results about linear functions and the uniqueness of the coefficients.

Degree elevation can be repeated. In each step we construct a new polyhedron whose faces are triangles with corners on the previous polyhedron. Our next objective is to prove, as we did in the one-dimensional case, that the functions representing these polyhedra converge to the common triangular patch they define. One of the facts we needed for proving this in the one-variable case was that a Lipschitz constant of a function is also a Lipschitz constant for a piecewise linear function representing the inscribed polygon resulting from the degree elevation construction. (L is a Lipschitz constant of f means: for any two points P, Q, $|f(P) - f(Q)| \le L|P - Q|$.) This is indeed rather obvious, not just for the inscribed polygons constructed in degree elevation, but for any inscribed polygon.

In the triangular case things are not quite so simple. If one uses ordinary distance in the definition of Lipschitz continuity for a function of two variables, a Lipschitz constant of a function need not be a Lipschitz constant for its control net. (In a discussion involving artificial kinds of distances, ordinary distance is referred to as *Euclidean* distance.) The reason for introducing triangular distance is that a Lipschitz constant for a function f is then also a Lipschitz constant for the control net formed using the values of f at the nodes of any triangulation of the domain triangle. Many convergence theorems in mathematics depend on devising a measure of distance which makes the problem tractable.

Before proceeding with our discussion of triangular Lipschitz continuity, we give an example when a Lipschitz constant L of a function f, defined by means of Euclidean distance, is not a Lipschitz constant of a function representing an inscribed polyhedron. Take the domain triangle $A = (0, 0)$, $B = (1, -1)$, $C = (1, 1)$ and the function $f(P) = |AP|$. For any two points P, Q, we have $|f(P)-f(Q)| = ||AP|-|AQ|| \le |PQ|$. Hence, based on Euclidean distance, 1 is a Lipschitz constant for the function $f(P)$. The linear function that interpolates the values at the three corners of the domain triangle is $B^1 f(x, y) = \sqrt{2}\,x$ and the smallest Lipschitz constant for the linear interpolant is $\sqrt{2}$. The next theorem asserts that such a thing can not happen if we use triangular distance in the definition of Lipschitz continuity and the sides of the triangles over which we interpolate are parallel to the sides of the domain triangle.

THEOREM 31.4. *Suppose $f(P)$ has triangular Lipschitz constant L_t. Let $Cn(P)$ be the function representing the control net of f corresponding to $\mathcal{T}^n$. Then L_t is also a triangular Lipschitz constant for the function $Cn(P)$.*

We formulate the first part of the proof as a lemma.

LEMMA. *Let L_t be a triangular Lipschitz constant for the function $f(P)$. Let $\triangle ABC$ be the domain triangle or, more generally, any triangle inside the domain triangle whose sides are parallel to the sides of the domain triangle. Let $l(P)$ be the linear function with the same values as f at A, B, and C. Then L_t is a triangular Lipschitz constant for the function $l(P)$.*

PROOF. We have to show that for any two points P, Q,

$$\frac{|l(P) - l(Q)|}{|PQ|_t} \le L_t. \tag{31.7}$$

The linearity of the function l implies that the quotient $|l(P) - l(Q)|/|PQ|_t$ depends only on the direction of the line through P and Q. (The language is ambiguous so let us state that here "direction" means the direction of the *undirected* line through P and Q. The quotient $|l(P)-l(Q)|/|PQ|_t$ is unchanged if P and Q are interchanged.)

If the segment PQ is parallel to one of the sides of $\triangle ABC$, say AB, then

$$\frac{|l(P) - l(Q)|}{|PQ|_t} = \frac{|l(A) - l(B)|}{|AB|_t} = \frac{|f(A) - f(B)|}{|AB|_t} \le L_t\,.$$

Consider, then, the case when PQ has any direction. We claim that there is a line segment parallel to PQ from a vertex of $\triangle ABC$ to a point on the opposite side of the triangle. To see this, draw the lines parallel to PQ through A, B, and C. For definiteness, let the line through A be the one which is between the other two. (It may also coincide with one of the others.) Then the point of intersection D of the line through A with the line BC is between B and C. We have to show that $|l(D) - l(A)|/|DA|_t \leq L_t$. Since the function l is linear and D is on the line segment BC, $l(D) - l(A)$ is between $l(B) - l(A)$ and $l(C) - l(A)$. Assume for definiteness that $|l(D) - l(A)| \leq |l(C) - l(A)|$. The last quantity is $|f(C) - f(A)|$. Because the sides of $\triangle ABC$ are parallel to the sides of the domain triangle, $|AD|_t = |AC|_t$ for all points D on the side BC. (It is this property that makes the triangular distance easier to use in the present context than the ordinary distance.) Putting our last two results together, we get $|l(D) - l(A)|/|DA|_t \leq |f(C) - f(A)|/|CA|_t \leq L_t$, and the proof of the lemma is completed.

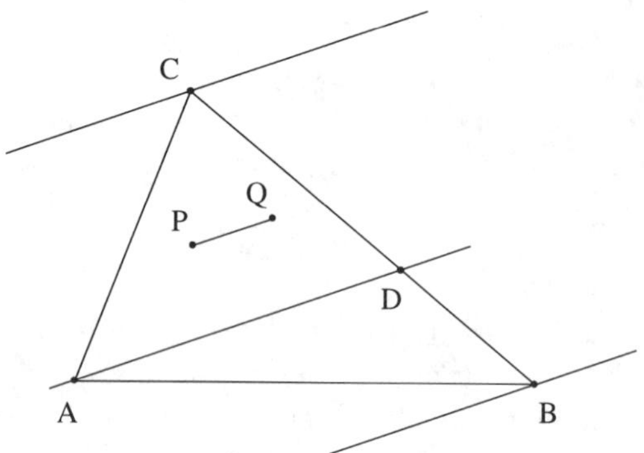

Figure 31.2. Proof of the lemma.

Let P and Q be two points in the domain triangle. The assertion of our theorem is

$$|Cn(P) - Cn(Q)| \leq L_t |PQ|_t. \tag{31.8}$$

The segment PQ passes through a certain number of triangles of $\mathcal{T}^n$. Let us denote the parts of the segment which lie in different subtriangles by PP_1, $P_1P_2, \ldots,$ P_kQ. Our lemma says that on each segment L_t is a Lipschitz constant for the linear function representing Cn on that triangle. Hence

$$\begin{aligned} |Cn(P) - Cn(Q)| &\leq |Cn(P) - Cn(P_1)| + |Cn(P_1) - Cn(P_2)| + \cdots \\ &\leq L_t(|PP_1|_t + |P_1P_2|_t + \cdots + |P_kQ|_t). \end{aligned}$$

Now we use the fact called to the reader's attention as Exercise 30.8, namely, that if the points P, P_1, $P_2, \ldots,$ Q lie on a straight line in this order, then

$|PP_1|_t + |P_1P_2|_t + \cdots + |P_kQ|_t = |PQ|_t$. Substituting this in the last inequality we get (31.8), and the proof of Theorem 31.4 is completed.

We are now ready to prove the following theorem of Gerald Farin:

THEOREM 31.5. *Let $Cn^0(P)$ be the control net based on the values of the function $f(P)$ at the nodes of T^n. The control nets obtained by repeated degree elevations starting with Cn^0 converge to $B^n f$.*

PROOF. Let $Cn^m(P)$ be the result of m degree elevations. The function $Cn^0(P)$ is piecewise linear, so it has a triangular Lipschitz constant L_t. By Theorem 31.4, L_t is a triangular Lipschitz constant for all the functions $Cn^m(P)$. We have $B^n f(P) = B^{n+m}Cn^m(P)$. By Theorem 30.1 we have

$$|Cn^m(P) - B^{n+m}Cn^m(P)| \leq \frac{4L_t}{(n+m)^{1/3}}. \tag{31.9}$$

Letting $m \to \infty$ in this formula proves Farin's theorem.

In Farin's excellent survey article "Triangular Bernstein-Bézier patches," *Computer Aided Geometric Design* 3 (1986), 83–127, the reader can find a proof based on explicit evaluation of the construction.

The triangular patches we have discussed so far are of functional form, i.e., we have a scalar function $f(P)$ defined on the domain triangle which represents the z-coordinate of the point of the patch lying directly above the point P of the domain triangle. As in the case of curves, a more flexible representation can be obtained if we define a vector function on the domain triangle. The surface then consists of all the points obtained by drawing the resulting vectors from the origin. A nonparametric representation of a surface by means of a scalar function $f(P)$ is a special case of a parametric representation; the vector function is $\overrightarrow{OP} + f(P)\hat{z}$. Many of our results for the nonparametric case can be extended without difficulty to the parametric case. We point out some of these items. We make changes in the notation, partly to indicate that we are now dealing with vectors and partly to make it easier for readers who wish to study the subject further from Gerald Farin's excellent book [Farin 93].

We introduce the notation $\mathbf{u}$ for the set of parameters (u, v, w) in the *parametric triangle*. In the nonparametric representation $z = f(P)$ of a Bézier patch the point $P = (u, v, w)$ of the domain triangle has a geometric meaning. It is the projection of the point of the patch onto the (x, y) plane, where the domain triangle is. In the parametric representation we do not have such a geometric association between the points of the patch and the corresponding points of the parametric triangle.

In the nonparametric case we started with a given function f and we constructed Bernstein polynomials to approximate it. The function values at the nodes of $\mathcal{T}^n$, interpolated linearly over each triangle of the triangulation, formed the control net. In the parametric case we are thinking more in terms of a design process of manipulating a control net to obtain a patch of a satisfactory shape, rather than constructing the shape in another way and then approximating it by a patch. In our discussion of parametric patches we are not going to introduce a vector analogue of the function f which was the function to be approximated in the nonparametric case. We consider the given quantities to be the vertices of a control net associated with $\mathcal{T}^n$.

Denote the *control point* assigned to the parameter set $(i/n, j/n, k/n)$ by $\mathbf{b}_{i,j,k}$; here $i + j + k = n$. The control net is the polyhedron obtained by linear interpolation over the triangles of the subdivided parameter triangle. The parametric Bézier triangular patch is defined by

$$\mathbf{b}^n(\mathbf{u}) = (uE_1 + vE_2 + wE_3)^n \mathbf{b}_{0,0,0} \tag{31.10}$$

where $\mathbf{u} = (u, v, w)$ is the set of barycentric coordinates of the point. Putting $w = 0$ and $v = 1 - u$ into (31.10), the right-hand side becomes

$$(uE_1 + (1-u)E_2)^n \, \mathbf{b}_{0,0,0} = \sum_{i=0}^{n} \binom{n}{i} u^i (1-u)^{n-i} \mathbf{b}_{i,n-j,0}.$$

This means that the boundary curves of the triangular patch are Bézier curves whose control polygons are the corresponding boundaries of the control net. This consideration enables us to deduce the first two of the following basic properties of a Bézier triangular patch from those of a Bézier curve; the third property follows from the fact that (31.10) is a convex combination of the control points.

1. The triangular patch interpolates its control net at the three corners.
2. A triangle of the control net which has a vertex at a corner of the net is tangential to the triangular patch at the corner.
3. The Bézier triangular patch lies entirely in the convex hull of its control net.

Consider the special case $n = 1$,

$$\mathbf{b}^1(P) = u\mathbf{b}_{1,0,0} + v\mathbf{b}_{0,1,0} + w\mathbf{b}_{0,0,1}. \tag{31.11}$$

In this case the triangular patch coincides with its control net, the triangle determined by three corners. The point in (31.11) is the point in the triangle $\mathbf{b}_{1,0,0}\mathbf{b}_{0,1,0}\mathbf{b}_{0,0,1}$ with barycentric coordinates (u, v, w). The equation (31.11) provides a formula for linear interpolation. Denote by $P_{0,1,1}$ the midpoint of the line segment $P_{0,1,0}P_{0,0,1}$. The barycentric coordinate u is then the ratio of length $P_{0,1,1}P$ to length $P_{0,1,1}P_{1,0,0}$, with similar statements for v and w. Thus, the three subtriangles with P as a vertex have areas u, v, and w times the area of the domain triangle.

As with Bézier curves, there are two geometrical constructions for the Bézier triangular patch. De Casteljau's produces one point of the patch exactly by a finite sequence of linear interpolations while degree elevation produces a sequence of polyhedra which slowly approximate the entire surface.

The value $\mathbf{b}^n(\mathbf{u})$ of a Bernstein triangular patch at any point is a linear combination of the control points. The operator formula (31.10) succinctly summarizes this. However, the formula is also confusing because it appears to involve members of the array $\mathbf{b}_{i,j,k}$ with index sums $< n$, and these are undefined and irrelevant to the final result.

Let us view the action of each factor $(uE_1 + vE_2 + wE_3)$ in (31.10) more closely. We have been using three variables u, v, and w and three indices i, j, and k to describe functions with two independent variables and the two-index arrays associated with them, in order to make the notation reflect the triangular symmetry of the situation. However, the effort to keep the symmetry in sight led to the appearance in our formulas of undefined quantities such as $\mathbf{b}_{0,0,0}$. The notation gives center stage to the equality of the three vertices of the triangle but forces us, if we want to understand the details of the computations, to track down what quantities are meaningful at each stage.

The following topics will be easier to understand if we let go of the symmetry and use instead a notation that does not bring in array elements with the wrong index-sum. In the operator formula (31.10), we can pull out the factor E_3 and get

$$\begin{aligned} \mathbf{b}^n(\mathbf{u}) &= (uE_1E_3^{-1} + vE_2E_3^{-1} + wI)^n E_3^n \mathbf{b}_{0,0,0} \\ &= (u\widetilde{E}_1 + v\widetilde{E}_2 + wI)^n \mathbf{b}_{0,0,n} \end{aligned} \tag{31.12}$$

where we wrote $E_1E_3^{-1} = \widetilde{E}_1$ and $E_2E_3^{-1} = \widetilde{E}_2$. These operators do not change the sums of the indexes and we could have omitted the third index since it is $n - i - j$ but we do not wish to change notation more than we need to.

We define

$$\mathbf{b}^1_{i,j,k} = (u\widetilde{E}_1 + v\widetilde{E}_2 + wI)\mathbf{b}_{i,j,k+1}, \tag{31.13}$$

$i, j, k \geq 0$, $i + j + k = n - 1$.

When we introduced triangulations, we called attention to the fact that there are two kinds of triangles in a triangulation: upright and inverted. The array $\mathbf{b}^1_{i,j,k}$ obtained by one application of the operator $(u\widetilde{E}_1 + v\widetilde{E}_2 + wI)$ to the array consisting of the vertices $\mathbf{b}_{i,j,k}$ of the control net forms an array consisting of the weighted average with weights u, v, w of the vertices of the upright triangles in this polyhedron, i.e., the triangles which are images of the upright triangles in $\mathcal{T}^n$. This is a triangular net with only n entries along the outside edges, one less than the control net itself.

We denote by $\mathbf{b}^l$ the result of repeating the above process l times:

$$\mathbf{b}^l_{i,j,k} = (u\widetilde{E}_1 + v\widetilde{E}_2 + wI)^l \mathbf{b}_{i,j,k+l}, \tag{31.14}$$

$i,\ j,\ k \geq 0,\ i+j+k = n-l$.

The three layers of inscribed polyhedra formed in space by the linear interpolations for a patch of degree 3 are shown in Fig. 31.3, following [Farin '93]. The "inverted" triangles were left transparent in all layers to make the diagram clearer. The left, right, and top corners are the points with parameters $(1,0,0)$, $(0,1,0)$ and $(0,0,1)$. The parameter values are $(u,v,w) = (0.22, 0.28, 0.50)$. The patch itself would have been too difficult to show.

The process is completed with the array $\mathbf{b}^n$, which has the single element

$$\mathbf{b}^n_{0,0,0} = (u\widetilde{E}_1 + v\widetilde{E}_2 + wI)^n \mathbf{b}_{0,0,n} = \mathbf{b}^n(u,v,w). \tag{31.15}$$

In Fig. 31.3 this element is the dot in the middle triangle. This is *de Casteljau's algorithm* for computing a point of a Bézier patch.

The *degree elevation* algorithm, which in the limit gives the entire surface, carries over from the nonparametric case without change; the derivation is valid for vectors, or alternatively one can apply the scalar result to the components of the vectors. We state the result in Farin's notation.

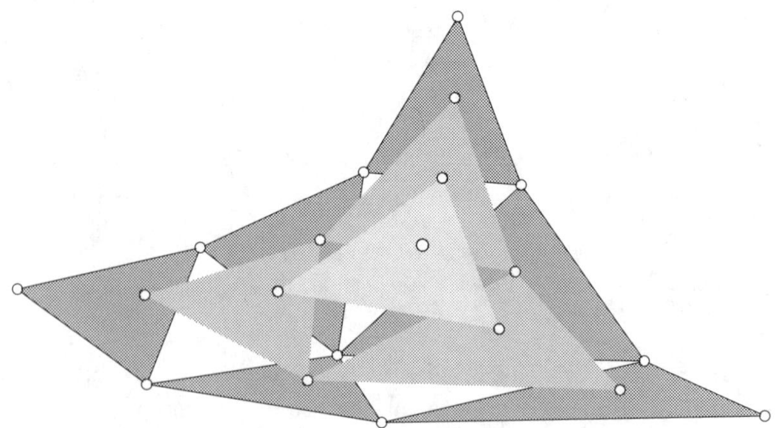

Figure 31.3. De Casteljau's iteration of linear interpolations.

The patch of degree n defined by the array $\{\mathbf{b}_{i,j,k}\}$ *where* $i + j + k = n$*, is the same as the patch of degree* $n + 1$ *defined by*

$$\mathbf{b}^{(1)}_{i,j,k} = \frac{i}{n+1}\mathbf{b}_{i-1,j,k} + \frac{j}{n+1}\mathbf{b}_{i,j-1,k} + \frac{k}{n+1}\mathbf{b}_{i,j,k-1}, \tag{31.16}$$

$i + j + k = n + 1$.

The notation is similar to what we used for the first set of linear interpolations in the de Casteljau algorithm; we distinguish the results of the present interpolation process from that of de Casteljau by putting in parentheses the superscript that counts the number of iterations. Note that this time the linear interpolations are performed over the "inverted" triangles of the parametric triangle. As we have noted before, when one of the indexes is 0, the corresponding weight in (31.16) is 0, and we are interpolating over a boundary segment of the net. We get a net with one more row of triangles than the original one. We illustrated the process in Fig. 31.4; we left the faces of the elevated control net $\{\mathbf{b}^{(1)}\}$ transparent.

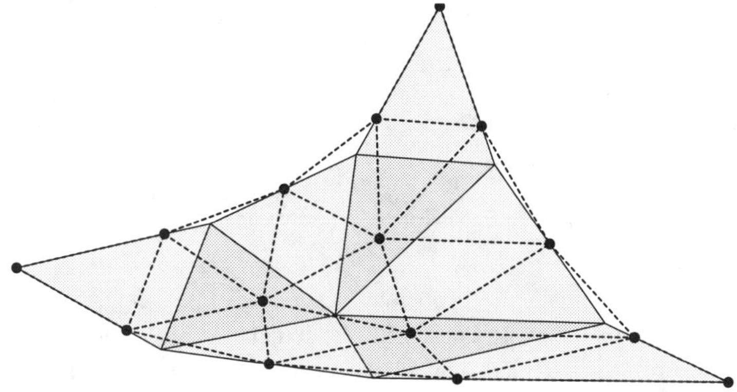

Figure 31.4. Control net and degree elevated control net.

An elevated control net is associated with the new set of control points $\{\mathbf{b}^{(1)}_{i,j,k}\}$. The process can be repeated to obtain $\{\mathbf{b}^{(2)}_{i,j,k}\}$, etc. By applying theorem 31.5 to each of the components of the vector-valued functions representing the control nets, we see that the sequence of the elevated control nets converges to the common Bézier triangular patch which is defined by each of them.

We presented the subdivision of Bézier curves in the chapter on that topic, and we close this chapter with a discussion of the operation with the same name and function for triangular patches. Before we start, we should tell the reader that although this topic is important to practical users of Bézier patches, it will not be used in the remaining part of this chapter, which we shall devote to Bézier patches. Our reason for discussing subdivision is to present the surprising fact that the de Casteljau arrays, which arose in the simpler context of computing individual points of a patch, also furnish a solution to the subdivision problem.

As in the case of curves, one purpose of subdivision is to be able to change a part of a patch but leave another part unchanged. The first step is to compute the control net of part of a patch, and that is how far we are going to pursue the subject. Consider an nth degree Bézier patch $\mathbf{b}(u, v, w)$ on a parametric triangle ABC, given by its control net $\mathbf{b}_{i,j,k}$. We have a point $D : (U, V, W)$ inside the triangle (it could be on one of the sides, or even outside), as in Fig. 31.5. We want to find the control net with respect to the parametric triangle ABD which produces the given patch.

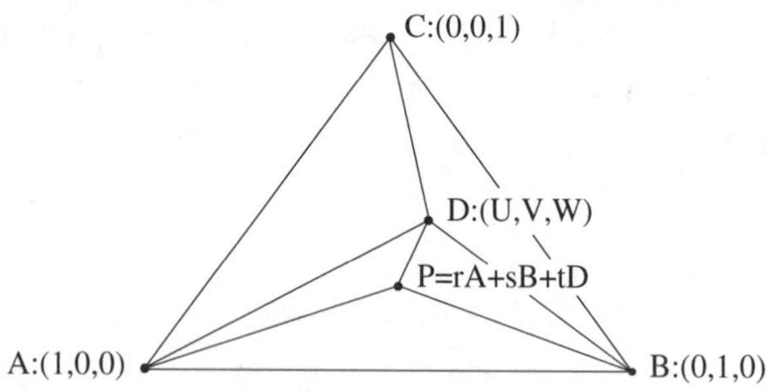

Figure 31.5. Subdivision of a parametric triangle.

Perhaps we should point out an obvious fact one may lose sight of. Our attention was focused on function values inside the domain triangle or the parametric triangle. However, the Bézier patches are defined by polynomials and thus they are defined outside the parametric triangle too. Thus, what we really wish to do is to generate the same patch as before but with a control net that extends only over part of the parametric triangle.

Let us denote barycentric coordinates of a point P with respect to $\triangle ABD$ by r, s, t. The control net of the patch $\mathbf{b}(u, v, w)$ with respect to the parametric triangle ABC is the array of coefficients of $B^n_{i,j,k}(r, s, t)$ in the expression of $\mathbf{b}(P)$ in terms of r, s, t. The coefficients can be found in a mechanical way by expressing u, v, w in terms of r, s, t, substituting in $\mathbf{b}(u, v, w)$ and rearranging. By examining the process more closely we shall find that this array of coefficients, i.e., the control net of the patch with $\triangle ABD$ as the parametric triangle, is in fact given by de Casteljau's algorithm.

The barycentric coordinates (u, v, w) of P with respect to $\triangle ABC$ are obtained from $P = rA + sB + t(UA + VB + WC)$. We get $(u, v, w) = (r, s, 0) +$

$t(U, V, W)$. Substitute this in the operator formula (31.12) for the patch:

$$\mathbf{b}(P) = (r\widetilde{E}_1 + s\widetilde{E}_2 + t(U\widetilde{E}_1 + V\widetilde{E}_2 + WI))^n \mathbf{b}_{0,0,n}. \tag{31.17}$$

Here the term which contains $r^i s^j t^k$ can be written as $B^n_{i,j,k}(r, s, t)$ times

$$\begin{aligned} \widetilde{E}_1^i \widetilde{E}_2^j (U\widetilde{E}_1 + V\widetilde{E}_2 + WI)^k \mathbf{b}_{0,0,n} &= (U\widetilde{E}_1 + V\widetilde{E}_2 + WI)^k \widetilde{E}_1^i \widetilde{E}_2^j \mathbf{b}_{0,0,n} \\ &= (U\widetilde{E}_1 + V\widetilde{E}_2 + WI)^k \mathbf{b}_{i,j,n-i-j}. \end{aligned} \tag{31.18}$$

Using the equation $n - i - j = k$ and recalling the definition (31.14) we see that the coefficients of $B^n_{i,j,k}(r, s, t)$ in (31.17) are the quantities $\mathbf{b}^k_{i,j,0}$ which occurred in the de Casteljau algorithm. To sum up, we have obtained the following result of Gerald Farin: *Let $\mathbf{b}_{i,j,k}$ be the control net of a parametric patch with ABC as parametric triangle. The control net for generating the same patch using ABD as parametric triangle is $\mathbf{b}^k_{i,j,0}$, $i + j + k = n$.*

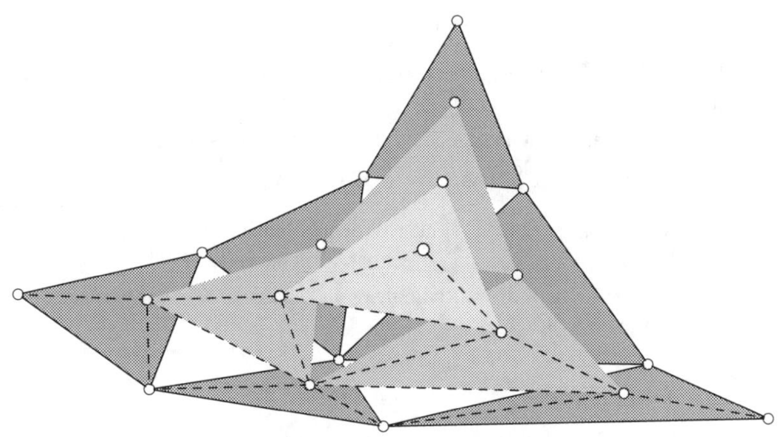

Figure 31.6. Control net with respect to $\triangle ABD$.

In Fig. 31.6 you can see the configuration we used to illustrate de Casteljau's algorithm. Now the point near the center of the topmost triangle is the vertex of the subdivision triangle. We marked with dashed lines the control net with respect to this triangle. The inverted triangles of this control net are shown transparent. The upright triangles of the dashed control net are parts of the the upright triangles in the bottom rows of the successive layers of de Casteljau triangles.

We noted earlier that a corner triangle of a patch is tangential to the patch at the corner. Applying this to the subdivided control net we get:

The last triangle in the de Casteljau construction is tangent to the patch at the point being constructed.

A special way to subdivide the parametric triangle is to cut it into two triangles instead of three. We show this in Fig. 31.7.

Exercises

31.1 Each node of the subdivision $\mathcal{T}^n$ of a triangle ABC is associated with a number such that the sum of two numbers on the endpoints of one diagonal is the same as the sum of two numbers on the endpoints of the other diagonal of any parallelogram formed by two adjacent subtriangles. Suppose numbers a, b, and c are associated with A, B, and C respectively. Find the sum of the numbers associated with all nodes. (A problem from the first round selection examination of Chinese team of IMO, 1987.)

31.2 Familiarize yourself with the de Casteljau algorithm by using it to evaluate $B^3(1/2, 1/4, 1/4)$, given the coefficient array

$$\begin{array}{ccccccc} & & & 4 & & & \\ & & -1 & & 2 & & \\ & 3 & & 5 & & 0 & \\ -5 & & 0 & & -2 & & 1 \end{array}$$

where the upper, left, and right corner elements are the coefficients of $B^3_{3,0,0}$, $B^3_{0,3,0}$ and $B^3_{0,0,3}$ respectively.

31.3 a) Show by algebraic computation that the center of similarity of the domain triangle and the upright triangle $T^n_{i,j,k}$ $(i + j + k = n - 1;\ i, j, k \geq 0)$ of $\mathcal{T}^n$ is the vertex $P^{n-1}_{i,j,k}$ of $\mathcal{T}^{n-1}$.

b) Show that the center of similarity of the domain triangle and the inverted triangle $T^n_{i,j,k}$ $(i + j + k = n + 1;\ i, j, k \geq 1)$ of $\mathcal{T}^n$ is the vertex $P^{n+1}_{i,j,k}$ of $\mathcal{T}^{n+1}$.

31.4 State and prove the Kelisky-Rivlin theorem (Theorem 26.6) for Bernstein triangular polynomials.

CHAPTER THIRTY-TWO

Convexity of Patches

This chapter is devoted mostly to the rather specialized questions of extending to triangular patches the theorems that a Bernstein approximant of a convex function is convex, and a Bézier curve with a convex control polygon is convex. It turns out that the first of these theorems can be extended to triangular patches in a suitable formulation but the analogue of the second theorem is false. We are going to prove that if a *nonparametric* control net is convex, the net obtained from it by degree elevation is also convex. From this, the convexity theorem will follow, as it did for curves.

In the second part of the chapter we discuss the formulas for derivatives of triangular patches. With the help of these we will be able to give a shorter, algebraic proof of the convexity theorem.

We already know from Theorem 31.2 that if the function $f(P)$ is convex then the restriction of the polynomial $B^n f(u, v, w)$ to a line segment which is parallel to any one of the sides of the domain triangle is a convex function in the domain triangle. One might think that this implies that the polynomial is a convex function in the domain triangle, but it does not. An example is the function $f(P) = \sqrt{x^2 + y^2}$ in the domain triangle $(0, 0)$, $(1, 0)$, $(0, 1)$. An exercise in Chapter 30 was to compute $B^2 f(P)$ in terms of x and y. We have $B^2 f(x, y) = x + y - (2 - \sqrt{2})xy$. This function is linear along lines $x = \text{const}$ and along lines $y = \text{const}$. The third set of lines parallel to a side of the domain triangle is the set of lines $y = \text{const} - x$. Along these lines f is given by $(2 - \sqrt{2})x^2 + \text{terms linear in } x$, which is convex, as it should be. However, along the line $x = y$ the quadratic term is $-(2 - \sqrt{2})x^2$, so the function is concave along this line.

Let us view figures showing the above function, control net, and triangular patch. Figure 32.1 shows the part of the cone $z = \sqrt{x^2 + y^2}$ which is above the domain triangle and the control net based on the triangulation $\mathcal{T}^2$. One can see that the control net is not convex.

Figure 32.2 shows the same control net but the surface we drew this time is the quadratic triangular patch $B^2 f(P)$. The only information about f which enters the definition of this patch are the function values that span the control net. Note that 5 of the 6 vertices of the control net lie on the plane $z = x + y$; the sixth

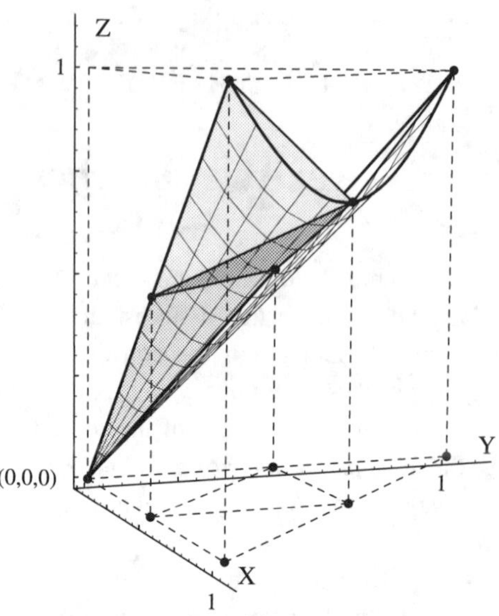

Figure 32.1. Cone $z = \sqrt{x^2 + y^2}$ and its Bézier net for $n = 2$.

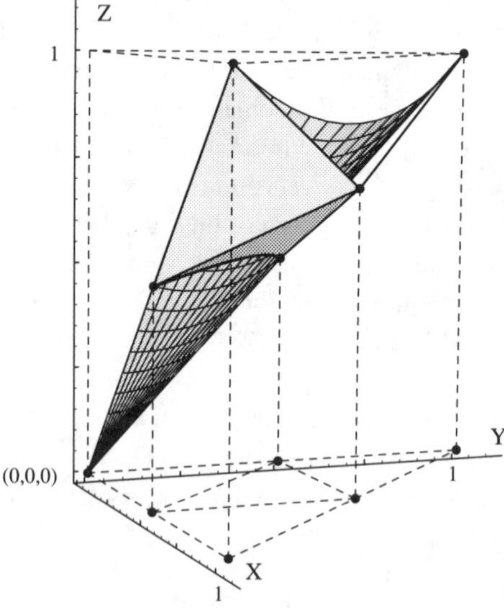

Figure 32.2. The net in the last figure and the Bézier patch.

vertex, $(0.5, 0.5, \sqrt{0.5})$ is below this plane. If the sixth vertex were also on the plane $z = x + y$, the quadratic patch would be just $z = x + y$. Even as it is, the patch is equal to $z = x + y$ along two edges, but it is pulled down along the edge $x + y = 1$, and this sagging causes it to be somewhat concave above lines coming from the origin. The concavity is slight and can not be clearly seen in the figure. We have already verified it by computation; one can also infer it from properties of triangular patches, as follows.

We know that the edges of the patch are represented by the one-variable Bernstein polynomials determined by the control points along those edges. Hence the line segments from $(0, 0, 0)$ to $(1, 0, 1)$ and from $(0, 0, 0)$ to $(0, 1, 1)$ are on the patch. Since the patch is the graph of a quadratic polynomial in x and y, it has a well-defined tangent plane at all points. The tangent plane at the origin must contain the two line segments we just mentioned, and hence it is the plane through these two lines. Thus the lines marked on the patch in Fig. 32.2 where the patch intersects vertical planes through the origin are tangential to the plane $z = x + y$ at the origin. The patch ends where it reaches the plane $x + y = 1$; here these curves are below the plane $z = x + y$. Hence they must be bending downwards in at least some places along the way, making the surface non-convex.

We see that, in contrast to the 1-dimensional case, the control net produced by a convex function of two variables need not be convex. Since the only information about the function affecting the Bézier patch is the shape of the control net, we see why a Bézier patch of a convex function may not be convex. For Bézier curves we were able to prove by degree elevation that a convex control polygon produces a convex Bézier curve. We will be able to generalize that argument to Bézier patches, although only in the nonparametric case.

THEOREM 32.1 [CHANG & DAVIS '84]. *If the control net of a nonparametric triangular patch is convex, then the patch is convex.*

PROOF. Let $cn(\mathbf{u})$ be the function whose graph is the convex control net of the patch. In other words, $cn(\mathbf{u})$ is a convex function and it is linear on each subtriangle of $\mathcal{T}^n$. Let $de(\mathbf{u})$ be the degree elevated control net obtained from $cn(\mathbf{u})$. This means, $de(\mathbf{u})$ is equal to $cn(\mathbf{u})$ at the vertices of $\mathcal{T}^{n+1}$ and is linear on each triangle of $\mathcal{T}^{n+1}$. We want to show that the function $de(\mathbf{u})$ is also convex. We have shown in the previous chapter that the polyhedrons obtained by repeated degree elevation converge to the triangular patch. It will follow that our triangular patch is the limit of convex functions and therefore convex.

We are going to use the following. *A piecewise linear function defined on a convex region is convex if and only if all the dihedral angles (angles between planes) of adjacent faces of its graph are convex downwards, i.e., measured from above, they are* $\leq 180°$.

We recall (Chapter 31, Exercise 3) the way $\mathcal{T}^n$ and $\mathcal{T}^{n+1}$ are related. The inverted triangles of $\mathcal{T}^n$ each contain one interior vertex of $\mathcal{T}^{n+1}$ and the upright triangles contain none. An upright triangle of $\mathcal{T}^{n+1}$ contains a vertex of $\mathcal{T}^n$ and an inverted triangle of $\mathcal{T}^{n+1}$ contains no such vertex. In each case, the three vertices of a triangle in one of the triangulations are in three triangles of the other triangulation which surround an "empty" triangle. In both triangulations each interior edge separates an upright and an inverted triangle. Also, in each triangulation every interior edge separates an empty and a non-empty triangle.

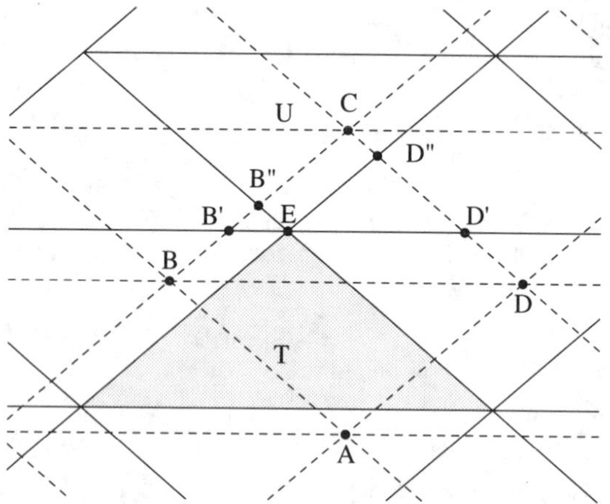

Figure 32.3. Adjacent triangles of the $n + 1$-triangulation.

Let BD be any interior edge of the $n+1$-triangulation, see Fig. 32.3. We want to show that the angle between two faces of the degree-elevated control net that meet above BD is convex downwards. Let BAD be the inverted and empty one of the two triangles of $\mathcal{T}^{n+1}$ with side BD and let BDC be the full one. We have to show that the dihedral angle of the graph of de above BD is convex. This is the case if the chord of the control net connecting the points $(A, de(A))$ and $(C, de(C))$ lies above or intersects the edge of the net from $(B, de(B))$ to $(D, de(D))$. Since $ABCD$ is a parallelogram, the midpoints of these two line segments cross the same vertical line at their midpoints. Thus the chord is on or above the edge above BD if

$$cn(A) + cn(C) \geq cn(B) + cn(D). \tag{32.1}$$

We have to prove (32.1).

Let T be the (upright) triangle of $\mathcal{T}^n$ surrounded by the triangles containing points A, B, and D, and let U be the inverted triangle of $\mathcal{T}^n$ containing the point C. The definitions of the other points we have marked can be read off Fig. 32.3.

The convexity of a control net and the elevated net obtained from it is unaffected if we subtract a linear function from the function of which it is the graph. The discussion is simpler if the function cn is 0 on the triangle T. We can reduce the general problem to this one by subtracting from cn the linear function which represents it in T. The corresponding degree elevated function is diminished by the same linear function. We do not introduce new names for the resulting functions; instead, we assume that cn has the value 0 on the triangle T.

Since cn is convex and equal to 0 on T, it is nonnegative everywhere. Indeed, suppose there is a point P of the domain triangle such that $cn(P) < 0$. Consider the chord from the center of T to the point of the control net under P. The part of this chord which is under the triangle T would be beneath the graph of cn, in contradiction with the convexity of the control net. In particular, $cn(A) > 0$ and thus (32.1) will be proved if we can show

$$cn(C) \geq cn(B) + cn(D). \tag{32.1a}$$

Let us examine the values of cn along $BB'B''$. In the triangle containing B, cn is linear. Also, cn is 0 along the common side of this triangle with T, and BB' is parallel to this side, so cn is constant on the line segment BB'. Since cn is convex, it is nondecreasing on $B'B''$. Thus we get

$$cn(B'') \geq cn(B) \quad \text{and similarly} \quad cn(D'') \geq cn(D).$$

The function cn is linear in U. Hence the sum of its values at two opposite corners of the parallelogram $ED''CB''$ is the same as the sum of the values at the other two corners. Since $cn(E) = 0$, this gives

$$cn(C) = cn(B'') + cn(D'') \geq cn(B) + cn(D)$$

which is (32.1a). The proof of Theorem 32.1 is completed.

Before we stated Theorem 27.4, we explained that the meaning of the term "convex", when applied to a curve, is somewhat different depending on whether we think of the curve as the graph of a function or a geometric object. The same holds for surfaces. When we were discussing nonparametric surfaces, "convex" meant the graph of a convex function. Parametric representations are used when the geometry is our primary interest. Hence, for the purposes of what follows, we are going to use the three-dimensional version of the alternative definition of a convex curve that we introduced before Theorem 27.4:

A surface is convex if it lies on the boundary of its convex hull.

The proof of Theorem 32.1 consisted of showing that degree elevation preserves convexity of a nonparametric triangular patch control net. Considering how trivial this is for curves, both in the nonparametric and the parametric case, the proof of Theorem 32.1 seems a bit involved. There is a good reason for this; the matter is really more delicate for surfaces than it is for curves, and for parametric control nets degree elevation need not preserve convexity. As a result our proof that a convex control net defines a convex patch does not carry over to the parametric case. Indeed, a convex parametric control net can define a nonconvex triangular patch. An example was given by Chen Falai.

The control net (see Fig. 32.4) consists of four triangles. The central triangle is the isosceles triangle with vertices $(0,\ 2,\ 0)$, $(0,\ -2,\ 0)$ and $(2,\ 0,\ 0)$. The three other triangles are vertical isosceles triangles of height 2 whose bottoms are the sides of the central triangle The tips of the isosceles triangles are $(0,\ 0,\ 2)$, $(1,\ -1,\ 2)$ and $(1, 1, 2)$. Keep in mind that *Mathematica*, which we used to make the figures in this chapter, draws the axes along edges of a box surrounding the object, not through the origin.

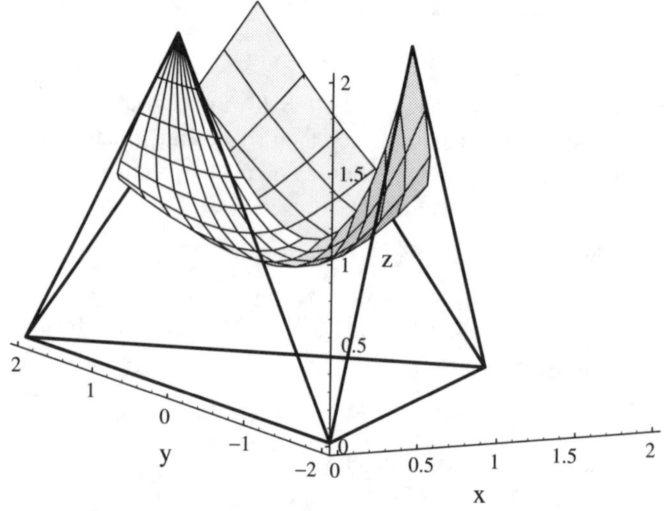

Figure 32.4. Convex net which generates a nonconvex parametric patch.

The table below shows the assignment of points in space to points in the parameter triangle. The triples in the first row are barycentric coordinates in the parameter triangle, while those in the second row are rectangular coordinates in space.

(u, v, w)	$(1, 0, 0)$	$\left(\frac{1}{2}, \frac{1}{2}, 0\right)$	$(0, 1, 0)$	$\left(0, \frac{1}{2}, \frac{1}{2}\right)$	$(0, 0, 1)$	$\left(\frac{1}{2}, 0, \frac{1}{2}\right)$
(x, y, z)	$(0, 0, 2)$	$(0, -2, 0)$	$(1, -1, 2)$	$(2, 0, 0)$	$(1, 1, 2)$	$(0, 2, 0)$

Using the table we can write down the parametric formulas for the quadratic triangular patch:

(32.2) $x = v^2 + 4vw + w^2 = (v+w)^2 + 2vw = (1-u)^2 + 2vw,$

(32.3) $y = -4uv - v^2 + w^2 + 4uw = (w-v)(1+3u),$

(32.4) $z = 2u^2 + 2v^2 + 2w^2.$

It is not clear from Fig. 32.4 that the patch is not convex; on the contrary, it looks convex. However, Fig. 32.5 which shows the corner near the tip (0, 0, 2) greatly stretched in the x-direction, reveals that the surface is not convex near the corners. The numbers on the axis labeled x indicate the degree of stretching. The axes are drawn along edges of a box containing the part of this corner located in $z \geq 1.6$. The z-axis, not drawn, is the vertical through the tip of the surface.

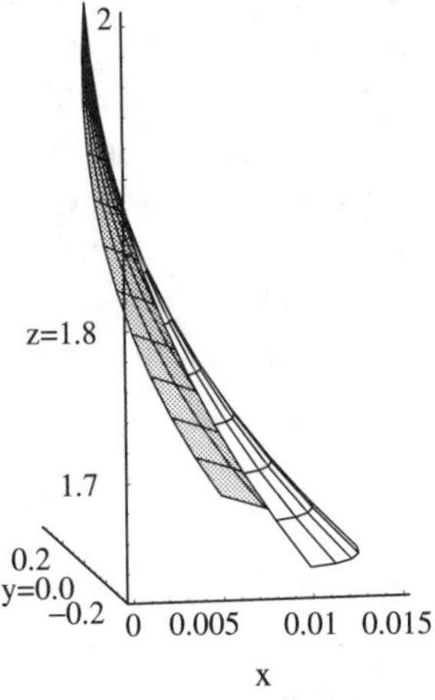

Figure 32.5. A corner of the patch, enlarged and stretched.

The figure can reasonably be accepted as a proof of nonconvexity of the corner, if one has checked that the right formula was put into *Mathematica* and one is satisfied that none of the bugs in *Mathematica 2.2* would cause it to draw such a figure incorrectly.

A more traditional approach is to use the figure only as a guide for checking the nonconvexity by computation. The figure shows that near the corner (0, 0, 2) the chord connecting two points on the edge of the surface, symmetrical with respect to the plane $y = 0$, is beneath the surface. One could verify numerically or algebraically that, say, the midpoint of one such chord is beneath the surface.

The geometric proof of Theorem 32.1 we presented above gives an intuitive insight of what causes the theorem to be true. However, one should not expect to obtain most facts of mathematics by visualizing the quantities involved. The backbone of mathematics is the formula. We end this chapter by showing how Theorem 32.1 can be proved by computation. We start by obtaining formulas for derivatives of triangular patches which are similar to formula (26.11) for the derivative of a Bernstein polynomial. Following the procedure we used there, we start with the operator formula for the triangular patch:

$$z = B^n f(P) = (uE_1 + vE_2 + wE_3)^n f_{0,0,0}. \tag{32.5}$$

We again note that the variables $u,\ v,\ w$ occur only in the operator, so we can differentiate the whole expression by differentiating the operator only.

The partial derivative of a function $F(u, v, w)$ with respect to, say, u is its rate of change when u changes and v and w are kept fixed. We appear to have a problem here. Our $u,\ v,\ w$ are barycentric coordinates, $u + v + w = 1$. The last equation prevents us from changing u while v and w are kept fixed. Thus, for a function $f(P)$ defined on the domain triangle, the expression $\partial f(u, v, w)/\partial u$ is undefined. However, the Bernstein polynomials are defined for all values of $u,\ v$, and w, although only the values with $u + v + w = 1$ are of interest in the geometrical discussion.

As we discussed in the one-variable case, we can use the power rule to differentiate the operator. We get for the first partials

$$\frac{\partial}{\partial u} B^n f(u, v, w) = n(uE_1 + vE_2 + wE_3)^{n-1} E_1 f_{0,0,0}\,,$$
$$\frac{\partial}{\partial v} B^n f(u, v, w) = n(uE_1 + vE_2 + wE_3)^{n-1} E_2 f_{0,0,0}\,,$$
$$\frac{\partial}{\partial w} B^n f(u, v, w) = n(uE_1 + vE_2 + wE_3)^{n-1} E_3 f_{0,0,0}\,.$$

These formulas have a puzzling feature. We expect derivatives to be obtained by applying a difference operator to the data, and that is how it was in the 1-variable case. Yet in the formulas we just obtained, we apply only shifting operators to the data. How can this be?

The answer lies in the fact that, although we have three independent variables in our formula for the Bézier patch, the only values meaningful in our work are the ones with $u + v + w = 1$; the rest are just an algebraic convenience. Thus, if we want to find, say, the rate of change of $B^n f(P)$ along a line $w =$ const, we have to take into account that if we increase u by a small amount, we must decrease v by the same amount. Hence the resulting expression will have the shifting operators E_1 and E_2 with equal and opposite coefficients, and this combination is a difference operator.

We shall also need the expressions for the second partials:

$$\frac{\partial^2}{\partial u^2} B^n f(u, v, w) = n(n-1)(uE_1 + vE_2 + wE_3)^{n-2} E_1^2 f_{0,0,0},$$

$$\frac{\partial^2}{\partial u \partial v} B^n f(u, v, w) = n(n-1)(uE_1 + vE_2 + wE_3)^{n-2} E_1 E_2 f_{0,0,0},$$

and so on.

As we noted, the rates of change meaningful for Bézier patches are the ones that pertain to moving along the plane $u + v + w = 1$. So let us consider a line segment $\mathbf{P}(t) = (u_0 + \xi t,\ v_0 + \eta t,\ w_0 + \zeta t)$, where t ranges over some interval. The condition that $u + v + w = 1$ implies

$$\xi + \eta + \zeta = 0. \tag{32.6}$$

Let $z = B^n f(\mathbf{P}(t))$. Taking the first derivative of z with respect to t, we get by the chain rule

$$\frac{dz}{dt} = \frac{\partial B^n f}{\partial u}\xi + \frac{\partial B^n f}{\partial v}\eta + \frac{\partial B^n f}{\partial w}\zeta.$$

Hence

$$\frac{dz}{dt} = n(\xi E_1 + \eta E_2 + \zeta E_3)(uE_1 + vE_2 + wE_3)^{n-1} f_{0,0,0}. \tag{32.7}$$

Similarly, for the second derivative, we have

$$\frac{d^2 z}{dt^2} = n(n-1)(\xi E_1 + \eta E_2 + \zeta E_3)^2 (uE_1 + vE_2 + wE_3)^{n-2} f_{0,0,0}. \tag{32.8}$$

We are now ready to show how one can prove Theorem 32.1 by algebraic computation. We use that a function of two variables defined on a convex domain is convex if its restriction to every straight line in its domain of definition is convex. This is just a restatement of the definition of a convex function of two variables. A twice differentiable function of one variable is convex if its second derivative is nonnegative. Hence, we need to show that if the control net is convex, the second derivative of the patch, given by (32.8), is nonnegative in every direction.

In our proof by degree elevation we have already used the fact that a non-parametric control net, i.e, a control net whose vertices lie vertically above the nodes of a triangulation of the domain triangle, is convex if and only if all its dihedral angles measured from above are $\leq 180°$. This will be so if and only if the sum of the ordinates at two ends of an interior edge of the polyhedron is at most equal to the sum of the ordinates at the other two vertices of the faces joined along that edge, because the midpoint of the chord connecting the tips of the two triangles is on the same vertical line as the midpoint of the edge. We formulate this condition in terms of difference operators.

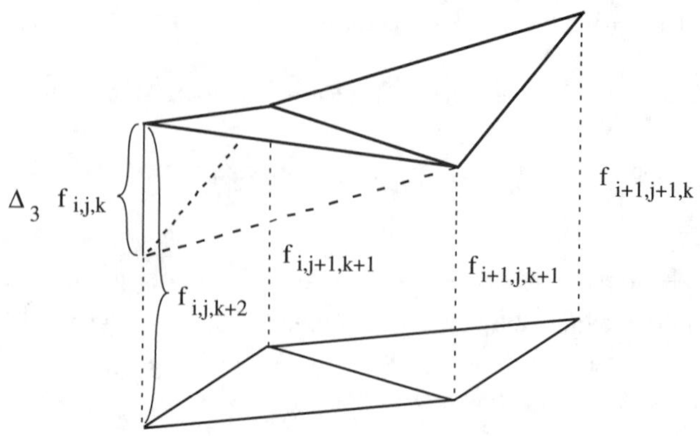

Figure 32.6. Measure of bending at an edge.

The lower half of Fig. 32.6 shows two adjacent triangles of a triangulation of the domain triangle. Together, they form a parallelogram. Above these two triangles we have two faces of a convex control net. The edge we have drawn is above a line $k = \text{const}$.

The figure consisting of the dotted triangle on the left and the triangle on the right is a parallelogram. Hence the sums of the ordinates for the two pairs of opposite vertices are the same. This gives us the following formula for the quantity marked as $\Delta_3 f_{i,j,k}$, which is nonnegative if the dihedral angle between the two triangles is convex downward:

$$\begin{aligned}\Delta_3 f_{i,j,k} &= f_{i,j,k+2} - f_{i,j+1,k+1} - f_{i+1,j,k+1} + f_{i+1,j+1,k} \\ &= \left(E_3^2 - E_3 E_2 - E_3 E_1 + E_2 E_1\right) f i, j, k \\ &= (E_3 - E_2)(E_3 - E_1) f_{i,j,k} \quad i, j, k \geq 0, \quad i + j + k = n - 2.\end{aligned}$$

By cyclic permutation of the indices we get formulas for the operators Δ_1 and Δ_2 which express the rise of the tip of one triangle above the plane of the other in case the edge joining the triangles is in the directions $i = \text{const}$ or $j = \text{const}$:

$$\Delta_2 = (E_2 - E_3)(E_2 - E_1), \qquad \Delta_3 = (E_3 - E_1)(E_3 - E_2).$$

The convexity of the control net over the triangulation T_n is expressed by

$$(32.9) \qquad \Delta_1 f_{i,j,k} \geq 0, \qquad \Delta_2 f_{i,j,k} \geq 0, \qquad \Delta_3 f_{i,j,k} \geq 0,$$

for $i, j, k \geq 0,\ i + j + k = n - 2$.

We stop here for a moment to think about what we have learned so far. The formula for second derivatives of a triangular patch, (32.8), contains all 6 products of the three operators $E_1,\ E_2,\ E_3$. On the other hand, if we know the equation of just one face of a control net and all values of

$$\Delta_1 f_{i,j,k},\ \Delta_2 f_{i,j,k},\ \Delta_3 f_{i,j,k} \tag{32.10}$$

with $i \geq 0,\ j \geq 0,\ k \geq 0,\ i+j+k = n-2$, we can determine the ordinates of all vertices of the control net. The formulas for the ordinates are linear. Thus the quantities (32.10) determine the control net up to a linear function. In particular, they determine all the second derivatives of the Bézier patch. (Actually, we do not need all the values (32.10), only enough of them to get from any face to any other face; this means that there are $(n-1)(n-2)/2$ relations among the quantities (32.10).)

We have the information (32.9) about the quantities $\Delta_h f_{i,j,k}$ and we know that the second derivative (32.8) we wish to prove positive can be expressed in terms of the Δs. It remains to produce such an expression so that we can read off that the second derivative is nonnegative.

First we express the operator $\xi E_1 + \eta E_2 + \zeta E_3$ in terms of the difference operators $E_1 - E_2,\ E_2 - E_3,\ E_3 - E_1$. We listed all three of these operators just for the sake of symmetry; each can be expressed in terms of the other two. Subtracting $0 = \xi E_3 + \eta E_3 + \zeta E_3$ from our operator we get

$$\xi E_1 + \eta E_2 + \zeta E_3 = \xi(E_1 - E_3) + \eta(E_2 - E_3). \tag{32.11}$$

We want to express products of two difference operators in terms of the Δ_i. The products of two different difference operators are by definition the Δ_i. For the squares we have

$$\begin{aligned} (E_1 - E_3)^2 &= (E_1 - E_3)(E_1 - E_2) + (E_1 - E_3)(E_2 - E_3) \\ &= \Delta_1 + \Delta_3. \end{aligned} \tag{32.12}$$

and similar formulas for the other two squares.

Squaring (32.11) and substituting these results we get

$$\begin{aligned} (\xi E_1 + \eta E_2 + \zeta E_3)^2 &= \xi^2(\Delta_1 + \Delta_3) + \eta^2(\Delta_1 + \Delta_3) + 2\xi\eta\Delta_3 \\ &= \xi^2\Delta_1 + \eta^2\Delta_2 + \zeta^2\Delta_3. \end{aligned}$$

Substitute this into (32.8):

$$\frac{d^2 z}{dt^2} = n(n-1)\big(\xi^2\Delta_1 + \eta^2\Delta_2 + \zeta^2\Delta_3\big)(uE_1 + vE_2 + wE_3)^{n-2} f_{0,0,0}.$$

Expand the second operator. By (32.9), applying the first operator to the result gives only nonnegative terms. Hence, the triangular patch is convex.

Exercises

32.1 Use operator algebra to show that if

$$\Delta_1 f_{i,j,k} = \Delta_2 f_{i,j,k} = \Delta_3 f_{i,j,k} = 0 \quad \text{for } i,\ j,\ k \geq 0,\ i + j + k = n - 2,$$

then the Bernstein triangular polynomial reduces to

$$z = u f_{n,0,0} + v f_{0,n,0} + w f_{0,0,n}.$$

(The problem is the same as Exercise 31.1 but here an algebraic solution is required.)

32.2 Algebraic demonstration of the nonconvexity of the corner, left out of the MS: The boundary curve $v = 0$ of the patch is given by the formulas

$$x = (1-u)^2, \quad y = (1-u)(1+3u), \quad z = u^2 + (1-u)^2. \tag{32.13}$$

The boundary curve $w = 0$ is given by the same expressions except that the sign of y is reversed. This reflects the symmetry of our configuration in the xz-plane. Hence the midpoint of a chord connecting a point of the boundary curve (32.13) and its mirror image in the plane $y = 0$ is

$$x = (1-u)^2, \quad y = 0, \quad z = u^2 + (1-u)^2. \tag{32.14}$$

32.3 We will show that when x is small, this point is below the surface, which shows that the surface is not convex. The intersection of the patch with the xz-plane is parametrized by $1 \geq u \geq 0$, $v = w = \frac{1}{2}(1-u)$. The formulas for the curve are

$$x = \tfrac{3}{2}(1-u)^2, \quad y = 0, \quad z = u^2 + \tfrac{1}{2}(1-u)^2. \tag{32.15}$$

If we express u in terms of x in (32.14), we get $z = 1 - 2\sqrt{x} + 2x$. A similar computation for the curve (32.15) gives $z = 1 - 2\sqrt{2/3}\sqrt{x} + x$. For small values of x the first curve is indeed below the second one, and the nonconvexity at the corners is verified.

APPENDIX A

Approximation

Taylor's Theorem, found in books on calculus, enables one to give polynomial approximations to functions in a routine manner which has even been incorporated in some elaborate computer algebra programs. However, we wish to make as much of this book as possible to be accessible to readers who have not had calculus yet. It turns out that the approximations we need can be derived using only algebra and geometry.

We introduce, for use in this section, a special convention about the symbols $\theta, \theta_1, \dots$. The convention is that these symbols stand for numbers ≤ 1 in absolute value. They may depend on all the variables that occur in the formula.

The identity

$$\frac{1}{1+x} = 1 - x + \frac{x^2}{1+x}.$$

gives the approximation, good for small values of x,

$$\frac{1}{1+x} = 1 - x + 2\theta x^2 \quad \text{for } |x| < \frac{1}{2}. \tag{A.1}$$

Next we derive an approximation for $\frac{1}{\sqrt{1+x}}$. We have

$$\left(1 - \frac{1}{2}x\right)^2 (1+x) = 1 - x^2\left(\frac{3}{4} - \frac{1}{4}x\right) = 1 - \theta x^2 \quad \text{for } |x| \leq 1.$$

We take the positive square root of this formula and use that the positive square root of a number is not farther from 1 than the number itself:

$$\left(1 - \frac{1}{2}x\right)\sqrt{1+x} = 1 - \theta_1 x^2 \quad \text{for } |x| \leq 1$$

or

$$\frac{1}{\sqrt{1+x}} = 1 - \frac{1}{2}x + \frac{\theta_1 x^2}{\sqrt{1+x}} = 1 - \frac{1}{2}x + 2\theta_2 x^2 \quad \text{for } |x| \leq \frac{1}{2}. \tag{A.2}$$

This approximation, like the previous one, is good only when x is very small since in that case the term containing the unspecified factor θ_1 is much smaller than the other terms.

In our discussion of iterations of the function $\sin x$ in Chapter 20 we need the estimate

$$x - \sin x \sim \tfrac{1}{6}x^3 \quad \text{as } x \to 0. \tag{A.3}$$

(We remind the reader that angles are measured in radians in this discussion. If we measured them in degrees, the formulas would be cluttered with πs and 180s.) The formula (A.3) is also an easy consequence of Taylor's theorem; we give here a derivation based on trigonometry.

We start with the relation

$$\sin x \sim x \quad \text{as } x \to 0. \tag{A.4}$$

This relation is rooted in the following facts. $2\sin x$ is the length of the chord of an arc of length $2x$ on the unit circle. The perimeter of a circle is the limit of the lengths of inscribed polygons as we take polygons with shorter and shorter sides. Hence the ratio of the length of a chord and the corresponding part of a circle must approach 1 as the chord length aproaches 0.

Next we derive an identity connecting the functions x and $\sin x$, using the method Archimedes devised for such computations.

In Fig. A.1, the quantity $\frac{1}{2}x$ is the area of the circular sector OAE while $\frac{1}{2}\sin x$ is the area of the triangle OAE. The difference is the area of the crescent ACE. We fill the crescent with triangles and we express the areas of the triangles in terms of sines of angles obtained by repeatedly halving x.

The first triangle we draw in the crescent is ECA. Its base is $2\sin(x/2)$. The equal sides have length $2\sin(x/4)$. The angles at the base have measure $x/4$. Hence the area of this triangle is $2\sin(x/2)\sin^2(x/4)$. Next we draw the triangles ABC, CDA, each of which has area $2\sin(x/4)\sin^2(x/8)$, etc. Since $x - \sin x$ is twice the area of the crescent, we have

$$x - \sin x = 4\sin\frac{x}{2}\sin^2\frac{x}{4} + 8\sin\frac{x}{4}\sin^2\frac{x}{8} + 16\sin\frac{x}{8}\sin^2\frac{x}{16} + \dots. \tag{A.5}$$

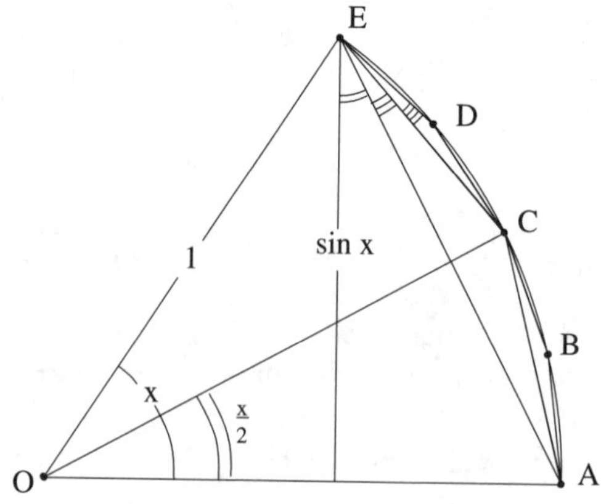

Figure A.1. Filling $x - \sin x$ by triangles.

If x is small then by (A.4) we may replace all the sin functions on the right-hand side of (A.5) by their arguments and change each term by only a small percentage. Moreover, for small x, all those terms have the same sign so their sum also changes by only a small percentage. (If a sum constains terms of opposite signs and we change the summands, the sum can change by a greater percentage than any of the summands.) Replacing the sin functions by their arguments gives the geometric series

$$\frac{x^3}{8} + \frac{x^3}{32} + \frac{x^3}{128} + \cdots = \frac{1}{6}x^3 .$$

This completes the proof of (A.3).

APPENDIX B

Limits and Continuity

Let $x_1, x_2, \ldots$ be an infinite sequence of real numbers. The sequence *converges to* x, or $\lim_{n\to\infty} x_n = x$, means that if n is large, x_n is close to x. More precisely: let I be any interval with center x. No matter how short I is, if we go out far enough in our sequence, all terms from there on are in I. Denoting the length of the interval I by 2ϵ and a term beyond which all terms lie in I by x_N, we get the customary definition of convergence:

DEFINITION. $\lim_{n\to\infty} x_n = x$ or $x_n \to x$ means that, given any $\epsilon > 0$, there is an integer $N = N(\epsilon)$ (i.e., N depends on ϵ) such that $|x_n - x| < \epsilon$ for $n \geq N(\epsilon)$. The number x is the *limit* of the sequence $\{x_n\}$.

The term "convergent sequence" implies that the sequence is infinite, but not that the terms are different; $1, 1, 1, \ldots$ is convergent.

EXAMPLE. Let $x_n = n/(n+1)$, $n = 1, 2, 3, \ldots$. Show that $\lim_{n\to\infty} x_n = 1$.

Given any $\epsilon > 0$, we show that $N(\epsilon) = \lceil 1/\epsilon \rceil$ has the property specified above. Since $n \geq N = \lceil 1/\epsilon \rceil$, $n + 1 > n \geq 1/\epsilon$ and

$$\left| \frac{n}{n+1} - 1 \right| = \frac{1}{n+1} < \epsilon.$$

It suffices to give *some* $N(\epsilon)$ in such proofs; we need not specify the smallest possible $N(\epsilon)$.

EXAMPLE. Let $0 < c < 1$. Show that $\lim_{n\to\infty} c^n = 0$.

Put $1/c = 1 + d$. Then $d > 0$. From the binomial theorem

$$\frac{1}{c^n} = (1+d)^n = 1 + nd + \frac{n(n-1)}{2}d^2 + \cdots > nd,$$

so $c^n < 1/(nd)$. Thus for $n \geq N(\epsilon) = \lceil 1/(d\epsilon) \rceil$, we have $c^n < \epsilon$.

We can think of real numbers as points on the number line. The definition of convergence also applies to sequences of points $\{P_n\}$ in two or more dimensions, with $|x_n - x|$ replaced by the distance $|P_n P|$.

DEFINITION. A set $\mathcal{S}$ is *closed* if it contains the limit of every convergent sequence whose terms belong to $\mathcal{S}$.

EXAMPLE. The set of nonnegative numbers is closed. The set of positive numbers is not closed because some sequences of positive numbers have limit 0, which is missing from the set. The set of numbers $\{x \mid 0 \leq x \leq 1\}$ is closed (a *closed interval*). The set of rational numbers is not closed, because a sequence of rational numbers can converge to an irrational number. A finite set (i.e., a set with finitely many elements) is closed. The only convergent sequences one can form out of elements of a finite set are sequences whose entries are all the same from a certain point on.

We introduce a term which is convenient in discussing the topics at hand.

DEFINITION. A *neighborhood* of a point x is any set $\mathcal{N}$ which contains all points at a distance $<\delta$ from x, for some $\delta > 0$. The quantity δ need not be the same for different neighborhoods $\mathcal{N}$.

In nontechnical language, "neighborhood" suggests a relatively small area. In mathematics, too, we think of a neighborhood as a small set, but it would be hard to include that in the definition, and it turns out to be unnecessary in the contexts in which the term is used. As an example, we reformulate the definition of convergence in terms of neighborhoods:

The sequence $\{x_n\}$ converges to x means that every neighborhood of x contains all except a finite number of the terms of the sequence.

This typical usage shows why we did not have to define a neighborhood to be a small set. What we are thinking is "every neighborhood, no matter how small", but there is no need to say "no matter how small" since, what is true for every neighborhood is true for arbitrarily small ones.

DEFINITION. We say that P is an *interior point* of a set of points $\mathcal{S}$ if $\mathcal{S}$ contains some neighborhood of P.

DEFINITION. A set $\mathcal{O}$ is *open* if every point of $\mathcal{O}$ is an interior point of $\mathcal{O}$. Note that a set which is not closed is not necessarily open.

If we peel an apple, the object we get still has a surface where molecules which are part of the peeled apple are in contact with air molecules which are not part of the apple. In the imaginary world of mathematics things can be different. If we remove the endpoints of a closed interval, such as $\{x \mid 0 \leq x \leq 1\}$, we get the *open interval* $\{x \mid 0 < x < 1\}$ which is bounded but has only interior points.

A common notation for a closed interval $\{x \mid a \leq x \leq b\}$ is $[a, b]$; if an endpoint is missing we use a parenthesis instead of a bracket. For instance, the interval $\{x \mid 0 \leq x < 1\}$is denoted by $[0, 1)$.

DEFINITION. P is a *boundary point* of a set $\mathcal{S}$ if every neighborhood of P contains points in $\mathcal{S}$ and points not in $\mathcal{S}$. Note that a point can be a boundary point of $\mathcal{S}$ without belonging to $\mathcal{S}$.

DEFINITION. The *closure* $\overline{\mathcal{S}}$ of a set $\mathcal{S}$ is the union of $\mathcal{S}$ and the set of its boundary points.

THEOREM 1. *The closure $\overline{\mathcal{S}}$ of $\mathcal{S}$ is closed. Every closed set containing $\mathcal{S}$ contains $\overline{\mathcal{S}}$.*

PROOF. Let $P_1, P_2, \ldots$ be a sequence of points in $\overline{\mathcal{S}}$ such that $P_n \to P$. We have to show $P \in \overline{\mathcal{S}}$, i.e., that P is either a point of $\mathcal{S}$ or a boundary point of $\mathcal{S}$. This amounts to showing that for any $\epsilon > 0$, there is a point Q of $\mathcal{S}$ at distance $<\epsilon$ from P. Let n be such that $|PP_n| < \epsilon/2$. Since P_n is a point or a boundary point of $\mathcal{S}$, there is a point $Q \in \mathcal{S}$ at distance $<\epsilon/2$ from it, and hence $|PQ| < \epsilon$. Thus $\overline{\mathcal{S}}$ is closed. The proof of the second half of Theorem 1 is an easy exercise.

In a discussion involving sets of points, all the points under consideration are on a line, a plane, in a three-dimensional space or possibly a space of more dimensions. The set of all points is the *universal set* for the given context. For example, when we discuss real numbers in geometric terms, the universal set is the real number line.

DEFINITION. The *complement* of a set $\mathcal{S}$ is the set of those points of the universal set which are not in $\mathcal{S}$.

THEOREM 2. *The complement of a closed set is open and the complement of an open set is closed.*

PROOF. Let $\mathcal{F}$ be a closed set and let $\mathcal{F}^c$ be the complement of $\mathcal{F}$. Let $P \in \mathcal{F}^c$. We show that for some positive integer n, the set of points $N_n : \{Q \mid |QP| < 1/n\}$ consists entirely of points of $\mathcal{F}^c$. Otherwise, for each n there would be a point $P_n \in \mathcal{F}$ such that $|P_nP| < 1/n$. But then we would have $\lim_{n\to\infty} P_n = P$ and since $\mathcal{F}$ is closed, we would have $P \in \mathcal{F}$. Thus, $\mathcal{F}^c$ must be open.

To prove the second part of the theorem, suppose $\mathcal{O}$ is an open set. Let $\mathcal{O}^c$ be the complement of $\mathcal{O}$. By the definition of open sets, no point in $\mathcal{O}$ can be the limit of a convergent sequence whose elements all belong to $\mathcal{O}^c$. This means that any convergent sequence with all elements in $\mathcal{O}^c$ has its limit in $\mathcal{O}^c$. Hence $\mathcal{O}^c$ is a closed set.

It is important to note that whether a set is open depends not only on the set but also on the universal set under discussion. A segment of the real number line without its endpoints is an open subset of the set of real numbers. However, a line segment is not an open set if the universal set consists of the points of a plane because an open set in the plane has to contain a disk around every one of its elements. In contrast to open sets, a closed set such as a closed interval will continue to satisfy the definition of "closed" even if we regard it as a subset of a higher dimensional space.

We next present some simple but basic facts about bounded sequences and sets.

DEFINITION. A sequence or a set of real numbers is said to be *bounded above* if there is a number M such that every element of the set is $\leq M$. M is an *upper bound* of the set. *Bounded below* and *lower bound* are defined similarly, and *bounded* means bounded both above and below. In more dimensions, we say a set of points is bounded if the set of the distances of its points from some point O is bounded.

THEOREM 3. *A convergent sequence is bounded.*

PROOF. Let the sequence $\{x_n\}$ converge to x as $n \to \infty$. For $\epsilon = 1$ we can find a positive integer N such that $|x_n - x| < 1$ for all $n \geq N$. This implies

$$|x_n| \leq |x| + |x_n - x| < |x| + 1$$

for $n \geq N$. Let $M = \max\{|x_1|, \ldots, |x_{N-1}|, |x| + 1\}$. We see that M can serve as a bound for the sequence $\{|x_n|\}$.

THEOREM 4 (THE LEAST UPPER BOUND PROPERTY). *Let S be a set of real numbers which is bounded above. Then S has an upper bound L that is smaller than any other upper bound of S. L is called the least upper bound of the set. Similarly, if S is bounded from below, it has a lower bound G that is greater than any other lower bound, and G is called the greatest lower bound.*

Remark. Is there anything to be proved here? Since the set is bounded above, there are numbers that are upper bounds of it, so can we not just say that L is the smallest of these numbers? The problem is that we don't know yet whether there is a smallest one among these upper bounds. The set of upper bounds could be like the set of positive numbers, which has no smallest element.

PROOF. To avoid introducing extra notation, assume all elements of $\mathcal{S}$ are <10 but some are >0. Think of the elements of the set as written in decimal form. Let a_0 be the largest digit that occurs to the left of the decimal point. Let $\mathcal{S}^0$ be the subset of $\mathcal{S}$ consisting of those numbers in $\mathcal{S}$ that have a_0 to the left of the decimal point. Let a_1 be the largest digit occurring just behind the decimal point in numbers of $\mathcal{S}^0$. Let $\mathcal{S}^1$ be the set consisting of those numbers in $\mathcal{S}^0$ that have the form $a_0 \cdot a_1 \cdots$. Repeat this construction over and over. The number $L = a_0 \cdot a_1 a_2 \cdots$ is an upper bound of $\mathcal{S}$. We have to check that if $L' < L$ then L' is not an upper bound of $\mathcal{S}$. Since $L' < L$, for some integer n, $L' < a_0 \cdot a_1 a_2 \cdots a_n$. From our construction of the digits, we know that $\mathcal{S}$ has elements which are greater than or equal to $a_0 \cdot a_1 a_2 \cdots a_n$, so L' is not an upper bound of $\mathcal{S}$.

DEFINITION. The sequence $\{x_1, x_2, \ldots\}$ is said to be *increasing* if it satisfies $x_n \leq x_{n+1}$ for $n = 1, 2, \ldots$. If $x_n < x_{n+1}$ holds for $n = 1, 2, \ldots$, then $\{x_n\}$ is called *strictly increasing. Decreasing* and *strictly decreasing* sequences are defined similarly.

Increasing and decreasing sequences are called *monotone* sequences.

THEOREM 5. *A bounded monotone sequence converges.*

PROOF. Let the sequence $\{x_n\}$ be increasing and bounded above; let L be its least upper bound. For any $\epsilon > 0$, $L - \epsilon$ is not an upper bound and there exists a term, say x_N, such that $x_N > L - \epsilon$. By virtue of the increasing property, we see that $x_n \geq x_N > L - \epsilon$ for $n \geq N$. This means that the definition of $\lim_{n\to\infty} x_n = L$ is satisfied.

DEFINITION. A *subsequence* of a sequence is a sequence obtained by selecting some (possibly all) of the terms of the original sequence without changing their original order.

It is clear from the above definitions that if a sequence converges to x, then its infinite subsequences also converge to x.

The following little fact is useful in proving the important Theorem 7.

THEOREM 6. *Every infinite sequence $\{x_n\}$ of real numbers contains an infinite monotone subsequence.*

PROOF. Let $\mathcal{S}$ be the set of indices of those terms of the sequence, if any, which are smaller than all subsequent terms. If $\mathcal{S}$ is infinite, with indices

$i_1 < i_2 < i_3 < \cdots$, then $x_{i_1} < x_{i_2} < x_{i_3} < \cdots$ is strictly increasing. Otherwise, $\mathcal{S}$ is finite. In that case, there is a least index i_1, such that for all $i \geq i_1$ we have $i \notin S$. Since $i_1 \notin S$, there exists $i_2 > i_1$ such that $x_{i_1} \geq x_{i_2}$. And $i_2 \notin S$ so there exists $i_3 > i_2$ such that $x_{i_2} \geq x_{i_3}$, and so forth. The sequence $x_{i_1}, x_{i_2}, x_{i_3}, \ldots$ is monotone decreasing, completing the proof.

THEOREM 7 (BOLZANO AND WEIERSTRASS' THEOREM). *Every bounded infinite sequence has a convergent subsequence.*

PROOF. By Theorem 6 the infinite sequence contains a monotone subsequence. By Theorem 5, the subsequence converges.

DEFINITION. A set of points is said to be *compact* if it is closed and bounded.

EXAMPLE. Any finite closed interval is compact. Any set consisting of finitely many points is compact. A circle is a compact set and so is the set of points which are on or inside a circle (a *closed disk*).

It is easy to show that a compact set of numbers contains its least upper bound and its greatest lower bound.

Our next topic is continuity of functions. Roughly speaking, a function f is *continuous at* x if changing x by a small amount causes only a small change in $f(x)$. The familiar functions, with the exception of the greatest integer function $\lfloor x \rfloor$, are continuous at every point where they are defined. The function $\lfloor x \rfloor$ is discontinuous at $x = 1$ for example, because if we decrease the value of x from 1, no matter how slightly, $\lfloor x \rfloor$ jumps from 1 to 0.

The points where $f(x)$ is undefined are disregarded in the mathematical definition of continuity. The function $1/x$ is continuous at all points where it is defined and hence it qualifies as a continuous function. What distinguishes it from the functions we usually think of as continuous is that it is undefined at $x = 0$ and there is no way to define $f(0)$ so as to make the function continuous.

We can reformulate the definition of continuity of f at x_0 as follows. f is continuous at x_0 means: If we take any neighborhood $\mathcal{N}$ of $f(x_0)$, no matter how small, $f(x) \in \mathcal{N}$ if x is close enough to x_0.

We get the customary definition of continuity if, in the above, we give a more precise meaning of "small" in terms of a number ϵ, and to "close enough" in terms of a number δ:

DEFINITION. Let f be a function defined on a set $\mathcal{S}$. Let x_0 be a point in $\mathcal{S}$. The function f is said to be *continuous at* x_0 if, for any $\epsilon > 0$, there exists a number $\delta > 0$ such that if $|x - x_0| < \delta$ and $x \in S$,

$$|f(x) - f(x_0)| < \epsilon.$$

The function f is said to be continuous on $\mathcal{S}$ if f is continuous at every point of $\mathcal{S}$.

Continuity at x can also be defined in terms of sequences, as follows. f is continuous at a point $x \in S$ if for every sequence $\{x_n\}$ of points of $\mathcal{S}$ such that $x_n \to x$ we have

$$\lim_{n\to\infty} f(x_n) = f(x). \tag{B.1}$$

The proof that continuity implies (B.1) is very easy. We prove the converse, that if (B.1) holds for all sequences which converge to x then f is continuous at x.

Consider the sequence of ever smaller neighborhoods $\mathcal{N}_n = \left(x - \frac{1}{n},\, x + \frac{1}{n}\right)$ of x. Let

$$\epsilon_n = \min\left(1,\, \tfrac{1}{2} \operatorname*{l.u.b.}_{\xi\in\mathcal{N}_n} |f(\xi) - f(x)|\right)$$

(In case the least upper bound does not exist, ignore that term.) In any case, there is a number $x_n \in \mathcal{N}_n$ such that $|f(x_n) - f(x)| \geq \epsilon_n$. The sequence $\{x_n\}$ converges to x. By (B.1) $|f(x_n) - f(x)| \to 0$ and thus $\epsilon_n \to 0$. Since $|f(\xi) - f(x)| < 2\epsilon_n$ for $\xi \in \mathcal{N}_n$, f satisfies the definition of continuity at x.

Another unexciting but very useful theorem is the following.

THEOREM 8. *If a function f is continuous at every point of a compact set $\mathcal{S}$, then there is a point $x_{\max} \in \mathcal{S}$ such that for all $x \in \mathcal{S}$, $f(x) \leq f(x_{\max})$. There is also a point $x_{\min}$ such that $f(x) \geq f(x_{\min})$ for all $x \in \mathcal{S}$.*

This theorem tells us that on a compact set a continuous function is bounded and assumes its least upper bound and its greatest lower bound. It might help to look at an example of what can happen on a noncompact set. Consider $f(x) = \tan x$ on the set $\mathcal{S} = \{x | 0 < x < \pi/2\}$. The function is continuous at every point of $\mathcal{S}$. It is bounded from below but does not assume its greatest lower bound, and it is not bounded from above.

PROOF. We prove that f has an upper bound on $\mathcal{S}$. Assume the contrary. Then there is a sequence of points $x_1, x_2, \ldots$ such that $f(x_n) > n$. Since $\mathcal{S}$ is compact, the sequence $\{x_n\}$ has a subsequence $x_{k_1}, x_{k_2}, \ldots$ which converges to some point $x^* \in \mathcal{S}$. The continuity of f requires that

$$\lim_{n\to\infty} f(x_{k_n}) = f(x^*)$$

which is violated here, so f must indeed be bounded on $\mathcal{S}$. (Some books allow ∞ as a function value and as a limit of sequences but we adhere to the opposite convention.)

Let L be the least upper bound of the values of $f(x)$ on $\mathcal{S}$. Let x_n be a point such that $f(x_n) > L - 1/n$. Then we can take a convergent subsequence of $\{x_n\}$ and show as above that the limit x^* satisfies $f(x^*) = L$.

THEOREM 9 (INTERMEDIATE VALUE THEOREM). *Let f be continuous on a closed interval $[a, b]$ and let y be any number in the interval $[f(a), f(b)]$. There exists a number c, $a \leq b$, such that $f(c) = y$.* See Fig. B.1.

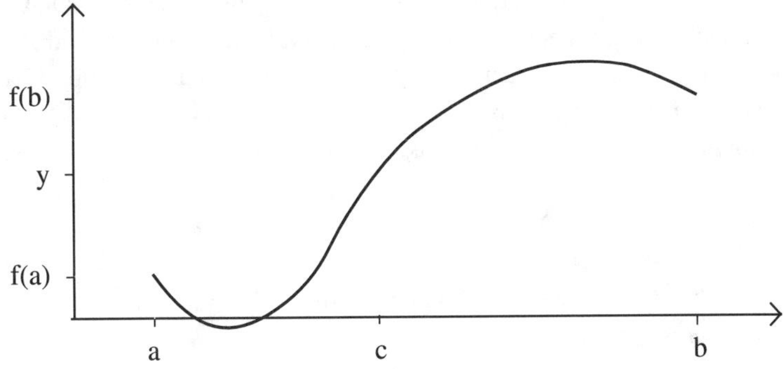

Figure B.1. Intermediate Value Theorem.

PROOF. If $f(a) = y$ or $f(b) = y$ then we can take $c = a$ or $c = b$. If neither of these equations holds, we may suppose without loss of generality[1] $f(a) < y$ and $f(b) > y$. Let c be the least upper bound (l.u.b.) of the set $\mathcal{N} = \{x \mid f(x) < y\}$. We claim that $f(c) = y$. Indeed, if we had $f(c) < y$, then by the continuity of f we would have $f(x) < y$ for values of x slightly greater than c. Similarly, if we had $f(c) > y$, then the values of f would be $> y$ in some interval with c in the middle, and the left end of that interval would be an upper bound of the set $\mathcal{N}$ smaller than c.

Here is a down-to-earth problem which can be solved using the intermediate value theorem for continuous functions. Tom started to climb Mount Timpanogos at 6:00 am and reached the top at 6:00 pm. The next morning, he started his descent at 6:00 am, following the same path, and arrived at the same place where he started the previous day. Show that there is a point on the path where his watch read the same time on those two different days.

[1]This phrase means, we are considering one of the possibilities but the others can be handled in the same way.

In this form the problem is a little confusing. As is often the case, a reformulation of the problem can lead to an easier solution. Our problem is essentially the same as the following. Suppose that Tom starts to climb and Brenda begins to descend the mountain at 6:00 a.m. on the *same* day. It is evident that they must meet, and their watches and positions will agree. The mathematical foundation for this problem is the intermediate value theorem: the sign of the directed distance separating them changes between 6:00 a.m. and the time when the first person reaches his or her destination.

Combining Theorems 8 and 9, we can say that a real-valued continuous function of a real variable maps a finite closed interval onto a finite closed interval. The same argument can be applied in a more general situation to prove that a continuous function maps a compact set onto a compact set.

An *infinite series* of real numbers u_n is an infinite sum

$$\sum_{i=1}^{\infty} u_i = u_1 + u_2 + u_3 + \cdots.$$

The u_i are the *terms* of the series and the quantities $S_n = u_1 + u_2 + \cdots + u_n$ are its *partial sums.*

An example of a series is $3/10 + 3/100 + 3/1000 + \cdots$, more familiar as $0.333\ldots$.

DEFINITION. If the sequence of partial sums S_n converges to a real number $\mathcal{S}$, then the infinite series is said to converge to the sum $\mathcal{S}$ and we write $u_1 + u_2 + u_3 + \cdots = S$. If S_n does not converge, the series is said to *diverge.*

EXAMPLE. If $|q| < 1$ then the *geometric series* converges and we have a simple formula for its sum:

$$\sum_{i=0}^{\infty} q^i = \frac{1}{1-q}.$$

PROOF. The nth partial sum of the series is

$$1 + q + q^2 + \cdots + q^{n-1} = \frac{1-q^n}{1-q}.$$

Since $|q| < 1$, we have $q^n \to 0$ as n tends to infinity. The statement is proved.

It follows that the series $3/10 + 3/100 + 3/1000 + \cdots$ converges to $1/3$, confirming a fact we have known since grade school.

So far each u_i was assumed to be a real number. We shall also consider series of functions $u_1(x) + u_2(x) + \cdots$. The set of points at which the series converges is called the *region of convergence* of the series, denoted by $\mathcal{D}$. The series of functions defines a function $S(x)$, its *sum function*, on $\mathcal{D}$. For example, let $u_i(x) = x^i$ for $i = 0, 1, 2, \ldots$, each of which is defined on $(-\infty, \infty)$, but $\mathcal{D}$ $=(-1, 1)$. If $x \in (-1, 1)$, we have $1 + x + x^2 + \cdots = \frac{1}{1-x}$. Note that if we used this formula unthinkingly to the sum of the series for $x = 2$, we would get the nonsensical result $1 + 2 + 4 + 8 + \cdots = -1$.

Let $S_n(x) = u_1(x) + u_2(x) + \cdots + u_n(x)$. We have $\lim_{n\to\infty} S_n(x) = S(x)$ for all $x \in \mathcal{D}$. By the definition of limit, we can say that for any $x \in \mathcal{D}$ and arbitrarily given $\epsilon > 0$ there exists a positive integer $N(\epsilon, x)$ such that we have

$$|S(x) - S_n(x)| < \epsilon \quad \text{for all } n > N(\epsilon, x). \tag{B.2}$$

The notation $N(\epsilon, x)$ stresses that N depends on x as well as ϵ.

THEOREM 10. *A sum of continuous functions is continuous.*

PROOF. Let $h(x) = f(x) + g(x)$ and suppose f and g are continuous at x_0. Let $\epsilon > 0$ be given. By continuity of f, there is a $\delta_1 > 0$ such that $|f(x) - f(x_0)| < \epsilon/2$ in $|x - x_0| < \delta_1$ and by the continuity of g, there is a $\delta_2 > 0$ such that $|g(x) - g(x_0)| < \epsilon/2$ in $|x - x_0| < \delta_2$. Let $\delta = \min(\delta_1, \delta_2)$. Then $|h(x) - h(x_0)| < \epsilon$ if $|x - x_0| < \delta$, which shows that h is continuous. The case of more than 2 summands follows similarly.

In his text *Cours d'analyse de l'Ecole Polytechnique* (1821), which raised the standard of rigor in calculus, Augustin Louis Cauchy (1798–1857) asserted that a convergent series of continuous functions is also continuous. However, Niels Henrik Abel (1802–1829) pointed out that this is not always true. For instance $u_1(x) = x$ and $u_i(x) = x^i - x^{i-1}$ for $i = 2, 3, \ldots$, are continuous on the the set of real numbers. The nth partial sum is $S_n(x) = x^n$. Hence the region of convergence is $\mathcal{R} = (-1, 1]$. However, the sum $S(x)$ of the series is discontinuous at the point $x = 1$ of $\mathcal{D}$ since $S(x) = 0$ for $-1 < x < 1$ and $S(1) = 1$.

Abel's observation highlighted the fact that in problems involving sequences of functions it is worth considering definitions of convergence other than the obvious one of convergence at every point of some set. This has grown into a very fruitful endeavor. Abel introduced the notion of *uniform* convergence to replace Cauchy's erroneous assertion.

DEFINITION. The series $u_1(x)+u_2(x)+\cdots$ is said to be uniformly convergent in a region $\mathcal{R}$ if for all ϵ there is an $N(\epsilon)$, such that $|S(x) - S_n(x)| < \epsilon$ for all $n > N(\epsilon, x)$ holds for all $x \in \mathcal{R}$. In other words, (B.2) can be satisfied for each ϵ by one N which is good for all points of $\mathcal{R}$.

EXAMPLE. Take the series in the last example but restrict x to the interval $\mathcal{R}: 0 \leq x \leq \frac{1}{2}$. Now $S(x) = 0$ everywhere on $\mathcal{R}$, and $|S(x) - S_n(x)| = x^{n+1} \leq 1/2^{n+1}$. Hence, for any ϵ, we can take $N(\epsilon)$ to be an integer such that $1/2^N < \epsilon$ and (B.2) will hold for all $x \in \mathcal{R}$.

THEOREM 11. *Suppose the functions $u_1(x)$, $u_2(x)$, ... are continuous and the series $u_1(x) + u_2(x) + \cdots$ converges uniformly in an interval I. Then the sum $S(x)$ of the series is continuous in I.*

PROOF. Let x_0 be any point in I and let $\epsilon > 0$ be given. We have to show that $|S(x) - S(x_0)| < \epsilon$ in some neighborhood of x_0.

Uniform convergence implies that there is a positive integer n such that

$$|S(x) - S_n(x)| \leq \epsilon/3$$

for all $x \in I$, where $S_n(x)$ denotes the nth partial sum of our series. The partial sum $S_n(x)$ is a finite sum of continuous functions and hence it is continuous in I. So there is a $\delta > 0$ such that for $x \in I$ and $|x - x_0| < \delta$ we have

$$|S_n(x) - S_n(x_0)| < \epsilon/3.$$

Thus for $x \in I$ and $|x - x_0| < \delta$ it follows that

$$\begin{aligned} |S(x) - S(x_0)| &\leq |S(x) - S_n(x)| + |S_n(x) - S_n(x_0)| + |S_n(x_0) - S(x_0)| \\ &< \epsilon/3 + \epsilon/3 + \epsilon/3 = \epsilon. \end{aligned}$$

Hence $S(x)$ is continuous at x_0.

The following is a simple criterion for uniform convergence:

THEOREM 12 (WEIERSTRASS). *If $|u_i(x)| \leq a_i$ for $x \in I$ and $i = 1, 2, \ldots$ and $a_1 + a_2 + \cdots = A$ is a convergent series of constants, then the series $u_1 + u_2 + \cdots$ converges uniformly in I.*

PROOF. Let T_n denote the *tail* $a_{n+1} + a_{n+2} + \cdots$ of the series $a_1 + a_2 + \cdots$. Convergence means that $\lim_{n\to\infty} T_n = 0$. Thus for arbitrarily given $\epsilon > 0$ there exists a positive integer $N(\epsilon)$ such that for $n > N$, we have $T_n < \epsilon$. It follows from the hypothesis that $|S(x) - S_n(x)| = |u_{n+1}(x) + u_{n+2}(x) + \cdots| \leq T_n < \epsilon$ for $n > N$ and $x \in I$, i.e., we have uniform convergence.

Most of what we have done so far can be described as a close examination of simple concepts. We hope it clarified the meaning of words such as continuous and convergent, but the reader probably did not find surprising anything we stated in this appendix so far. In contrast, the next theorem tells us something rather unexpected.

THEOREM 13 (H. E. HEINE, 1821–1881 AND EMILE BOREL, 1871–1956). *Suppose a family $\mathcal{O}$ of infinitely many open sets covers a finite closed interval I, or, more generally, any compact (closed and bounded) set I in one or more dimensions. Then I can also be covered with finitely many sets of $\mathcal{O}$.*

PROOF. We formulate the proof when I is a closed interval, but the same idea can be used in the general case. The proof is by assuming the assertion of the theorem to be false and deriving a contradiction. Suppose it is impossible to cover I with finitely many sets of $\mathcal{O}$. Then at least one of the two closed half-intervals from the midpoint of I to the ends can not be covered by finitely many sets of $\mathcal{O}$. We can continue to bisect and obtain an infinite sequence of closed intervals, each half as long as the previous one, which can not be covered by finitely many sets of $\mathcal{O}$.

Let P be a limit point of a sequence of endpoints of these intervals. Possibly P is an endpoint of I but even then, P is in I because I is a closed set. P is covered by an open set O_P of $\mathcal{O}$. The set O_P by itself covers all intervals whose endpoints are close enough to P. This contradicts the definition of P, completing the proof of the Heine-Borel theorem.

We apply the Heine-Borel theorem to prove the following which is due essentially to Heine; historically, it preceded the Heine-Borel theorem. One should not think that it took until the middle of the 19th century for mathematicians to uncover the simple fact expressed by Theorem 14; what Heine gets credit for is recognizing that a proof ought to be given, and thereby starting to create some concepts which turned out to be basic for the later development of mathematics.

THEOREM 14. *Let $f(x)$ be continuous on a finite closed interval, which for definiteness we take to be* $[0, 1]$. *Given any ϵ, there is a piecewise linear function $p(x)$ defined on* $[0, 1]$ *such that*

$$|f(x) - p(x)| < \epsilon \quad \text{on } 0 \le x \le 1. \tag{B.3}$$

PROOF. Let us say, for the purpose of this proof, that the *spread* of a function f is $< \epsilon$ on an interval I if $x \in I$ and $y \in I$ imply $|f(x) - f(y)| < \epsilon$.

LEMMA. *Suppose the spread of f is $< \epsilon$ on I. Let $ch(x)$ be a function whose graph is a chord with endpoints $(a, f(a)$ and $(b, f(b))$, where $a < b$, $a \in I$, $b \in I$. Then*

$$|f(x) - ch(x)| < \epsilon \quad \text{on } a \le x \le b. \tag{B.4}$$

PROOF. The value $ch(x)$ of the chord function is between its values $f(a)$ and $f(b)$ at the ends of the chord. By assumption $|f(x) - f(a)| < \epsilon$ and $|f(x) - f(b)| < \epsilon$, and these two inequalities imply (B.4).

Continuity of f implies that each $x \in [0, 1]$ the center of some open interval $\mathcal{O}_x$ such that the spread of f on $\mathcal{O}_x$ is $< \epsilon$. Let $\mathcal{O}'_x$ be the middle third of $\mathcal{O}_x$. The Heine-Borel Theorem tells us that finitely many of the $\mathcal{O}'_x$, say $\mathcal{O}'_1, \ldots, \mathcal{O}'_m$, cover $[0, 1]$. Let δ be the length of the shortest of these intervals. Let $[a, b]$ be any interval of length $< \delta$. Let k be such that $a \in \mathcal{O}'_k$. Then the entire interval $[a, b]$ is contained in $\mathcal{O}_k$ and hence the spread of f on $[a, b]$ is $< \epsilon$. Hence by the Lemma we obtain the graph of a function $p(x)$ satisfying Theorem 14 by taking any sequence of abscissas $0 = x_0 < x_1 < \cdots < x_n = 1$ such that $x_{i+1} - x_i < \delta$ and drawing the chords through the points $(x_0, f(x_0))$, $(x_1, f(x_1)), \ldots$.

APPENDIX C

Convexity

We have collected in this appendix several facts about convexity. They are more or less obvious from the geometric point of view but it is useful to list them and to see how the properties can be proved rigorously, instead of referring to what is obvious from one's mental image of a convex set. In a rigorous presentation points are pairs or triples or n-tuples of numbers and all assertions have to be derived from the properties of real numbers. This is what our proofs aim to do. We still use geometric language but the terms we use can be easily expressed in terms of real numbers and sets. For instance, if we have two points in the plane, $P = (a, b)$ and $Q = (c, d)$, "the line segment PQ" is the set of points $\{(x, y) : x = (1 - t)a + tc,\ y = (1 - t)b + td,\ 0 \le t \le 1\}$.

A set $\mathcal{S}$ of points is said to be *convex* if, whenever $P \in \mathcal{S}$ and $Q \in \mathcal{S}$, all points of the line segment PQ are in $\mathcal{S}$.

A function $f(x)$ is said to be *convex* on an interval I if for all $a \in I,\ b \in I$, every point of the chord connecting $(a, f(a))$ and $(b, f(b))$ is on or above the graph of $f(x)$. Expressing this in algebraic language: f is convex in I means that for any $a,\ b \in I$,

$$f(\lambda a + \mu b) \le \lambda f(a) + \mu f(b) \quad \text{if } \lambda \ge 0, \quad \mu \ge 0, \quad \lambda + \mu = 1.$$

If every chord of a curve segment lies on or above the curve then the line segment connecting any two points which are on or above the curve is also on or above the curve. Thus we have

THEOREM 1. *Convexity of the function $f(x)$ on an interval I is equivalent to convexity of the set of points lying on or above the graph of the function in I, i.e., the set $\{(x, y) \mid x \in I,\ y \ge f(x)\}$.*

The definition of convexity for sets applies without change in any number of dimensions. The definition of convexity is applied to functions of two or more variables with the change that the domain of the independent variables be a convex set. As in the one-dimensional case, a function is convex if and only if the set of points lying on or above its graph is convex.

THEOREM 2. *If f is differentiable on an interval I and the derivative is nondecreasing then f is convex on I.*

PROOF. Let the abscissas of a chord be a and b, $a < b$. We have to show that for any value c between a and b, the point $(c, f(c))$ lies on or below the chord from $(a, f(a))$ to $(b, f(b))$. To simplify the formulas we take c to be 0. (Rewriting the proof with a general value of c may help the reader understand it.) We have to show

$$f(0) \leq \frac{bf(a) - af(b)}{b - a} \quad \text{or} \quad 0 \leq b(f(a) - f(0)) - a(f(b) - f(0)).$$

By the mean value theorem $f(a) - f(0) = af'(\xi_a)$, $f(b) - f(0) = bf'(\xi_b)$ where $\xi_a \in (a, 0)$ and $x_b \in (0, b)$. The inequality we have to prove becomes $0 \leq ab(f'(\xi_a) - f'(\xi_b))$. In the expression on the right side, a is negative and b is positive, and since f' is increasing and $\xi_a < \xi_b$, the last factor is nonpositive. This completes the proof.

THEOREM 3. *The intersection of any collection of convex sets is convex.*

This is an immediate consequence of the definition of convexity. Note that there may be an infinite number of sets in the collection.

THEOREM 4. *Let $f(x)$, $g(x)$ be functions of a scalar or vector variable x which are convex on a set D. Then $h(x) = \max(f(x), g(x))$ is also convex on D.*

PROOF. The set of points on or above the graph of h is the intersection of the sets of points above the graphs of f and g and hence it is convex.

THEOREM 5. *Let $\mathcal{S}$ be a closed connected*[1] *set of points in any number of dimensions. Suppose there is a $\delta > 0$ such that if P and Q are two points of $\mathcal{S}$ less than δ apart, then all points of the line segment PQ are in $\mathcal{S}$. Then $\mathcal{S}$ is convex.*

Roughly speaking, the theorem says that if all small pieces of $\mathcal{S}$ are convex then $\mathcal{S}$ is convex.

PROOF. There are several mathematical concepts which formalize the intuitive notion of the set $\mathcal{S}$ being connected, i.e., that it consists of one piece. Clearly, any definition of connectedness must imply the following: If P and Q are points of $\mathcal{S}$, and $\epsilon > 0$ is an arbitrarily small number, there is an open polygon[2] with

[1] We discuss the meaning of this term in the proof.

[2] An *open polygon* is the union of a finite sequence of line segments $A_0A_1, A_1A_2, \ldots, A_{n-1}A_n$, where the endpoints A_0 and A_n may be different. A *closed* polygon is one in which $A_n = A_0$; a polygon without an adjective is usually understood to be closed. In Appendix B we discuss open and closed sets; the two uses of this pair of words are unrelated.

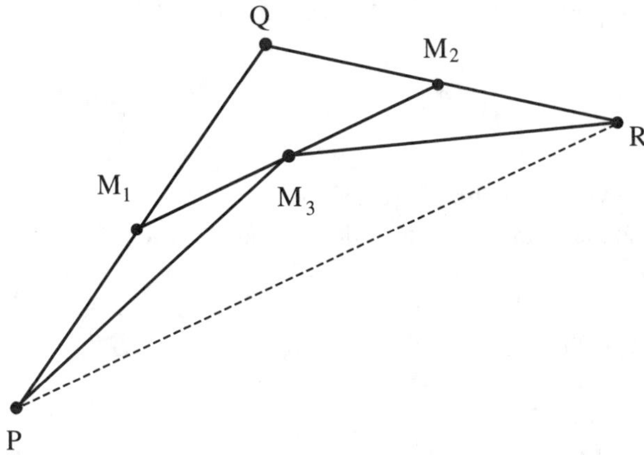

Figure C.1. Proof of the lemma.

endpoints P and Q, *vertices* in $\mathcal{S}$ and side lengths $< \epsilon$. This is all we need. All other mathematical definitions of connectedness require more.

LEMMA. *Under the hypotheses of Theorem 5, if P, Q, R are points such that*

$$|PQ| < \delta, \qquad |QR| < \delta \tag{C.1}$$

then $\mathcal{S}$ contains the line segment PR.

PROOF OF THE LEMMA By hypothesis the line segments PQ and QR are in $\mathcal{S}$. Let M_1, M_2, M_3 be the midpoints of PQ, QR, and M_1M_2, see Fig. C.1. By (C.1) and the triangle inequality $|PM_1| < \delta/2$, $|M_1M_3| < \delta/2$. Hence $|PM_3| < \delta$ and thus the entire line segment PM_3 is in $\mathcal{S}$. Similarly, the line segment M_3R is also in $\mathcal{S}$. The height of $\triangle PM_3R$ is half the height of $\triangle PQR$. The construction can be repeated. We get triangles of base PR and ever smaller heights whose two sides are in $\mathcal{S}$. The tops of these triangles approach the base and since $\mathcal{S}$ is a closed set, the base PR is also in $\mathcal{S}$. The lemma is proved.

Applying the lemma with 2δ, 4δ, etc. in place of δ, we get that for line segments of any length, if $\mathcal{S}$ contains PQ and QR then $\mathcal{S}$ contains PR. From this it is easy to derive the convexity of $\mathcal{S}$. Let A, B be any two points in $\mathcal{S}$. They can be connected by an open polygon with vertices in $\mathcal{S}$ and sides $<\delta$. Hence the sides of the polygon are also in $\mathcal{S}$. If two consecutive sides PQ and QR of a polygon are in $\mathcal{S}$ then so is PR. Thus we can replace pairs of consecutive sides of the open polygon by line segments which are also in $\mathcal{S}$ until we conclude that the line segment AB is in $\mathcal{S}$.

Consider n points $P_1, \ldots, P_n$. A *convex combination* of these points is any point of the form

$$\text{(C.2)} \qquad \begin{aligned} &t_1 P_1 + \cdots + t_n P_n, \\ &n \geq 1, \ t_1 \geq 0, \ldots, t_n \geq 0, \quad \text{and} \quad t_1 + \cdots + t_n = 1. \end{aligned}$$

We can rephrase the definition of a convex set in the form: a set $\mathcal{S}$ is convex means that if two points P and Q are in $\mathcal{S}$ then $\mathcal{S}$ contains every convex combination of P and Q.

THEOREM 6. *For all positive integers n, a convex set $\mathcal{S}$ contains the convex combinations of any n of its points.*

PROOF. We prove this by induction. For $n = 1$ there is nothing to prove and for $n = 2$, the statement is the definition of convexity. Suppose $n = 3$. We may assume that the coefficient t_1 in (C.2) is not 1, since otherwise we are back in the case $n = 1$. Then

$$t_1 P_1 + t_2 P_2 + t_3 P_3 = t_1 P_1 + (t_2 + t_3)\left(\frac{t_2}{t_2 + t_3} P_2 + \frac{t_3}{t_2 + t_3} P_2\right).$$

The point in the parentheses on the right side is in $\mathcal{S}$, and the entire right side is a convex combination of it and P_3, and hence it is in $\mathcal{S}$ also. The same procedure can be used to go up to any number n of points.

THEOREM 7. *Let $\mathcal{S}$ be a set of points in a space of any number of dimensions and let $\mathcal{H}$ be the set of convex combinations of finite subsets of $\mathcal{S}$. The set $\mathcal{H}$ is convex.*

PROOF. We have to show that if $\mathcal{H}$ contains the points P and Q then it contains any convex combination R of P and Q. The points P and Q are convex combinations of points in the union of the sets on which P and Q are based. To prove Theorem 7 one has to show that a convex combination of convex combinations of points of a set is itself a convex combination. This involves a few lines of algebra which we leave to the reader.

The set $\mathcal{H}$ is the *convex hull* of the set $\mathcal{S}$. Every convex set containing $\mathcal{S}$ contains $\mathcal{H}$. We can say $\mathcal{H}$ is the smallest convex set containing $\mathcal{S}$.

In a space of n dimensions it suffices to take the convex combinations of all subsets of $n + 1$ points instead of all finite subsets in Theorem 7. (Caratheodory's theorem.)

THEOREM 8. *If the set $\mathcal{S}$ is convex then its closure $\mathcal{C}$ and its interior $\mathcal{I}$ are also convex.*

PROOF. Let P, Q be points in $\mathcal{C}$ and let $R = \lambda P + \mu Q$ be a convex combination. The definition of closure is that there exist sequences of points $\{P_i\}$, $\{Q_i\}$ in $\mathcal{C}$ such that $P_i \to P$, $q_i \to Q$. By convexity of $\mathcal{C}$ the points $R_i = \lambda P_i + \mu Q_i$ are also in $\mathcal{C}$ and $R_i \to R$. Hence R is in the closure of $\mathcal{C}$. The convexity of $\mathcal{I}$ is equally easy to show, and we leave it as an exercise.

THEOREM 9. *Let $\mathcal{C}$ be a convex set in 3 dimensions. If we project $\mathcal{C}$ onto a plane by parallel or central projection the image is convex.*

This is obvious and we could have mentioned it at the beginning, but only now are we coming to a point where we need it. Another fact which we need now is

THEOREM 10. *If a convex set $\mathcal{C}$ has a nonempty interior $\mathcal{I}$ then the boundary of $\mathcal{I}$ is the boundary of $\mathcal{C}$.*

PROOF. An example of a set in the plane with an interior whose boundary includes points other than those of the boundary of the interior is a disk with a line segment attached. Roughly speaking, we have to prove that a convex set with a nonempty interior has no hairs.

Let B be a boundary point of $\mathcal{C}$. Arbitrarily close to B there is a point $P \in \mathcal{C}$. By assumption $\mathcal{C}$ contains a disk and by convexity it contains the points of every triangle with vertex P and base in the disk. Hence there are interior points arbitrarily close to P and hence to B. Thus every boundary point B of $\mathcal{C}$ is a boundary point of $\mathcal{I}$. To prove the inclusion the other way, let P be a boundary point of $\mathcal{I}$. Since $\mathcal{I} \subseteq \mathcal{C}$, P is either a boundary point or an interior point of $\mathcal{C}$. But P can not be an interior point of $\mathcal{C}$ because then it would be an interior point of $\mathcal{I}$ also.

DEFINITION. In $\begin{smallmatrix}\text{two}\\ \text{three}\end{smallmatrix}$ dimensions, we say a $\begin{smallmatrix}\text{line}\\ \text{plane}\end{smallmatrix}$ *divides* a set if there are points of the set on both sides of the $\begin{smallmatrix}\text{line}\\ \text{plane}\end{smallmatrix}$.

THEOREM 11. *If the interior $\mathcal{I}$ of a convex set $\mathcal{C}$ is not empty, a line or plane divides $\mathcal{C}$ if and only if it divides $\mathcal{I}$.*

PROOF. Clearly if a line or plane divides $\mathcal{I}$, it divides $\mathcal{C}$. If $\mathcal{I}$ is in one of the closed halfplanes or halfspaces defined by a line or plane l then so is the closure of I and hence, by Theorem 10, so is $\mathcal{C}$ also.

DEFINITION. Let B be a boundary point of a convex set $\mathcal{C}$ in the plane. A line l through B is a *line of support at B* if it does not divide $\mathcal{C}$. In three dimensions a *plane of support* is defined similarly.

THEOREM 12. *In* $\begin{smallmatrix}\text{two}\\ \text{three}\end{smallmatrix}$ *dimensions there is a* $\begin{smallmatrix}\text{line}\\ \text{plane}\end{smallmatrix}$ *of support through every boundary point P of a convex set $\mathcal{C}$.*

PROOF. Consider first the two-dimensional case. If the interior $\mathcal{I}$ of $\mathcal{C}$ is the empty set, then $\mathcal{C}$ is a line segment and that line is a support line at every boundary point. Let us therefore consider the case when $\mathcal{I}$ is not empty. We prove that through any point P which is not an interior point there is a line which does not divide $\mathcal{C}$. (However, such a line is not called a support line if P is not on the boundary of $\mathcal{C}$.)

The half-lines which end at P and go through a point of $\mathcal{I}$ fill a sector with vertex P. The angle of this sector is $\leq 180°$ since P is not in $\mathcal{I}$ and hence not in the interior of a triangle with all three vertices in $\mathcal{I}$. Hence there is a line through P which does not divide $\mathcal{I}$ and hence by theorem 11 it does not divide $\mathcal{C}$. This proves theorem 12 in the two-dimensional case.

Consider now the case of three dimensions. If the entire set $\mathcal{C}$ lies in a plane, then that plane is a support plane at P. Thus it suffices to consider the case when there are four points in $\mathcal{C}$ which are not coplanar. By convexity the tetrahedron spanned by these points is in $\mathcal{C}$, hence $\mathcal{C}$ has a nonempty interior $\mathcal{I}$. By Theorem 11 it suffices to prove that $\mathcal{I}$ has a support plane through P.

Let Π be any plane through P. If Π does not intersect $\mathcal{I}$ then it is a support plane. Otherwise $\Pi \cap \mathcal{I}$ is a convex set which does not contain P. Hence there is a line l in Π through P which is disjoint from $\mathcal{I}$.

Take a plane $\Pi^{\perp}$ perpendicular to l and project $\mathcal{I}$ by perpendicular projection on $\Pi^{\perp}$. Let L be the point into which l projects. The image of $\mathcal{I}^{\perp}$ is a convex set not containing L. Hence there is a line l' through L which does not divide $\mathcal{I}^{\perp}$. Hence $\mathcal{I}$ is in one of the two closed halfspaces defined by the plane of the lines l and l', and the proof of Theorem 12 is complete.

THEOREM 13. *Let $f(x)$ be convex on an interval I. Let $l(x)$ be a linear function and suppose $f(x) = l(x)$ on a nonvanishing interval $J \subseteq I$. Then $f(x) \geq l(x)$* on *I. The same conclusion can be drawn if f and l are functions of two variables and J is a nonempty open set.*

PROOF. Let P be a point on the graph of f whose abscissa is in J and let Q be any other point of the graph. The chord PQ is on or above the graph. Hence it is on or above the line $y = l(x)$ near P. Therefore the entire chord is on or above the line $y = l(x)$. The same argument can be used in more dimensions.

DEFINITION. A *piecewise linear* function f of one or several variables is a continuous function whose domain of definition is partitioned into a finite number of subregions in each of which $f(x)$ is equal to a linear function.

THEOREM 14. *A convex piecewise linear function has the form*

$$f(x) = \max(l_1(x), \dots l_k(x)) \tag{C.3}$$

where the l_i are linear functions.

PROOF. Theorem 13 tells us that a convex piecewise linear function does have this form. Conversely, if $f(x)$ has the form (C.3) then the set of points on or above the graph is an intersection of closed halfspaces and hence it is convex, which means that the function $f(x)$ is convex.

THEOREM 15. *Let $f(x)$ be a piecewise linear function of one variable, defined on the interval $I = [a_0, a_n]$. Let $a_0, a_1, \dots, a_n$ be the endpoints of the successive intervals where f is linear and let $l_i(x)$ be the linear function representing $f(x)$ in $[a_{i-1}, a_i]$. Suppose each of the angles between successive segments is convex downward, i.e., $f(x)$ is convex in each interval $[a_i, a_{i+2}]$. Then $f(x)$ is convex on the interval $[a_0, a_n]$.*

PROOF. This theorem is similar to Theorem 2 and can be proved similarly. We can also deduce it from Theorem 5, as follows. Let $\delta = \min_{i=0,\dots,n-1}|a_{i+1} - a_i|$. Then any subinterval $J \subseteq I$ of length $< \delta$ is contained in one of the intervals $[a_i, a_{i+2}]$. Hence the graph of the restriction of f to J is convex and therefore the set of points which are on or above the graph of f and have abscissas differing by $< \delta$ is convex. By Theorem 5 we conclude that the set of points above the entire graph is convex and hence so is the function f.

Consider a (continuous) piecewise linear function $f(x_1, x_2)$ defined on the interior and boundary of a convex polygon R. The domain boundary between adjacent linear functions is a line, so R is partitioned into polygons which are each the domain of a single linear function. Thus the graph of f is a polyhedron Π.

THEOREM 16. *Let $f(x_1, x_2)$ be a piecewise linear function defined on a convex region R and let Π be its graph. Suppose the dihedral angle formed by any two adjacent faces of Π is convex downwards. This means, if R_1 and R_2 are regions where f is represented by the linear functions l_1 resp. l_2, then $\max(l_1, l_2)$ represents f in $R_1 \cup R_2$. Then f is convex on R.*

PROOF. Let P and Q be two points on the graph of f. The intersection of Π with the vertical plane through the chord PQ is an open polygon Π_v. Its vertices are at the intersections of the vertical plane through PQ with the edges or vertices of the graph of f. Since the angles at the edges of Π are convex downwards, the corresponding angles of Π_v are also convex downwards. Hence, if the chord PQ does not pass above a vertex of Π then the angles of Π_v are all convex downwards by the assumption of our theorem and Π_v is convex by Theorem 17. Hence the chord PQ is on or above Π. If the chord PQ passes over a vertex of Π, it can be approximated by chords which do not pass over a vertex of Π, and we can still arrive at the same conclusion.

THEOREM 17. *Suppose f is convex on an interval I and $A = (a, y_a)$, $B = (b, y_b)$ are points on the graph of f. Let $y = l(x)$ be the equation of the line through A and B. Then if x is in I but outside the interval (a, b), $l(x) \le f(x)$.*

PROOF. Suppose $a < b < c$ and $f(c) < l(c)$. By convexity this implies $f(b) < l(b)$ because the graph of the curve is below the chord through A and $(c, f(c))$. But $l(b) = f(b)$, so we have a contradiction.

THEOREM 18. *If f is convex and the graph of g is obtained by replacing the arc of the graph of f from $C = (c, f(c))$ to $D = (d, f(d))$ by the chord, then g is also convex.*

PROOF. By Theorem 17, the part of the graph of f which is not between C and D is on or above the line CD. Hence the region above the graph of g is the intersection of the region above the graph of f and the upper half plane bounded by the line through C and D. This is the intersection of two convex sets and hence it is convex. By Theorem 1, the function g is therefore convex.

THEOREM 19. *Let the function f be defined and continuous on a closed interval I. Suppose that if $L(x)$ is any linear function, $f(x) - L(x)$ has at most two sign changes in I. Then f is either convex or concave on I. (*Concave *means that every chord lies below the graph.)*

PROOF. We first show that if AB is any chord, the curve segment between A and B is either entirely on or below the chord or entirely on or above it. Suppose the contrary. Then we have a point C of the arc which is above the chord and a point D which is below it. Assume without loss of generality that the order of the points on the curve is A, C, D, B, as shown on the left-hand side in Fig. C.2. Let $y = L(x)$ be the equation of a line which passes between the chord and the curve at both C and D. Such a line will be above the chord and hence above the curve at A, below it at C, above it at D and below it at B, hence $f(x) - L(x)$ would have at least three sign changes.

We have established that our curve is not crossed by its chords but we still have to exclude the possibility that some chords lie entirely on or above the curve

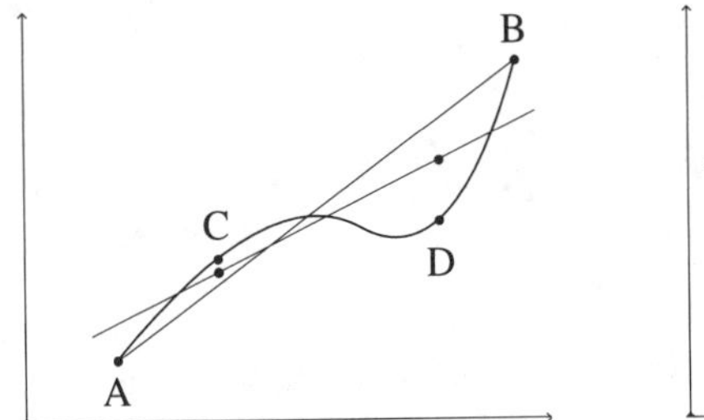

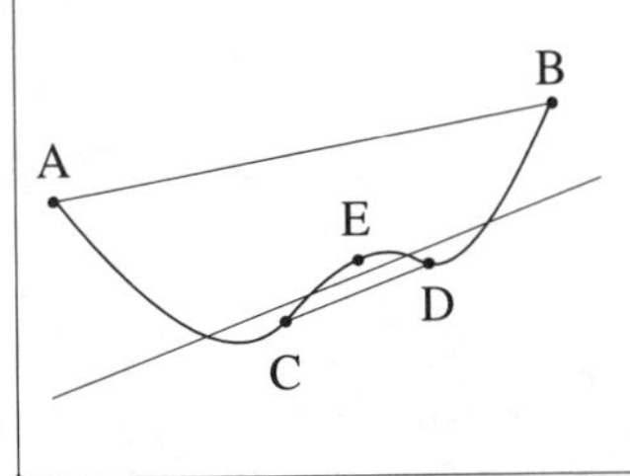

Figure C.2. Proof of Theorem 19.

while others lie entirely on or below it. Take AB to be the chord lying above the entire interval I. Assume without loss of generality that the chord AB lies on or above the curve. We show that any other chord CD also lies on or above the curve. Suppose, on the contrary, that the curve segment CD has some point E above the chord CD, as shown on the right side of Fig. C.2. Now let $y = L(x)$ be the equation of a line l parallel to and above the chord CD, but below E. Since l is below E which is below the chord AB, $f(x) - L(x)$ is positive at either A or B. It is negative at C, positive at E and negative at D. This implies at least three sign changes, which is not allowed. Hence either every chord is on or above the curve or every chord is on or below it, as claimed.

References

Baker, H. F. (1942), "A remark on Polygons", *J. London Math. Society*, **17** 162–164.

Barnhill, R. E. (1985), "Surfaces in CAGD: A survey with new results", *Computer Aided Geometric Design*, **2** 1–17.

Chang, G. Z. (1982), "Proving Pedoe's Inequality by complex number computation", *Amer. Math. Monthly*, **89** 692.

Chang, G. Z. (1982), "A Proof of a Theorem of Douglas and Neumann by Circulant Matrices", *Houston J. Math.*, **8** 15–18.

Chang, G. Z. and Davis, P. J. (1983), "Iterative processes in elementary geometry", *Amer. Math. Monthly*, **90** 421–431.

Chang, G. Z. and Davis, P. J. (1983), "A Circulant formulation of the Napoleon-Douglas-Neumann Theorem", *Linear Algebra and its Appl.*, **54** 87–95.

Chang, G. Z. and Davis, P. J. (1984), "The convexity of Bernstein polynomials over triangles", *J. Approx. Theory*, **40** 11–28.

Chang, G. Z. (1985), "Planar metric inequalities derived from the Vandermonde determinant", *Amer. Math. Monthly*, **92** 495–499.

Chang, G. Z. and Feng Y. Y. (1986), "Limit of iterates for Bernstein polynomials defined on higher dimensional domains", *Kexue Tongbao*, **31** 157–160.

Coxeter, H. S. M. (1969), *Introduction to Geometry*, John Wiley & Sons.

Davis, P. J. (1975), *Interpolation and Approximation*, Dover Publications, Inc., New York.

Davis, P. J. (1979), *Circulant Matrices*, John Wiley & Sons.

Demir, Haseyin (1986), "Incircles within", *Math. Magazine*, **59**, no. 2, 77–83.

Demir, H. (1990), Problem 1327, *Math. Magazine*, **63** 275.

Demir, H. (1991), Problem 1371, *Math. Magazine*, **64** 132.

Devaney, R. L. (1986), *An Introduction to Chaotic Dynamical Systems*, The Benjamin Cummings Publishing Co.

Devaney, R. L. (1990), *Chaos, Fractals, and Dynamics*, Addison-Wesley Publishing Co.

Do Carmo, E. P. (1976), *Differential Geometry of Curves and Surfaces*, Prentice-Hall, Inc.

Douglas, J. (1940), "Geometry of polygons in the complex plane", *J. Math. Phys.*, **19** 93–130.

Douglas, J. (1940), "On linear polygon transformations", *Bull. Amer. Math. Soc.*, **46** 551–560.

Engel, Arthur (1993), *Exploring Mathematics with your Computer*, New Math. Lib., Math. Assoc. of Amer.

Farin, Gerald (1993), *Curves and Surfaces for Computer Aided Geometric Design*, Academic Press.

Furno, A. L. (1981), "Cycles of differences of integers". *J. of Number Theory* **13** 255–264.

Greitzer, S. L. (1978), *International Mathematical Olympiads 1959–1977*, New Math. Lib., Math. Assoc. of Amer.

Hardy, G. H., Littewood, J. E. and Pólya, G. (1952), *Inequalities*, Cambridge Univ. Press.

Hirschman, I. I. and Widder, D. V. (1955), *The Convolution Transform,* Princeton Univ. Press.

Hoschek, Josef and Lasser, Dieter (1993), *Fundamentals of Computer Aided Geometric Design*, A K Peters.

Kay, D. C. (1969), *College Geometry*, Holt, Rinehart & Winston, Inc.

Kelisky, R. P. and Rivlin, T. J., "Iterates of Bernstein polynomials", *Pacific J. of Math.*, **21** 511–520.

Kelly, L. M., ed., *Lecture Notes in Math.* Springer-Verlag.

Kelly, P. J. and Merriell, D (1964), "Concentric polygons", *Amer. Math. Monthly*, **71** 37–41.

Klamkin, M. S. (1986), *International Mathematical Olympiads 1978–1985*, New Math. Lib., Math. Assoc. of Amer.

Klamkin, M. S. (1988), *USA Mathematical Olympiads 1972–1986*, New Math. Lib., Math. Assoc. of Amer.

Klamkin, M. S. (1984), Solution to Problem 83–5 in *SIAM Review*, 275–276.

Kosmak, Ladislav (1960), "A note on Bernstein polynomials of convex functions", *Math. (Cluj)* **2**, 281–282.

Larson, L. C. (1983), *Problem-Solving Through Problems*, Springer-Verlag.

Li, T. Y. and Yorke, J. A. (1975), "Period three implies chaos", *Amer. Math. Monthly*, **82** 985–992.

Li, Y. S. and Qi, D. X. (1979), *Splines Methods*, (in Chinese), Science Press, Beijing.

Liu, D. Y. (1982), "A theorem on the convexity of planar Bézier curves of degree n" (in Chinese), *Chinese Annals of Math.*, **4** 158–165.

Lorentz, G. G. (1986), *Bernstein Polynomials*, 2nd ed., Chelsea Publishing Co.

Melzak, Z. A. (1973), *Companion to Concrete Mathematics*, John Wiley & Sons. The reader should be warned that this book is full of errors.

Neumann, B. H. (1941), "Some remarks on polygons", *J. London Math. Society*, **16** 230–245.

Neumann, B. H. (1942), "A remark on polygons", *J. London Math. Society*, **17** 165–166.

Newman, D. J. (1982), *A Problem Seminar*, Springer-Verlag.

Pedoe, D. (1943), "An inequality for two triangles", *Proc. Cambridge Philos. Soc.*, **38** 397.

Pedoe, D. (1957), *Circles*, Permagon Press.

Pedoe, Daniel (1970), *Geometry—A Comprehensive Course*, Dover Publications.

Propp, James (1983), "Problem Proposal", *Math. Magazine*, **56**.

Sagher, Yoram (1988), "What Pythagoras could have done", *Amer. Math. Monthly*, **95** 117.

Schoenberg, I. J. (1973), "Cardinal Spline Interpolation", *CBMS Conf. Series in Appl. Math.*, No.12, SIAM, Philadelphia, Pa., vi+125 pp.

Sederberg, T. W. (1984), "Planar piecewise algebraic curves", *Computer Aided Geometric Design* **1** 241–255.

Sederberg, T. W. and Chang, G. Z. (1990), "Best linear common divisors for approximate degree reduction", *Computer Aided Design*, **25** 163–168.

Stefan, P. (1977), "A theorem of Sarkovskii on the existence of periodic orbits of continuous endomorphisms of the real line", *Comm. Math. Phys.*, **54** 237–248.

Stolarsky, K. B. (1971), "Cubic triangle inequalities", *Amer. Math. Monthly*, **78** 879–881.

Stolarsky (1975), Problem 1, in *The Geometry of Metric and Linear Spaces*.

Su, B. C., and Liu, D. Y. (1989), *Computational Geometry –Curve & Surface Modeling*, Academic Press.

Zhang, J. Z. (1987), "From plain facts to an amazing theorem", (in Chinese), *The Nature*, 532–536.

Zhang, J. Z. and Li, H. (1991), *Real Iterations* (in Chinese), Hunan Educational Publishing House.

Ziegler, Z. (1968), "Linear approximation and generalized convexity", *J. Approx. Theory*, **1** 420–433.

Zippin, Leo (1962) *Uses of Infinity*, New Math. Lib. vol. 7, Math. Assoc. of Amer.

Hints and Solutions

1.1 Here we have the iterative function: $f(x) = x/3 + 10$. We solve the equation $f^9(x) = g$ for x. From Problem 1.1, we get

$$f^9(x) = \left(\frac{1}{3}\right)^9 x + 5\frac{3^9 - 1}{3^8}.$$

Thus

$$x = 3^9 g - 15(3^9 - 1).$$

1.2 By mathematical induction we can show that $f^n(x) = \frac{x}{1+nx}$, for $n = 1, 2, 3, \ldots$.

1.3 For $n = 1, 2, 3, \ldots$, we have

$$f^n(x) = \frac{x}{\sqrt{1 - nx^2}}.$$

1.4 Among the distributions in which no four monkeys have the same number of coconuts, the distribution

$$0, 0, 0, 1, 1, 1, \ldots, 32, 32, 32, 33$$

minimizes the number of coconuts. The minimum is $3(1 + 2 + \cdots + 32) + 33 = 1617$, which is greater than 1600.

1.5 Let x be the number of apples initially on the beach. After the first monkey left, the number of apples on the beach is $f(x) = 4(x - 1)/5$ which can be rewritten as

$$f(x) + 4 = \tfrac{4}{5}(x + 4) = \tfrac{4}{5}(f^0(x) + 4).$$

Iteratively, we have

$$f^5(x) + 4 = \tfrac{4}{5}(f^4(x) + 4) = \left(\tfrac{4}{5}\right)^2 (f^3(x) + 4) = \cdots = \left(\tfrac{4}{5}\right)^5 (x + 4).$$

Hence,

$$f^5(x) = \left(\tfrac{4}{5}\right)^5 (x + 4) - 4.$$

In order to have a positive integer $f^5(x)$, $x + 4$ must be a multiple of 5^5. So the least positive integer value of x is $5^5 - 4 = 3121$. Consequently, the least number of apples left after all five monkeys visited the beach is 1020.

1.6 If n is an odd number, then we have obviously $f(n) = 2$ and $L_n = 1$. Consider even n. First we show that $f(n)$ can be expressed as $f(n) = p^m$ where p is a prime. If this is not true, then we can write $f(n) = st$ in which $s(>1)$ and $t(>1)$ are relatively prime. From $s|n$, $t|n$ we have $st|n$, a contradiction to the fact $f(n) = st$.

Now we can write $f(n) = p^m$ for even n. If p is odd then $L_n = 2$; if $p = 2$ then $L_n = 1$ for $m = 1$ and $L(n) = 3$ for $m = 2, 3, \ldots$.

2.2 Use the arithmetic mean–geometric mean inequality $A \geq G$.

2.3 Use the arithmetic mean–harmonic mean inequality $A \geq H$.

2.4 Suppose that in n weeks the prices of sugar are $p_1, p_2, \ldots, p_n$ dollars/pound. Then Brenda pays on the average $A := (p_1+p_2+\cdots+p_n)/n$ for one pound of sugar and Tom pays $H := n/(\frac{1}{p_1}+\frac{1}{p_2}+\cdots+\frac{1}{p_n})$ for one pound of sugar. If $p_1, p_2, \ldots, p_n$ are not all the same, then $A > H$.

In finance, the same idea is called *dollar averaging*. If we invest a fixed amount in a stock each month, the average price per share we will have paid will be less than if we buy a fixed number of shares each month. There is however a downside to dollar averaging: if the stock price declines steadily, we lose more with dollar averaging.

2.5 By induction, from $0 < a_2 \le a_1(1-a_1)$ follows $a_1 < 1$ and $a_2 \le \frac{1}{4} < \frac{1}{2}$. Assume that $a_n < 1/n$ for some $n \ge 2$. We have

$$\begin{aligned} a_{n+1} \le a_n(1-a_n) &= -\left(a_n - \frac{1}{2}\right)^2 + \frac{1}{4} \\ &\le -\left(\frac{1}{n} - \frac{1}{2}\right)^2 + \frac{1}{4} = \frac{1}{n}\left(1 - \frac{1}{n}\right) = \frac{n-1}{n^2} < \frac{1}{n+1}. \end{aligned}$$

2.6 By the arithmetic mean–geometric mean inequality we have

$$\frac{a_1^2}{a_2} + a_2 \ge 2a_1, \qquad \frac{a_2^2}{a_3} + a_3 \ge 2a_2, \quad \ldots, \qquad \frac{a_n^2}{a_1} + a_1 \ge 2a_n.$$

Adding the above inequalities gives the desired result.

2.7 First of all, by the arithmetic mean–geometric mean we have

$$(1+a_1)(1+a_2)\cdots(1+a_n) \le \left(1 + \frac{S}{n}\right)^n = \sum_{i=0}^{n} \binom{n}{i}\left(\frac{S}{n}\right)^i.$$

Then the desired result comes from

$$\binom{n}{i}\frac{1}{n^i} = \frac{n!}{i!(n-i)!}\frac{1}{n^i} = \frac{1}{i!}\frac{(n-i+1)\ldots n}{n^i} \le \frac{1}{i!},$$

for $i = 0, 1, 2, \ldots, n$.

2.8 We have $a_i \le a_{i+1} \le b_{i+1} \le b_i$. Bounded monotone sequences converge. Let $a_n \to A$ and $b_n \to B$. Taking the limit in the equation defining a_{i+1}, we get $A = \frac{1}{2}(A+B)$, hence $A = B$. It is easy to verify that $a_{i+1}b_{i+1} = a_i b_i$ and hence $a_i b_i = a_0 b_0$. Taking the limit, we get $A^2 = a_0 b_0$, as claimed.

3.1 It is clear that $c(a-b)^2 + 4abc = c(a+b)^2$. Hence the difference between the left-hand side and the right-hand side of the stated inequality becomes

$$\begin{aligned} &a[(b-c)^2 - a^2] + b[(c-a)^2 - b^2] + c[(a+b)^2 - c^2] \\ &= a(b-c+a)(b-c-a) + b(c-a+b)(c-a-b) \\ &\quad + c(a+b+c)(a+b-c). \end{aligned}$$

This can be written as the product $(b+c-a)(c+a-b)(a+b-c)$; because of the triangle inequality, each factor is positive.

3.2 After expanding $(a+b+c)^2$, we see that the desired inequalities become

$$ab + bc + ca \le a^2 + b^2 + c^2 < 2(ab + bc + ca).$$

The first inequality follows by adding $2ab \le a^2 + b^2$, $2bc \le b^2 + c^2$ and $2ca \le c^2 + a^2$; and the second inequality comes from

$$\begin{aligned} &2(ab + bc + ca) - (a^2 + b^2 + c^2) \\ &\quad = c(2a + 2b - c) - (a - b)^2 = 2c(a + b - c) - (a - b)^2 + c^2 \\ &\quad = 2c(a + b - c) + (c + b - a)(c + a - b) > 0. \end{aligned}$$

3.3 Let ABC be an isosceles triangle with the base BC. Through point A draw a line l parallel to BC. Any point A' on l makes a triangle $A'BC$ which has the same base and height of triangle ABC.

Extend the line BA to C' such that $BA = AC'$ and join $A'C'$. We see that $A'C = A'C'$. If $A' \neq A$ then in triangle $A'BC'$ we have

$$A'B + A'C = A'B + A'C' > BC' = 2AB = AB + AC$$

by the triangle inequality.

3.4 The problem is simpler than it may appear at first. In fact, the perimeter is always ≥ 4 and the sum of the diagonals is $\geq 2\sqrt{2}$ Let $PQRS$ be a convex quadrilateral with area 1. Think of the diagonal PR as fixed. By Ex. 3.3, if we move Q and S parallel to PR onto the perpendicular bisector of PR, the perimeter decreases and the sum of the diagonals clearly does not increase. So it suffices to prove the inequalities for the resulting rhomboid, i.e., if $|PQ| = |QR|$ and $|PS| = |RS|$.

We have $1 = |PQ||PS|\sin(\angle(PQ, PS)$, hence by the arithmetic-geometric mean inequality, $|PQ| + |PS| \geq 2$. Also, $1 = \frac{1}{2}|PR||QS|\sin(\angle(PR, QS)$, hence by the arithmetic-geometric mean inequality, $|PR| + |QS| = 2\sqrt{2}$. Equality in either inequality occurs only if the two lengths occurring are equal and the angle between them is a right angle, i.e., for a square.

5.1 To bring each cup upside down, each cup must be turned an odd number of times. Since the number of cups is odd, the total number of turns of cups has to be odd.

5.2 In the series 1,2,..., 1995, there are exactly 998 integers greater than or equal to 998. Hence among $a_{998}, a_{999}, \ldots, a_{1995}$ there is at least one $a_m \geq 998$. Thus $ma_m \geq (998)^2$.

5.3 If we add 1 to all the numbers in our list, the effect of the operation $\mathcal{O}$ is to multiply two unequal entries in the list. The question then becomes to characterize the numbers we get if we start with 2 and 3 and form products of unequal previously obtained numbers. It is clear that, in addition to 2 and 3, we get all numbers of the form $2^m 3^n$, $m \geq 1, n \geq 1$, and no others.

6.2 It suffices to prove this when there are three piles. If two of the piles have the same number of pennies then we can get 0 in one operation. Hence it suffices to consider the case when the initial numbers are x_0, y_0, z_0 with $x_0 < y_0 < z_0$. The assertion of the exercise will be proved by showing that as long as $x > 0$ one can produce a triad which has one pile with $r < x$ pennies, where r is the remainder in the division of y_0 by x_0: $y_0 = qx_0 + r, 0 \leq r < x_0$.

Let the binary representation of q be $q = a_0 + a_1 2 + a_2 2^2 + \cdots + a_k 2^k$, where $a_k = 1$. Now double x $k+1$ times. When $x = 2^i x_0$, subtract at the next doubling from y if $a_i = 1$ and from z if $a_i = 0$. Until the last doubling and subtraction has been completed, $x < 2^k x_0 \le y$ and until the doubling and subtraction before the last one have been completed, $x < 2^{k-1} x_0 < z$, so all these operations are permissible.

6.3 A set of $2n+1$ integers is "good" if the set has the property described in the hypothesis. We can assume $x_1, x_2, \ldots, x_{2n+1}$ are all nonnegative, otherwise we can add $-\min(x_1, \ldots, x_{2n+1})$ to each x_i to make a new set which is again good. Since the sum of $2n$ integers is always even, no matter which of the x_i is taken away, all of the a_i must have the same parity. Subtracting from each x_i the same number $\min(x_1, \ldots, x_{2n+1})$, we obtain a new set of nonnegative integers. The new set is also good and contains at least one zero. Hence the numbers in the set are all even. Dividing each member in the new set by 2, we obtain another new set which is good and contains at least one zero. We can repeat this procedure infinitely many times. This implies that the new set consists of zeros only; and the numbers of the original set are all the same.

It is obvious that the same proof can be applied to the case in which all x_i are rational numbers.

7.1 To simplify the notation, assume $\overline{x} = 0$. We have

$$A^{(j+1)} = \sum_i \frac{1}{2}\left|x_i^{(j)} + x_{i+1}^{(j)}\right| \le \sum_i \frac{1}{2}\left(\left|x_i^{(j)}\right| + \left|x_{i+1}^{(j)}\right|\right) = A^{(j)}.$$

If we take $x_i^{(0)}$ to be 0 for all odd indices and so that the mean is 0, then $A^{(1)} = A^{(0)}$.

For S we have the identity

$$S^{(j+1)} - S^{(j)} = -\sum_i \left(\frac{x_i^{(j)} - x_{i+1}^{(j)}}{2}\right)^2 \le 0. \tag{S.4}$$

Since $S^{(k)} \ge 0$ for all k, the left-hand side of (S.4) must approach 0. Hence $\lim_{j\to\infty} x_i^{(j)} - x_{i+1}^{(j)} = 0$ for all i. Since the arithmetic mean of the numbers is $\overline{x}$ for all j, it follows that they all approach $\overline{x}$.

8.1 The assertion is true for $n = 0$. Assume we have proved it for $n < 2^m$. For $n = 2^m + k$, $k = 0, 1, \ldots, 2_m - 1$ it follows from the fact that the rows of the mod 2 Pascal triangle for this range of exponents are made up of two copies of the Pascal triangle mod 2, with an inverted triangle of 0's between them. Thus the row with $n = 2^m + k$ has twice as many 1's as the row with $n = k$. Also, $d(2_m + k) = 1 + d(k)$ for $k < 2^m$, and thus the assertion is verified up to $n = 2m + 1 - 1$.

8.2 Suppose for some n, there are equal numbers of odd and even values of $\binom{n}{i}$, $i = 0, 1, \ldots, n$. Then by the previous exercise we have

$$n + 1 = 2^{d(n)} + 2^{d(n)} = 2^{d(n)+1},$$

and $n = 2^{d(n)+1} - 1$. But by Theorem 8.7 $\binom{2^m - i}{k}$ is odd for $i = 0, 1, \ldots, n$, which gives a contradiction.

8.3 We have

$$\binom{p^k}{i} = \frac{p^k}{i} \prod_{j=1}^{i-1} \frac{p^k - j}{j}.$$

In each of the fractions in the product we can cancel the highest power of p dividing j, since $j < p^k$ and hence p^k is divisible by that power. In the fraction in front of the product we can cancel the highest power of p which occurs in i, but in this fraction the numerator will remain divisible by p. Hence the integer $\binom{p^k}{i}$ is equal to a fraction whose numerator is divisible by p but whose denominator is not divisible by p, so by the unique factorization theorem $\binom{p^k}{i}$ must be divisible by p.

The second part of our statement says that in the row of the Pascal triangle above the one we just considered, none of the entries is divisible by p. The first entry is 1, so it is not divisible by p. Hence the second entry is not divisible by p either, because the sum of the first two entries is one of the entries in the next row which is divisible by p. We can go on to the third, fourth, ... entries in the same way, using the fact that if the sum of two numbers is divisible by p and one of the summands is not, then the other summand is not divisible by p either.

8.4 $c(341) = 1023$.

8.5 The only part of the proof which differs somewhat from that of Theorem 8.8 is the proof that the inequality always holds, since this can not be deduced from Theorem 8.10. Suppose that $1 < k < n - 1$. By (8.12), $T^k\mathbf{u}|_n = T^k\mathbf{u}|_{n-k} = 1$. The vector $T\mathbf{u} = \mathbf{w}$ does not contain two 1's whose indices differ by k, hence $T^{k-1}\mathbf{w} \neq \mathbf{w}$.

8.6 Let $M_i = \max(a_i, b_i)$ and $m_i = \min(a_i, b_i)$. We show that the set consisting of the numbers m_i is $\{1, 2, \ldots, n\}$. If $b_1 < a_1$ then we must have $b_i = n + 1 - i$ and $a_i = n - 1 + i$. Otherwise, by the monotonicity of the sequences, there exists a k, which could be 1 or n, such that

$$M_i = b_i \quad \text{and} \quad m_i = a_i \quad \text{for } i \leq k,$$

$$M_i = a_i \quad \text{and} \quad m_i = b_i \quad \text{for } i > k.$$

Then $m_i < m_k < M_k$ for $i \leq k$ because the a's form an increasing sequence and $b_k > m_i$ for $i > k$ because the b_i are decreasing. Thus $M_k = b_k$ is greater than all the m_i. A similar argument shows that $M_{k+1}a_{k+1}$ is also greater than all the m_i. The sequence $M_1, M_2, \ldots$ is decreasing down to M_k and increasing from M_{k+1} on. Hence all the M_i are greater than all the m_j, and this implies that the set of the m_i's is $\{1, 2, \ldots, n\}$ and the set of the M_i is the $\{n + 1, \ldots, 2n\}$. The sum of the differences is $\sum M_i - \sum m_i = \sum(n + i - i) = n^2$.

8.7 The number of black checkers after one transformation is the number of color changes as we go around the circle. This is even since we end with the same color as we started with. If $n/2$ is even, $4 \mid n$, as claimed.

8.8 We show that $n - 1$ transformations suffice to make all entries equal. The proof for part 1) is easy. For the first transformation, choose $\alpha = (a_1 + a_2)/2$; For the second transformation, take $\alpha = (a_1' + a_3')/2$, etc. One final transformation will annihilate the n-tuple.

We prove 2) by induction. Let $\{a_1, \ldots, a_n\}$ be a set of nonnegative numbers which require n transformations. We show how one can add a nonnegative number to the set so that the new set requires $n+1$ transformations.

We note that if α is outside the range

$$(*) \qquad [0,\ \max(a_1, \ldots, a_n) = A]$$

then the transformation can be combined with the next transformation so that a single step will achieve what we did in two steps. Thus in a sequence of transformations of minimal length, α_1 is in the range $(*)$. The corresponding range $[0, A']$ in the next step will be contained in $[0, A]$, etc. Take $a_{n+1} > nA$. If we apply a sequence of n transformations which changes all of $a_1, \ldots, a_n$ into 0, then a_{n+1} will be diminished by at most A at each step and hence will not become 0. Thus we need $n+1$ transformations to make this set into all 0's.

9.1 Suppose the integers x, y and z satisfy the equation $x^2+y^2+z^2 = 2xyz$. Note that $2xyz$ is even. Therefore, either x, y and z are all even or two of them are odd and the other is even. Consider the second case, say $x = 2k$, $y = 2m+1$ and $z = 2n+1$, where k, m and n are integers. Substituting these equalities into the equation, we get $4k^2+8p+2 = 4q$, where p and q are integers. This implies $2k^2+4p+1 = 2q$, an impossibility. Hence x, y and z are all even. Let $x = 2x_1$, $y = 2y_1$, and $z = 2z_1$. From the original equation, we obtain $x_1^2+y_1^2+z_1^2 = 4x_1y_1z_1$. We repeat the above reasoning and see that x_1, y_1, z_1 must be all even. Continuing the same argument, we find that each of the original three integers is divisible by 2 an endless number of times. Hence x, y and z must all be zero.

10.1 Each $i \in \{1, 2, \ldots, n\}$ appears in the matrix just n times, where n is an odd integer. By symmetry, i appears in elements off the main diagonal an even times. This implies that each i must appear on the main diagonal at least once. Since there are only n elements on the main diagonal, each i must be there exactly once.

11.1 Let $r_0 = 1$. Let $2\theta_n$ be the central angle subtended by the side of the nth regular polygon. It is clear that $\theta_n = \pi/(3\times 2^{n-1}i)$. We have $r_n = r_{n-1}\cos\theta_n$. Inductively

$$r_n = \cos\frac{\pi}{3}\cos\frac{\pi}{6}\cdots\cos\frac{\pi}{3\times 2^{n-1}}.$$

The formula $\cos\frac{1}{2}x = \sin x/2\sin\frac{1}{2}x$ gives

$$r_n = \frac{\sin\frac{\pi}{3}}{2^n\sin\frac{\pi}{3\times 2^{n-1}}}.$$

Let n tend to infinity. From the formula $\lim_{x\to 0}\sin x/x = 1$ we obtain

$$\lim_{n\to\infty} r_n = \frac{3\sqrt{3}}{4\pi}.$$

12.1 Taking AB as a side, we construct an equilateral triangle ABC', where C' is a point inside the square. Noting that $BC' = BD$, we can immediately infer that $\angle C'DE = 15^\circ$. Hence $C = C'$ and $\Delta ABC = \Delta ABC'$ is equilateral.

12.2 1) If ABC is equilateral then $\alpha = \beta = \gamma = 40^\circ$;

2) If ABC is right isosceles, with right angle at C, then $\alpha = \beta = 45^\circ$ and $\gamma = 30^\circ$.

12.3 Let $a = y + z$, $b = z + x$, and $c = x + y$ where $x > 0$, $y > 0$ and $z > 0$. The stated inequality becomes

$$(y+z)^2 + (z+x)^2 + (x+y)^2$$
$$\geq 4\sqrt{3xyz(x+y+z)} + (y-z)^2 + (z-x)^2 + (x-y)^2.$$

After simplification we get

$$xy + yz + zx \geq \sqrt{3xyz(x+y+z)}.$$

Squaring both sides gives

$$(xy + yz + zx)^2 \geq 3xyz(x+y+z).$$

Expanding the left side of the above inequality and rearranging, we obtain

$$x^2y^2 + y^2z^2 + z^2x^2 \geq x^2yz + xy^2z + xyz^2.$$

The last inequality is equivalent to

$$(xy - yz)^2 + (yz - zx)^2 + (zx - xy)^2 \geq 0.$$

13.1 Let the vertices of a parallelogram, in complex representation, be 0, z_1, $z_1 + z_2$, z_2. Then the squares of the diagonals are

$$|z_1 + z_2|^2 = (z_1 + z_2)(\overline{z}_1 + \overline{z}_2) = |z_1|^2 + |z_2|^2 + (\overline{z}_1 z_2 + z_1\overline{z}_2),$$

and

$$|z_1 - z_2|^2 = (z_1 - z_2)(\overline{z}_1 - \overline{z}_2) = |z_1|^2 + |z_2|^2 - (\overline{z}_1 z_2 + z_1\overline{z}_2).$$

By adding these two equalities, we get

$$|z_1 + z_2|^2 + |z_1 - z_2|^2 = 2(|z_1|^2 + |z_2|^2).$$

13.2 Triangle $z_1z_2z_3$ is equilateral if and only if $\Delta z_1z_2z_3 \sim \Delta z_2z_3z_1$. The direct similarity is equivalent to

$$\frac{z_2 - z_1}{z_3 - z_1} = \frac{z_3 - z_2}{z_1 - z_2}.$$

and this can easily be changed into the equality in the question.

13.3 Without loss of generality, we assume that $|z_1| = |z_2| = |z_3| = 1$. From $z_3 = -(z_1 + z_2)$ we have

$$1 = |z_3|^2 = z_3\overline{z}_3 = (z_1 + z_2)(\overline{z}_1 + \overline{z}_2) = 2 + (z_1\overline{z}_2 + \overline{z}_1 z_2).$$

Thus $z_1\overline{z}_2 + \overline{z}_1 z_2 = -1$. It follows that $|z_1 - z_2|^2 = (z_1 - z_2)(\overline{z}_1 - \overline{z}_2) = 2 - (z_1\overline{z}_2 + \overline{z}_1 z_2) = 3$. Similarly

$$|z_2 - z_3|^2 = |z_3 - z_1|^2 = 3.$$

Hence $|z_1 - z_2| = |z_2 - z_3| = |z_3 - z_1|$. The triangle $z_1z_2z_3$ is equilateral.

13.4 Let $|z_1|^2 = |z_2|^2 = |z_3|^2 = s$. Taking absolute values in $z_1(z_2 + z_3) = -z_2z_3$ we get $|z_2 + z + 3|^2 = s$. Hence

$$|z_2 - z_3|^2 = (z_2 - z_3)(\overline{z}_2 - \overline{z}_3)$$
$$= 2(z_2\overline{z}_2 + z_3\overline{z}_3) - (z_2 + z_3)(\overline{z}_2 + \overline{z}_3) = 4s - s$$

By symmetry, $|z_3 - z_1|^2 = |z_1 - z_2|^2 = 3s$ also, hence the triangle is equilateral.

13.5 It is easy to verify the following identity

$$(B-C)(x-A)^2+(C-A)(x-B)^2+(A-B)(x-C)^2$$
$$=-(B-C)(C-A)(A-B),$$

where A, B, C are complex numbers and x is a variable.

Since the left-hand side of the equality is a quadratic polynomial in x, just substitute $x=A$, $x=B$ and $x=C$ successively to see if the equality holds.

Putting $x=P$ into the equality and taking the absolute value in both sides gives the desired inequality.

13.6 Hint: Verify the identity

$$(B-C)(x-A)(y-A)+(C-A)(x-B)(y-B)+(A-B)(x-C)(y-C)$$
$$=(B-A)(C-B)(A-C).$$

13.7 Let the convex quadrilateral have A, B and C and D as complex representation for its vertices.

From the assumption we have

$$|A+B-(C+D)|+|B+C-(D+A)|=|A-B|+|B-C|+|C-D|+|D-A|.$$

We must have equality in both of the inequalities

$$|A+B-(C+D)|\le|D-A|+|B-C|,$$
$$|B+C-(D+A)|\le|A-B|+|C-D|.$$

This can be true only when $DA\parallel CB$ and $AB\parallel DC$. Hence $ABCD$ is a parallelogram.

13.8 The assumptions can be written equivalently in the algebraic expression

$$\frac{V-A}{U-A}=\frac{U-V}{B-V}=\frac{C-U}{V-U}.$$

It follows that

$$\frac{V-A}{U-A}=\frac{(V-A)+(U-V)+(C-U)}{(U-A)+(B-V)+(V-U)}=\frac{C-A}{B-A},$$

i.e., $\triangle AUV\sim\triangle ABC$.

13.9 By hypothesis, we have

$$\frac{z_3-z_1}{z_2-z_1}=\frac{w_3-w_1}{w_2-w_1}.$$

This implies that

$$\frac{z_3-z_1}{z_2-z_1}=\frac{(1-t)(z_3-z_1)+t(w_3-w_1)}{(1-t)(z_2-z_1)+t(w_2-w_1)}=\frac{u_3-u_1}{u_2-u_1}.$$

13.10 We apply the Neuberg-Pedoe inequality to triangles ABC and $B'C'A'$:

$$(b')^2(-a^2+b^2+c^2)+(c')^2(a^2-b^2+c^2)+(a')^2(a^2+b^2-c^2)\ge 16FF',$$

then do the same thing to triangles ABC and $C'A'B'$:

$$(c')^2(-a^2+b^2+c^2)+(a')^2(a^2-b^2+c^2)+(b')^2(a^2+b^2-c^2)\ge 16FF'.$$

Adding the last two inequalities side by side, and then dividing both sides of the resulting inequality by 2, the desired inequality follows. If equality holds, equality must hold in the last two inequalities. Thus, the triangles $B'C'A'$ and $C'A'B'$ are similar to ABC; hence ABC and $A'B'C'$ must be equilateral.

14.1 Let ABC be an arbitrary triangle with positive orientation. Let $u := e^{\frac{i\pi}{3}}$. Let $A'B'C'$ be the outer Napolean triangle of ABC. The fact that $\triangle BA'C$ is isosceles with an angle $2\pi/3$ at C' is expressed by $(C - A')u^2 = B - A'$. We solve for A' to obtain $(1 - u^2)A' = B - Cu^2$; similarly

$$(1 - u^2)B' = C - Au^2, \quad (1 - u^2)C' = A - Bu^2.$$

Using that $u^3 = -1$ and $u^2 = u - 1$, we get

$$\begin{aligned} u(1 - u^2)(B' - A') &= (C - B + (C - A)u^2)u = (C - B)u - (C - A) \\ &= (C - B)(1 + u^2) - (C - A) = (A - B) + (C - B)u^2 \\ &= (1 - u^2)(C' - A'). \end{aligned}$$

It follows that $u(B' - A') = C' - A'$, i.e., $\triangle A'B'C'$ is equilateral.

14.2 Multiplying the equation $|z_1 - z_0| = |z_1|$ by $|z| \neq 0$, we get $|1 + zz_0| = 1$. Since $z_0 \neq 0$, we have $|z + 1/z_0| = 1/|z_0|$. Hence the locus of the point z is the punctured circle of radius $1/|z_0|$ and center at $-1/z_0$, with the origin excluded.

14.3 Let z satisfy $|z| = 1$ and the given equation. Taking the absolute value on both sides of the equivalent equation $z^n(z - 1) = 1$, we have $|z - 1| = 1$. It follows that $z = e^{\pm\frac{\pi}{3}i}$. Hence $z - 1 = e^{\pm\frac{2\pi}{3}i}$. From $z^n(z - 1) = 1$ we get

$$e^{\pm(\frac{n+2}{3})\pi i} = 1.$$

We conclude that $\frac{n+2}{3}$ should be an integer multiple of 2.

14.4 Let $z = iu$, the original equation becomes

$$11u^{10} + 10u^9 + 10u + 11 = 0.$$

It suffices to show that $|u| = 1$. From the last equation we find

$$u^9 = -\frac{10u + 11}{11u + 10}.$$

Taking absolute value, we get

$$\begin{aligned} |u|^{18} &= \frac{|10u + 11|^2}{|11u + 10|^2} = \frac{(10u + 11)(10\overline{u} + 11)}{(11u + 10)(11\overline{u} + 10)} \\ &= 100\frac{|u|^2 + 121 + 110(u + \overline{u})}{121|u|^2 + 100 + 110(u + \overline{u})} = \frac{A}{B}. \end{aligned}$$

We see that

$$B - A = 21(|u|^2 - 1).$$

If $|u| > 1$ then $B > A$, making $|u|^{18} < 1$ a contradiction; if $|u| < 1$ then $B < A$, making $|u|^{18} > 1$, a contradiction. Thus $|u| = 1$.

14.5 Let **C** be the point of Π which is originally above C. The turn around A moves **C** to the location of the mirror image C' of C in the line AB. The turn around B moves **C** back to C. Next, consider the point **A** of Π, which is originally above A. The turn around A leaves it where it was, and the turn around B moves it to the mirror image A' of A in the line BC. Thus the effect of the two rotations on the line segment **CA** is to turn it counterclockwise by 2γ around the point C. But as long as the only permitted motion of Π is to slide it while keeping it parallel to the plane of the three points A, B, C, the position of two of its points determines the position of the whole plane Π.

15.1 Let $M = (\alpha, \beta, \gamma)$, the barycentric coordinate of M with respect to the domain triangle ABC. From Problem 15.1 we see that

$$A' = \left(0, \frac{\beta}{1-\alpha}, \frac{\gamma}{1-\alpha}\right).$$

Thus

$$[MBA'] = \frac{\alpha\gamma}{\beta+\gamma}[ABC],$$

and similar expressions for $[MCB']$ and $[MAC']$. By hypothesis we have

$$\frac{\beta\gamma}{\alpha+\beta} = \frac{\gamma\alpha}{\beta+\gamma} = \frac{\alpha\beta}{\gamma+\alpha},$$

or

$$\alpha(1-\gamma) = \beta(1-\alpha) = \gamma(1-\beta).$$

Without loss of generality, we assume that $\alpha \geq \beta \geq \gamma$. Then $\alpha(1-\gamma) = \gamma(1-\beta) \leq \gamma(1-\gamma)$, and so $\alpha \leq \gamma$. Thus $\alpha = \beta = \gamma = 1/3$ and M is the centroid of the triangle ABC.

15.2 Let the point P have barycentric coordinates (α, β, γ) with respect to the base triangle $P_1P_2P_3$. We see that

$$\frac{P_1P}{PQ_1} = \frac{P_1Q_1 - PQ_1}{PQ_1} = \frac{1}{\alpha} - 1 \quad \text{and similarly}$$

$$\frac{P_2P}{PQ_2} = \frac{1}{\beta} - 1, \quad \frac{P_3P}{PQ_3} = \frac{1}{\gamma} - 1.$$

If the three said numbers are all greater than 2, then $\alpha < \frac{1}{3}$, $\beta < \frac{1}{3}$, and $\gamma < \frac{1}{3}$, and so $\alpha + \beta + \gamma < 1$, a contradiction. In the same way we can show that at least one of those three numbers is greater than or equal to 2.

15.3 Take ABC as the base triangle. Using barycentric coordinates, we set

$$K = (0, 1-k, k), \quad L = (l, 0, 1-l),$$

$$M = (1-m, m, 0), \quad \text{where} \quad 0 < k, l, m < 1.$$

Without loss of generality, we assume $[ABC] = 1$. Hence

$$[AML] = m(1-l), \quad [BKM] = k(1-m), \quad [CLK] = l(1-k).$$

Multiplying the above three equations we get

$$[AML][BKM][CLK] = k(1-k)l(1-l)m(1-m).$$

Since $x(1-x) \leq \frac{1}{4}$ for $0 < x < 1$, at least one of the factors on the left-hand side must also be $\leq 1/4$.

15.5 We have $[MAB] + [MCD] = [MBC] + [MDA]$, both quantities being equal to half the area of the parallelogram. On the other hand, if q is the quotient of the geometric progression, the right-hand side is q times the left-hand side. Hence $q = 1$. Thus M is halfway between the sides AB and CD of the parallelogram and also halfway between BC and DA. Thus, M is the midpoint of the parallelogram.

15.6 The argument used in the previous problem again gives that M is the midpoint of the parallelogram.

15.7 Let ABC be the coordinate triangle and $A = (1, 0, 0)$, $B = (0, 1, 0)$, $C = (0, 0, 1)$. Let

$$D = (0, 1 - x, x), \quad E = (y, 0, 1 - y), \quad F = (1 - z, z, 0).$$

Then

$$U = \left(0, \frac{2-x}{2}, \frac{x}{2}\right), \qquad X = \left(0, \frac{1-x}{2}, \frac{1+x}{2}\right),$$
$$V = \left(\frac{y}{2}, 0, \frac{2-y}{2}\right), \qquad Y = \left(\frac{1+y}{2}, 0, \frac{1-y}{2}\right),$$
$$W = \left(\frac{2-z}{2}, \frac{z}{2}, 0\right), \qquad Z = \left(\frac{1-z}{2}, \frac{1+z}{2}, 0\right).$$

Without loss of generality, we assume $[ABC] = 1$. Thus

$$[UVW] = 1 - (x + y + z)/2 + (xy + yz + zx)/4,$$
$$[XYZ] = (1 + xy + yz + zx)/4,$$
$$[DEF] = 1 - (x + y + z) + (xy + yz + zx).$$

The required result is an immediate consequence of these formulas.

16.1 Direct similarity of the three triangles yields

$$\frac{C' - A}{B - A} = \frac{A' - B}{C - B} = \frac{B' - A}{C - A} = u.$$

Thus

$$A' = B + (C - B)u, \qquad B' = A + (C - A)u, \qquad C' = A + (B - A)u.$$

This gives $A' + C' = B + A + u(C - A) = B + B'$. We conclude that $A'B'C'B$ is a parallelogram.

16.2 We use that in the complex plane, rotating a vector by 90° is the same as multiplying it by i:

$$C = B + (B - A)i, \qquad D = A + (B - A)i,$$
$$B = E + (E - A)i = E(1 + i) - Ai, \qquad F = E - (C - E)i = E(1 + i) - Ci.$$

From the last two equations we get

$$\begin{aligned} F &= B + (A - C)i = B + Ai - Ci \\ &= B + Ai - (Bi - (B - A)) = 2B - (A + (B - A)i) = 2B - D. \end{aligned}$$

The equation $D + F = 2B$ implies that B bisects DF.

16.3 Let z_1, z_2, z_3, and z_4 be complex representations of four vertices of the parallelogram with positive orientation. The centers of the four externally erected squares are denoted by w_1, w_2, w_3, and w_4 successively. Thus

$$2w_1 = z_1(1+i) + z_2(1-i), \qquad 2w_2 = z_2(1+i) + z_3(1-i),$$
$$2w_3 = z_3(1+i) + z_4(1-i), \qquad 2w_4 = z_4(1+i) + z_1(1-i).$$

Since $z_1z_2z_3z_4$ is a parallelogram, we have $z_1+z_3 = z_2+z_4$. With the last equation in mind, we can easily show that $i(w_2-w_1) = w_3-w_2$ and $i(w_3-w_2) = w_4-w_3$. We conclude that $w_1w_2w_3w_4$ is a square.

16.4 As in the previous solution, we can write

$$X = [B(1+i) + C(1-i)]/2, \quad Y = [C(1+i) + A(1-i)]/2,$$
$$Z = [A(1+i) + B(1-i)]/2.$$

It follows that $i(X-A) = Y - Z$.

17.1 This is just the trigonometric formulation for the statement in Problem 17.2.

17.2 Let $f(z) = a_m z^m + \cdots + a_1 z + a_0$. Consider first the case $z = 0$. We have $f(0) = a_0$. We have to show that the arithmetic mean of the values of f taken over the points $z_1, z_2, \ldots, z_n$ is a_0. The vertices of a regular n-gon with center at the origin can be written as $z_j = u\omega^j$, where u is a complex constant and $\omega = e^{2\pi i/n}$. Substituting this into the formula for $f(z)$ and using formulas (17.1) we get the required result.

To get the result for an arbitrary value of z, note that for an arbitrary complex number ζ, $f(\zeta) = f((\zeta - z) + z)$. Let $\zeta - z = w$. Then $f(\zeta) = p(w)$, where p is a polynomial of degree m in w with coefficients depending on z. In terms of p, what we have to prove is that $p(0)$ is the arithmetic mean of the values of p at the vertices of a regular n-gon with center $w = 0$, which we have proved above.

18.1 Let the centers of D, O_1, and O_2 be C, C_1, and C_2. Denote by AB the diameter of D which is perpendicular to CC_1.

We have $AO_1 + BO_1 \geq AB = 2AO$; hence $AO_1 = BO_1 \geq AO$. This implies that neither A nor B is in D_1. Obviously, A and B cannot both be in D_2.

19.1 We use the notation of the examples in this chapter.

Let $u = 1 - 1/x$. Thus $x' = 1/(1-u^2)$. Hence $u' = 1 - 1/x' = g(u) = u^2$ and $g^n(u) = u^{2^n}$. Finally we have

$$f^n(x) = \frac{1}{1 - g^n(u)} = \frac{x^{2^n}}{x^{2^n} - (1-x)^{2^n}}.$$

19.2 Note that $f(x) = (x+1)^2 - 1$. Let $u = 1 + x$. Thus $x' = u^2 - 1$ and $u' = 1 + x' = u^2$. We have $g(u) = u^2$ and $g^n(u) = u^{2^n}$. It follows that

$$f^n(x) = g^n(u) - 1 = (1+x)^{2^n} - 1.$$

We have to find a function $m(u)$ such that $m^5(u) = u^2$. The answer is $m(u) = u^{\sqrt[5]{2}}$. Hence the desired function is

$$\phi(x) = m(u) - 1 = (1-x)^{\sqrt[5]{2}} - 1.$$

19.3 Assume $u = x^2$, we have

$$u' = g(u) = x'^2 = \frac{u}{u+c}.$$

The transformation can be further simplified by introducing the variable $v = \frac{1}{u}$. We have

$$v' = \frac{1}{u'} = \frac{u+c}{u} = 1 + cv.$$

The nth iterate of this linear transformation can be obtained by inspection, or from the formula in Chapter 1. It is

$$v_n = 1 + c + c^2 + \cdots + c^{n-1} + c^n v.$$

We express this result in terms of the original variable x. We have $v = 1/x^2$, $x_n = \sqrt{1/v_n}$ and hence

$$f^n(x) = \sqrt{g^n(u)} = \frac{x}{\sqrt{c^n + (1 + c + c^2 + \cdots + c^{n-1})x^2}}.$$

If one proceeds to sum the geometric series one gets two different formulas, the sum of the series being n when $c = 1$, and $(1 - c^n)/(1 - c)$ otherwise.

20.1 One way to solve this problem is to note that $b_n = 2a_n - 1$ satisfies the recursion formula $b_{n+1} = \frac{1}{2}(b_n + 1/b_n)$. This is the special case $a = 1$ of the formula for calculating $\sqrt{a}$ which we discussed in Chapter 1. It follows that if $b_0 > 0$, $b_n \to 1$. Hence, if $a_0 > \frac{1}{2}$ then $a_n \to 1$.

20.2 The first inequality follows easily by induction, since

$$f^{n+1}(1) = 2^{f^n(1)} < 2^{g^{n-1}(1)} < 3^{g^{n-1}(1)} = g^n(1)$$

if $f^n(1) < g^{n-1}(1)$.

To prove the second inequality, it suffices to prove that

$$g^{n-1}(1)/f^{n+1}(1) < 1/2$$

for all $n \geq 3$. This can also be proved by induction, starting from

$$g^2(1)/f^4(1) = 27/65536 < 1/2$$

and observing that from $g^{n-1}(1)/f^{n+1}(1) < 1/2$ it follows that

$$\frac{g^n(1)}{g^{n+2}(1)} = \frac{3^{g^{n-1}(1)}}{2^{f^{n+1}(1)}} < \frac{3^{f^{n+1}(1)/2}}{2^{f^{n+1}(1)}} = \left(\frac{\sqrt{3}}{2}\right)^{f^{n+1}(1)} < \left(\frac{\sqrt{3}}{2}\right)^{f^4(1)} < \frac{1}{2}.$$

22.1 It is easy to see that f^3 is linear on the inteval $[(i-1)/8, i/8]$ for $i = 1, 2, \ldots, 8$, and that f^3 assumes the values 1 and 0 alternatively at 0, 1/8, 2/8, 3/8, ..., 7/8, 1. A graphical solution shows that the line $y = x/2$ intersects $y = f^3(x)$ at eight points over the interval $[0, 1]$.

22.3 First of all we can show that

$$\cos^2 x + 2\sin x\cos x - \sin^2 x = \cos 2x + \sin 2x = \sqrt{2}\sin(2x + \pi/4).$$

For $A = B = 0$, $M = \sqrt{2}$ and hence the minimum of M is $\leq \sqrt{2}$. We have

$$2M \geq f\left(\frac{\pi}{8}\right) - f\left(\frac{5\pi}{8}\right) = 2\sqrt{2} - \frac{A\pi}{2}.$$

The inequality $M \le \sqrt{2}$ implies $A \ge 0$. Using $f(5\pi/8)$ and $f(9\pi/8)$ in a similar way we get $A \le 0$. Thus $A = 0$ and it is easy to see now that $B = 0$ also when $M \le \sqrt{2}$.

22.4 Note that $||\tau_n|| = 1/2^{n-1}$, the exercise becomes the Chebyshev Theorem.

22.5 The definition says that $T_n(t) = \cos n\theta$ and $\theta = \arccos t$. Thus

$$T_m(T_n(t)) = \cos(m \arccos T_n(t)) = \cos mn\theta = T_{mn}(t).$$

22.6 Hint: Use the trigonometric identity $\cos 2\phi = 2\cos^2\phi - 1$.

22.7 Hint: Recall the trigonometric identity

$$\cos(m+n)\theta + \cos(m-n)\theta = 2\cos m\theta \cos n\theta.$$

24.1 a) If the two components of $A\mathbf{x}$ have opposite signs, the components of $\mathbf{x}$ can not all have the same sign.

b) Take $a_{ij} = 1 + (-1)^{i+j}$, $x_1 = 1$, $x_2 = -1$.

24.2 Suppose $c_i \le 0$ for $i \le k$ and $c_i \ge 0$ for $i > k$. If we bring the negative terms to the right-hand side and divide by x^k we get

$$c_{k+1}x + c_{k+2}x^2 + \cdots + c_n x^{n-k} = -c_0 x^{-k} - c_1 x^{1-k} - \cdots - c_k.$$

On the left-hand side we have a function which increases from 0 to ∞ as x moves from 0 to ∞. On the right side we have a function which is positive and nonincreasing in $(0, \infty)$. These two curves must intersect at one point.

24.3 We can divide out factors x, so we may suppose that $c_0 \ne 0$. Then by Descartes' Rule the number of positive roots of $p(x)$ is at most $\mathcal{V}\{c_0, c_1, \ldots, c_{n-1}, c_n\}$. The number of negative roots of $p(x)$ is the number of positive roots of $p(-x)$, which is at most $\mathcal{V}\{c_0, -c_1, c_2, \ldots, (-1)^n c_n\}$. The sum of these two sign change totals is at most n because where there is a sign change in the first sequence, there is no sign change in the second one. Since there are n real roots, Descartes' upper bound for the number of roots must be attained for both the positive and the negative roots.

24.4 Divide $p(x)$ by the highest power of x that divides it, and consider the resulting polynomial as our $p(x)$. Then $c_0 \ne 0$ and we have to prove that no two consecutive coefficients are 0. The number of roots which are positive or negative is (see the solution of Exercise 2) at most $\mathcal{V}\{c_0, c_1, \ldots, c_{n-1}, 1\} + \mathcal{V}\{c_0, -c_1, c_2, \ldots, (-1)^n\}$. If two consecutive coefficients are 0, then this sum is at most $n-1$, contrary to the hypothesis that $p(x)$ has n real roots.

24.5 If X denotes the column vector with components $x_1, x_2, \ldots$ and $Y = SX$ then $y_1 = x_1$, $y_2 = x_1 + x_2$, $y_3 = x_1 + x_2 + x_3$, etc. If, say, x_1 is positive, then y_1 is also positive. If y_{i-1} is positive but y_i is negative then x_i must be negative. Similarly, if y changes to positive again, x must have changed sign before or at that stage, etc.

Note that the above is just another way of looking at the transformation we considered in the proof of Descartes' Rule.

24.6 If a row is removed, Bx is obtained from Ax by removing an element of Ax, hence Bx has no more sign changes than Ax. If the jth column of A is removed to form B, then $Bx = Ay$, where y is obtained from x by inserting a 0 as a new entry in the jth place. This does not change the number of sign changes.

24.7 The following examples demonstrate this:

$$\begin{bmatrix} 1 & 0 \\ 0 & 0 \end{bmatrix} + \begin{bmatrix} 0 & 0 \\ 0 & -1 \end{bmatrix}, \qquad \begin{bmatrix} 0 & 1 & 0 \\ 0 & 0 & 1 \\ 0 & 0 & 0 \end{bmatrix} + \begin{bmatrix} 0 & 0 & 0 \\ 0 & 0 & 0 \\ 1 & 0 & 0 \end{bmatrix}.$$

24.8 If A has nonnegative elements, $A\mathbf{x}$ has no sign change if $\mathbf{x}$ has none. We have to check if it is possible to have 2 sign changes in $A\mathbf{x}$ and only one in $\mathbf{x}$. If that can occur, then we can get the sign sequence $+, -, +$ in $A\mathbf{x}$ with at most one sign change in $\mathbf{x}$. The vectors $\mathbf{x}$ with one sign change satisfy either the inequalities

$$x_1 \ge 0 \quad x_3 \le 0$$

or the inequalities

$$x_1 \le 0 \quad x_3 \ge 0.$$

Thus we have to check whether the sign sequence $+, -, +$ in $A\mathbf{x}$ is compatible with either one of the last two pairs of inequalities.

24.9 We eliminate x from the system of inequalities (24.13) by subtracting multiples of the equation $x + 2y = u$. We get

$$-2y \ge -u, \quad y \ge 0, \quad -3y \ge 3 - 2u, \quad y \ge 4 - u.$$

or

$$0 \le y, \quad 4 - u \le y, \quad y \le \frac{1}{2}u, \quad y \le \frac{2}{3}u - 1. \tag{S.5}$$

This will have a solution y if all the lower bounds are $\le$ all the upper bounds:

$$0 \le \frac{1}{2}u, \quad 0 \le \frac{2}{3}u - 1, \quad 4 - u \le \frac{1}{2}u, \quad 4 - u \le \frac{2}{3}u - 1.$$

These inequalities simplify to

$$0 \le u, \quad \frac{3}{2} \le u, \quad \frac{8}{3} \le u, \quad 3 \le u.$$

The last inequality implies all the others, so the smallest value u can take is 3. Substituting this in (S.5) we get that $y = 1$ and then $u = x + 2y$ gives $x = 1$.

24.10 The number of minors and the expression we claim for it are both symmetric functions of m, and n, i.e., they are unchanged if m and n are interchanged. So we may assume that $m \ge n$.

We can specify a $k \times k$ minor M by giving the k row indices which are represented in M and the $n - k$ column indices which are *not* represented in M. Thus to each minor we have associated an n-element subset of the union of row and column indices of A. Conversely, to every such subset there corresponds a minor, with the exception of the subset which contains all n column indices and no row index.

24.11 To see that $\mathbf{v}$ is orthogonal to the jth column of A, construct a $(k+1) \times (k+1)$ matrix by adding a copy of the jth column of A as the $k + 1$st column and expand the resulting determinant with respect to the last column. To see that $\mathbf{v}$ is different from 0, construct a nonsingular $(k+1) \times (k+1)$ matrix by attaching to A a column independent of the columns of A and expand the determinant of this matrix with respect to the new column.

24.12 By assumption, **b** has two nonzero entries with the same sign and indices of opposite parity. To avoid introducing more notation, assume these are b_1 and b_2, and that they are positive. Take $a_i = (-1)^i$ for $i > 2$. Let $s = a_3b_3 + a_4b_4 + \cdots$. If $s \geq 0$, take $a_2 = 1$ and $a_1 = -(b_2 + s)/b_1$; if $s < 0$, take $a_1 = -1$ and $a_2 = (b_1 - s)/b_2$.

24.13 A vector **x** with k components has at most $k - 1$ sign changes. The v. d. property of A implies that $A\mathbf{x}$ is not strictly alternating for any **x**. Hence, by Ex. 24.12, the orthogonal complement of the column space of A is strictly alternating. A generating vector of the orthogonal complement is given in Ex. 24.11, and the fact that this is strictly alternating is equivalent to the statement in the exercise.

25.1 This is an easy exercise in calculus but can also be derived from the inequality of the arithmetic and geometric means, as follows. Let $h = n - k$. Then

$$h^k k^h B_k^n(x) = \binom{n}{k} (hx)^k (k - kx)^h$$

The n factors on the right side are nonnegative and their sum is kh for all values of x. By the inequality we mentioned, such a product is greatest when the factors are equal. That means $hx = k - kx$ or $x = x_k$, as claimed.

25.2 We have

$$\frac{B_{i+1}^n(x)}{B_i^n(x)} = \frac{x}{1-x}\,\frac{n-i}{i+1} = \frac{1-x_i}{1-x}\,\frac{x}{x_{i+1}}.$$

Substituting x_k for x we get the value

$$\frac{1-x_i}{1-x_k}\,\frac{x_k}{x_{i+1}}$$

for the ratio. For $i < k$ the first factor is >1 and the second factor is ≥ 1. For $i \geq k$ the first factor is ≤ 1 and the second factor is <1. Thus the sequence $B_0^n(x_k), B_1^n(x_k), B_2^n(x_k), \ldots$ increases until the lower index reaches the value k and decreases after that.

25.3 The ratio can be written as

$$\frac{(x+c)^k}{x^k}\,\frac{(1-x-c)^{n-k}}{(1-x)^{n-k}} = \left(1 + \frac{c}{x}\right)^k \left(1 - \frac{c}{1-x}\right)^{n-k}.$$

All the factors in the last expression are clearly decreasing in our interval.

25.4 Suppose $0 \leq x < y \leq x_k$. We have to show that $B_k^n(x) < B_k^n(y)$. It was shown in Exercise 25.3 that

$$\frac{B_k^n(y)}{B_k^n(x)} > \frac{B_k^n(x_k)}{B_k^n(x_k - (y - x))}.$$

We have shown that on the interval [0, 1], the maximum of $B_k^n(x)$ is at x_k. Thus the ratio on the right is >1.

25.5 By Exercise 25.4, all the terms in the sum are increasing on $[0, x_K]$. To see what happens on the rest of the interval, note that by (25.2) our sum is equal to $1 - \sum_{k=0}^{K-1} B_k^n(x)$.

Here, again by Exercise 25.4, all the subtracted terms are decreasing on $[x_k, 1]$.

25.6 This can be shown by a variant of the useful method of "summation by parts." We have

$$\begin{aligned} B^n f(x) &= y_0 B_0^n(x) + y_1 B_1^n(x) + y_2 B_2^n(x) + \cdots \\ &= y_0\left(B_0^n(x) + B_1^n(x) + B_2^n(x) + \cdots\right) \\ &\quad + (y_1 - y_0)\left(B_1^n(x) + B_2^n(x) + \cdots\right) \\ &\quad + (y_2 - y_1)\left(B_2^n(x) + \cdots\right) + \cdots + (y_n - y_{n-1})B_n^n(x). \end{aligned}$$

By assumption the factors $y_{i+1} - y_i$ are nonnegative, and the polynomials they multiply are increasing by the result of the previous exercise.

25.7 First we prove: Let $a < b < c$. If the function f is continuous on the interval $[a, c]$ and L is a Lipschitz constant for f in the subintervals $[a, b]$ and $[b, c]$ then L is a Lipschitz constant for f in the full interval $[a, c]$.

We have to show that if x and y are in $[a, b]$, $|f(y) - f(x)| \leq L|y - x|$. This is so by hypothesis if x and y are in the same subinterval. Otherwise, we may suppose that $x < b < y$. Then

$$|f(y)-f(x)| \leq |f(y)-f(b)|+|f(b)-f(x)| \leq L(|y-b|+|b-x|) = L|y-x|.$$

In the above, the interval $[a, c]$ was partitioned into two subintervals and we concluded that if f has Lipschitz constant L in each subinterval then L is a Lipschitz constant for the whole interval. If we have more than two subintervals, the same conclusion can still be reached, by fusing the subintervals one by one.

A piecewise linear function is obviously Lipschitz continuous on every interval where it is linear, with the absolute value of the slope as Lipschitz constant. If the number of line segments is finite then the largest one of these absolute values will be a Lipschitz constant for each of the subintervals and hence for the whole interval.

25.8 If $g(x) = f(x) - cx$, we have $f(x) - B^n f(x) = g(x) - B^n g(x)$ because a linear function cx passes through the operator B^n unchanged. If we take c to be the arithmetic mean of the l.u.b. and the g.l.b. of $(f(y) - f(x))/(y - x)$, $L_{\min}$ will be a Lipschitz constant for g. Theorem 25.6 then gives the required result.

26.1 The sequence $a_0, a_1, a_2, \ldots$ is an arithmetic progresion. The result follows from Theorem 26.2, 1).

26.2 We can compute the result of applying B^n to the function x^3 by using (26.7). We get

$$B^n\left(x^3\right) = \frac{1}{n^2}(x + 3(n-1)x^2 + (n^2 - 3n + 2)x^3).$$

from which the result follows.

26.3 $B_k^n(x)$ has x^k as a factor, hence for $k > i$ all the powers of x appearing in the polynomial $y_k B_k^n(x)$ have exponents $>i$.

26.4 The statement is that $B^n(f - g)$ is identically 0 only if $f - g$ vanishes at $x = x_0, x_1, \ldots, x_n$. In our previous notation, we have to prove that $B^n f$ is not identically 0 if at least one of the $y_k \neq 0$.

Let y_k be the first nonzero entry in the sequence $y_0, y_1, \ldots$. Since all subsequent terms in the Bernstein polynomial have a factor x^{k+1}, we have $B^n f(x) = x^k(y_k\binom{n}{k} + a_1 x + \cdots + a_{n-k}x^{n-k})$. For sufficiently small positive values of x both factors are different from 0.

26.5 It suffices to show that for $0 \le j \le n$, x^j can be so represented. That this is possible, and a formula for the representation, follow from the identity $x^j = x^j(x + (1-x))^{n-j}$.

26.6 We have $a = f(0) > 0$ and $c = f(1) > 0$. We write $f(x)$ as

$$f(x) = (\sqrt{a}(1-x) - \sqrt{c}x)^2 + 2(\sqrt{ac} + b)x(1-x).$$

We see that if in addition to $a > 0$ and $c > 0$ we also have $\sqrt{ac} + b > 0$, then $f(x)$ is > 0 on the interval $[0, 1]$. The quantity inside the square root is positive for $x = 0$ and negative for $x = 1$. Hence it vanishes somewhere inside the interval, and at that point $f(x)$ has the same sign as $\sqrt{ac}+b$. Thus a necessary and sufficient condition is that $a > 0,\ c > 0,\ b > -\sqrt{ac}$.

26.7 We can represent ax^2+bx+c in the form $\alpha(x+1)^2+2\beta(x+1)(1-x)+\gamma(1-x)^2$. Comparing coefficients gives

$$\alpha = \tfrac{1}{4}(a+b+c), \quad \beta = \tfrac{1}{4}(c-a), \quad \gamma = \tfrac{1}{4}(a-b+c).$$

We conclude, as in the previous exercise, that the set of inequalities $\alpha \ge 0,\ \gamma \ge 0, \beta \ge -\sqrt{\alpha\gamma}$ is a necessary and sufficient condition for the polynomial to be nonnegative in $[-1, 1]$. In the case when $\alpha\gamma = 0$, the necessity of the condition $b \ge 0$ requires a different justification but it is very simple to provide it.

26.8 Think of i as fixed for the rest of the proof. We will use j to denote variable indices. Although no function f is named in the problem, it will be easier to relate the proof to the proof of Theorem 26.3 if we think of the y_j as the values of some function $f : y_j = f(x_j)$.

The second differences of a linear function are all 0, and a linear function is unchanged by the operator B^n. Thus it suffices to prove (26.23) for any function which differs from f by a linear function.

Consider first the case $M = 0$. Then we have $\Delta^2 x_j \le 0$ or $\Delta x_j \ge \Delta x_{j+1}$, i.e., the sequence of differences $x_{j+1} - x_j$ is nonincreasing. Hence we can subtract from $f(x)$ a linear function to obtain a function $g(x)$ such that $g(x_0), g(x_1), \ldots, g(x_i)$ is a nondecreasing sequence, $g(x_i) = 0$ and $g(x_i), g(x_{i+1}), \ldots, g(x_n)$ is a nonincreasing sequence. It follows that $B^n g(x)$ is negative or 0 on the whole interval $[0, 1]$ and in particular, at x_i. This completes the proof of the case $M = 0$. The transition to the case $M \ne 0$ is the same as in the proof of Theorem 26.3.

26.9 The graph of $B^n f$ is obtained as a limit of degree elevations from the control polygon Π of $B^n f$. It is rather obvious that if we replace a segment AB of the graph of a convex function by the chord AB, the resulting curve is the graph of a convex function. A proof of this can be found in Appendix C. Consequently, if f is convex, the control polygon of $B^n f$ is the graph of a convex piecewise linear function. The construction of the degree elevated polygon consists of replacing parts of this graph by chords, so the result is again convex. We have shown that the graph of $B^n f$ is the limit of degree elevations so it, too, is convex.

26.10 The formula for the derivative of a Bernstein basis polynomial follows easily from the product rule and the power rule of differentiation.

To derive (26.13) for the derivative of $B^n f$:

$$\frac{d}{dx}\sum_{0}^{n} f_k B_k^n(x) = n\sum_{0}^{n} f_k\left(B_{k-1}^{n-1}(x) - B_k^{n-1}(x)\right) = \sum_{0}^{n-1}\frac{f_{k+1}-f_k}{1/n}\,B_k^{n-1}(x);$$

two terms of the expanded first sum are missing from the second sum but they are both equal to 0.

26.11 For the purpose of this proof, let the phrase "change of direction" mean: change from increasing to decreasing or from decreasing to increasing. We have to show that the number of changes of direction of $B^n f$ on $(0, 1)$ is at most equal to the number changes of direction of f. Clearly the number of changes of direction of a piecewise linear interpolant of f is not more than that of f. By Farin's Theorem 26.4 we can get $B^n f$ from f by an infinite sequence of piecewise linear interpolations. If $B^n f$ has c changes of direction then all close enough approximants of $B^n f$ also have at least c changes of direction. Hence c is at most equal to the number of changes of direction of f.

27.3 Let the cubic Bézier curve $\mathbf{P}(t)$ have control points $\mathbf{P}_0, \mathbf{P}_1, \mathbf{P}_2, \mathbf{P}_3$. Hence we have $\mathbf{P}_0 = \mathbf{P}(0) = (0, 0)$, $\mathbf{P}_3 = \mathbf{P}(1) = (3, 0)$. By $\mathbf{P}'(0) = 3(\mathbf{P}_1 - \mathbf{P}_0)$, we get $\mathbf{P}_1 = \mathbf{P}_0 + \mathbf{P}'(0)/3 = (0, 1)$. Similarly, from $\mathbf{P}''(0) = 6(\mathbf{P}_2 - 2\mathbf{P}_1 + \mathbf{P}_0)$, it follows that $\mathbf{P}_2 = (2, -2)$. Finally, we obtain $\mathbf{P}'(1) = 3(\mathbf{P}_3 - \mathbf{P}_2) = (3, 6)$.

27.4 Let (a, b) be any point. Then the highest power of t in both terms of the sum $r(t) = (x(t)-a)^2 + (y(t)-b)^2$ has a positive coefficient, hence $r(t)$ is a nonconstant polynomial. Let d be the degree of $r(t)$. Then $r(t)$ can not take on any value for more than d values of t. Thus the parametric curve is not an arc of a circle with center (a, b) for any point (a, b).

A rational representation of a circle can be obtained from any pair of polynomials $p(t), q(t)$ such that $p^2 + q^2$ is a perfect square. The simplest example is $p(t) = 2t$, $q(t) = 1 - t^2$, $p^2 + q^2 = (1 + t^2)^2$, which gives the representation

$$x(t) = \frac{2t}{1+t^2}, \quad y(t) = \frac{1-t^2}{1+t^2}.$$

27.5 We must have that $\mathbf{Q}_0 = \mathbf{P}_0 = (0, 0)$ and $\mathbf{Q}_2 = \mathbf{P}_4 = (6, 0)$. The tangent vector at $\mathbf{Q}_0$ must be parallel to $\mathbf{Q}_0\mathbf{Q}_1$ and $\mathbf{P}_0\mathbf{P}_1$, hence $\mathbf{Q}_1 = (h, h)$ for some real h. Similarly $\mathbf{Q}_1\mathbf{Q}_2$ must be parallel to $\mathbf{P}_3\mathbf{P}_4$. Thus $\mathbf{Q}_1 = (6, k)$ for some real k. We see that $h = k = 6$ and hence $\mathbf{Q}_1$ must be $(6, 6)$. We can easily verify that if we apply two degree elevations to these three control points, we do indeed get the five points given in the exercise.

28.1 One can easily verify that the unique solution of the system of equations which determine the spline is when all the m_j are equal to the slope m of the line. A more elegant way of arriving at the result is to note that the line is the unique curve which minimizes the integral of the square of the second derivative.

28.2 The second derivative of the cubic spline is 0 at both endpoints. The second derivative of a nonlinear cubic vanishes at only one point, and the second derivative of a nonlinear quadratic function does not vanish anywhere.

29.1 Take any function defined in the interval [0,1) whose integral over this interval is 0, and continue it forward and backward periodically.

29.2 We have

(S.6)
$$\frac{d}{dx}Sf(x) = f\left(x+\frac{1}{2}\right) - f\left(x-\frac{1}{2}\right).$$

Thus, if

(S.7)
$$f\left(x+\frac{1}{2}\right) - f\left(x-\frac{1}{2}\right) = f'(x)$$

and if $Sf(x) = f(x)$ at one point, say the origin, then $Sf(x) \equiv f(x)$. We can satisfy the last condition by taking f to be an odd function, not identically 0, in $\left[-\frac{1}{2}, \frac{1}{2}\right]$.

If we write (S.7) in the form

$$f\left(x+\frac{1}{2}\right) = f\left(x-\frac{1}{2}\right) + f'(x).$$

we see that it gives us $f(x)$ in the interval $\left(\frac{1}{2}, 1\right]$ from the values in $\left[-\frac{1}{2}, \frac{1}{2}\right]$. Once we have the function in the interval [0, 1] we can extend f to (1, 1.5], etc. We can go backwards and extend $f(x)$ to $-\infty$ also. However, there are two difficulties with this construction which we are now going to discuss.

The first one arises at the half-integer points, where the values used in the recursion jump back by 1/2. This can easily cause jumps in our function. To assure continuity at half-integer values, we require that the original function $f(x)$ defined on $[-1/2, 1/2]$ be 0 on an interval containing 0 in its interior and on intervals containing the two endpoints $-1/2$ and $1/2$ of the base interval. Then our continuation formula will give the value 0 near all half-integer values of x, and the extended function will be continuous where the recursion formula is switched.

The second point which requires attention is that in each new half-interval we are using the derivative of the function in the previous half-interval. In terms of the function on the base interval, this means forming higher and higher derivatives. Hence, to be able to extend the function to the entire real axis, the initial function must have infinitely many derivatives.

Another approach to the problem is try to find some simple function which is invariant under S. We have

$$\int_{x-1/2}^{x+1/2} e^{ct}\,dt = \frac{e^{c/2} - e^{-c/2}}{c} e^{cx}.$$

Thus $Se^{cx} = e^{cx}$ if

(S.8)
$$\frac{e^{c/2} - e^{-c/2}}{c} = 1.$$

It turns out that this equation has no real solution. Indeed one would not expect one; the functions e^{cx} are convex downward, and it is obvious from the geometry that for a nonlinear convex function $Sf(x) > f(x)$. However, the exponential function is defined for complex exponents by Euler's formula $e^{x+iy} = e^x(\cos y + i \sin y)$. It is easy to verify that the law of exponents remains valid with this definition. The natural logarithm of a complex number z with argument $\theta \in (-\pi, \pi]$ radians is $\ln z = \ln(|z|) + \theta$ (*principal value*), or any value differing from the principal value by an integer multiple of $2\pi i$. One can verify this by substituting in Euler's formula.

We used *Mathematica* to obtain $c \approx 1.95210902 + 1.74190298i$ as an approximate solution of (S.8). From this we can conclude that Se^{cx} is a function satisfying $Sf = f$. We did not contemplate complex functions when we posed the problem. However, if $Sf = f$ then the real and the imaginary parts of f must separately satisfy the equation. Hence $f(x) = e^{(1.95210902\ldots)x}\cos(1.74190298\ldots)x$ is a function invariant under S.

We glossed over a subtle point. One can easily confirm by computation that $c = 1.95210902 + 1.74190298i$ very nearly satisfies (S.8), but that does not logically imply that it is very close to a number which satisfies (S.8) exactly. In most instances, including the present one, this can be proved by means of the contracting mapping idea explained in Chapter 21. Most algorithms for solving nonlinear equations or systems of equations are based on the contracting mapping principle. Newton's method can be regarded as a way of transforming an equation into a form in which the contracting mappping method can be applied if we start near enough to a solution. The fact that the algorithm gives a result which satisfies the equations up to 6 or 8 digit accuracy makes it practically certain that the existence of an exact solution nearby can be demonstrated by using the contracting mapping principle.

29.3 The highest powers of x with nonzero coefficients in the polynomials

$$\text{(S.9)}\qquad \begin{aligned} x^k,\quad \Delta x^k &= (x+1)^k - x^k,\\ \Delta^2 x^k &= (x+2)^k - 2(x+1)^k + x^k, \ldots, \Delta^k x^k \end{aligned}$$

are $x^k, x^{k-1}, \ldots, x^0$. Every polynomial $p(x)$ of degree $\le k$ can be represented in exactly one way as a linear combination of the polynomials (S.9). Indeed, the coefficient of x^k in $p(x)$ will be the coeficient of x^k in the representation in terms of (S.9). Then the only way to get the right coefficient for x^{k-1} is to add the proper multiple of Δx^k, etc. In particular, a linear combination of the polynomials (S.9) is identically 0 only if all the coefficients are 0.

We have to prove that representation in terms of the polynomials (29.39) is also unique. The jth expression (S.9) contains exactly the first j functions x^k, $(x+1)^k$, $(x+2)^k, \ldots$ with nonzero coefficients. Hence if the a_j in the expression for $p(x)$ are not all 0, $p(x)$ can be represented by a combination of the (S.9) with coefficients which are not all 0, and we have seen that such a function is not identically 0.

29.4 The first $k-1$ derivatives of the function $\phi(x)$ are continuous at the nodes. Hence the difference between the two polynomials which represent ϕ on the intervals separated by $x_i = -\frac{1}{2}(k+1) + i$ is a polynomial of degree k which vanishes, together with all its derivatives of orders $\le k-1$, at $x = x_i$, so it has the form $a_i(x - x_i)_+^k$, where a_i is a constant. Hence

$$\text{(S.10)}\qquad \phi(x) = a_0(x - x_0)_+^k + a_1(x - x_1)_+^k \cdots + a_{k+1}(x - x_{k+1})_+^k$$

We have to show that the vector $[a_0, a_1, \ldots, a_{k+1}]$ is determined uniquely up to a constant factor.

For $x > x_{k+1}$ $\phi(x) \equiv 0$. In this range none of the terms in (S.10) are truncated, hence

$$\text{(S.11)}\qquad a_0(x - x_0)^k + a_1(x - x_1)^k + \cdots + a_{k+1}(x - x_{k+1})^k \equiv 0.$$

Take the $k+2$nd term to the right-hand side and use the result of Exercise 29.3,- which says that every polynomial of degree k has a unique representation in terms of the first $k+1$ terms of (S.11). This tells us that the coefficients a_j are determined once we know a_{k+1}, hence all polynomials with the properties ϕ differ only by constant factors, q.e.d.

29.5 The resulting system of $n+1$ equations for the remaining coefficient has a dominant diagonal.

29.6 Use the equations (29.33) to eliminate x_{-1} and x_{n+1} from (29.32). The result is a system of $n+1$ equations for $n+1$ variables with a dominant diagonal. Thus, there is always a unique solution for these equations. x_{-1} and x_{n+1} are then determined by (29.33).

29.7 We wish to apply Theorem 29.2 to approximate $f(x)$ by a function with continuous second derivative. Since this theorem assumes that $f(x)$ is defined for all x, extend f as a linear function with slope $\pm L$ for negative values of x until it reaches the x-axis and let it be 0 from then on. Do the same on the half-line $x > 1$.

We estimate the difference $B^n f(x) - f(x)$ by going from $f(x)$ to $B^n f(x)$ in three steps: first from $f(x)$ to its approximant $g_h(x) = S_h^2 f(x)$, then from g_h to its Bernstein polynomial, and finally from the Bernstein polynomial of g_h to the Bernstain polynomial of f. This is expressed by the inequality

$$\begin{aligned} &|B^n f(x) - f(x)| \\ \text{(S.12)}\qquad &\le |g_h(x) - f(x)| + |B^n g_h(x) - g_h(x)| + |B^n f(x) - B^n g_h(x)| \\ &\le |g_h(x) - f(x)| + |B^n g_h(x) - g_h(x)| + |B^n(f(x) - g_h(x))| \end{aligned}$$

By Theorem 29.2 the first term is $\le$ Lh/3. Since the Bernstein polynomial is a weighted mean of certain function values, the third term is also bounded by the same quantity. We discard some of the information in Theorem 25.4 by replacing $x(1-x)$ by its largest possible value, $\frac{1}{4}$. If M is an upper bound for $|g_h''(x)|$, then

$$|B^n g_h(x) - g_h(x)| \le \frac{M}{8n} \le \frac{L}{4nh},$$

where the last inequality follows from Theorem 25.2. Substituting these values in (S.13) we get

$$\text{(S.13)}\qquad |B^n f(x) - f(x)| \le \frac{2Lh}{3} + \frac{L}{4nh}$$

The last formula is true for any positive value of h. To get the best estimate, choose h so as to minimize the last sum in (S.13). The product of the two terms is $L^2/6n$ for all values of h. The sum is $\ge$ twice their geometric mean, i.e., it is $\ge \sqrt{2/3n}L$, and equality will hold if we choose h so as to make the two terms equal. This completes the solution of the exercise.

30.1 We do the 3-triangulation on the triangle ABC, see Fig. 30.3

In the figure, the centroid is denoted by G. Draw a line EF through G. If the line is parallel to one side of the triangle, the area difference is exactly one subtriangle in the 3-triangulation, its area is just one ninth of that of ABC. In the general case, let D be the intersection of EF and B_1B_2. It is clear that

$$\Delta B_1DG \cong \Delta C_1EG \quad \text{and that} \quad \Delta DB_2G \cong \Delta EA_1G.$$

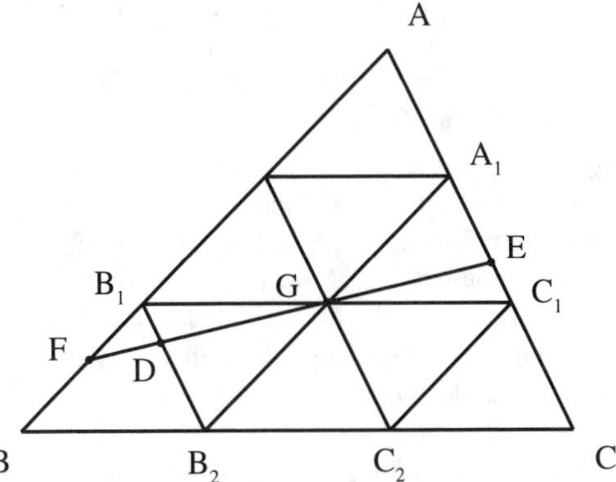

Figure 30.3. 3-triangulation.

Hence the difference is that between the areas of triangle B_1FD and the quadrilateral FBB_2D. Since these two figures form a subtriangle, the difference is at most the area of the subtriangle, one ninth of the area of triangle ABC.

30.2 If P and P' are on a line parallel to a side of the domain triangle, say on a line $w = const$, then it is easy to see that $|PP'| = |u' - u| = |v' - v| = |PP'|_t$. Thus, if the sides of a polygon $P_0P_1 \cdots P_n$ are parallel to the sides of the domain triangle, the length of the polygon is $|P_0P_1|_t + \cdots + |P_{n-1}P_n|_t \geq |P_0P_n|t$. To get a polygon for which equality holds, suppose $|u' - u| \leq |v' - v| \leq |w' - w|$. Let Q be the point whose first two coordinates are u' and v. Then $|PQ| = |u' - u|$ since the second coordinate is constant along PQ. Similarly, $|QP'| = |v' - v|$. Hence $|PQ| + |QP'| = |u' - u| + |v' - v| = |PP'|_t$.

30.3 We have by equation (30.6)

$$\begin{aligned}\big(B^n f\big)(P) &= \sum_{i+j+k=n} f(P_{i,j,k}) B^n_{i,j,k}(P) \\ &= \sum_{i+j+k=n} \phi(i/n) B^n_{i,j,k}(P) = \sum_{i=0}^{n} \phi(i/n) B^n_i(u).\end{aligned}$$

A linear function is a sum of linear functions of a single variable and hence is not changed by B^n. To reduce $B^n uv$ to the one-variable case, use $uv = \frac{1}{2}(u + v)^2 - u^2 - v^2 = (1 - w)^2 - u^2 - v^2$ and the formula $B^n x^2 = x^2 + \frac{1}{n}x(1 - x)$ for the Bernstein polynomial of a quadratic function.

30.4 Assume

(S.14) $\quad B^n f(u, v, w) = 0$ for $u \geq 0,\ v \geq 0,\ w \geq 0,\ u + v + w = 1.$

$B^n f$ is a homogeneous polynomial, i.e., each term has the same the total degree n. If we multiply all the variables by a factor t, the function value is multiplied by t^n. Hence (S.14) remains true even when the condition $u + v + w = 1$ is omitted.

$B^n f(u, v, w)$ is a polynomial of degree n in w. The coefficient of w^k is a polynomial $C_k(u, v)$ in u and v. For any fixed values of u and v with $u \geq 0$,

$v \ge 0,\ u + v < 1$, $B^n f(u, v, w)$ vanishes for more than n values of w. Hence all its coefficients are 0, i.e., $C_k(u, v) = 0$ for $u \ge 0,\ v \ge 0$. If we now expand $C_k(u, v)$ in powers of v and use the same argument, we see that all the coefficients in the polynomial $B^n f$ are 0.

30.5 Let $A = (x_a, y_a, z_a)$ and $B = (x_b, y_b, z_b)$ be two points on the cone represented by f. We have to show the points of the chord AB lie on or above the cone.

Project, from the origin, the upper half-space onto the plane $z = 1$. The formula for this transformation is $(x, y, z) \to (\frac{x}{z}, \frac{y}{z}, 1)$. The cone is projected onto the circle $C : \{(x, y, z) \mid x^2 + y^2 = 1,\ z = 1.\}$ The points of the upper halfspace which are above the cone, i.e., the points with $z > \sqrt{x^2 + y^2}$, are projected into the interior of the circle and points below it are projected into the exterior. A straight line is projected into a straight line, so the chord AB is projected into a chord of the circle C. Hence the projection of AB is inside C and thus AB is above the cone.

The reader might wonder about the following. The only reason for giving a proof of something which is as obvious as the object of this exercise is to practice the art of rigorous deduction. Yet, if one uses geometric images it difficult to be sure that one is not inadvertently using something that has not been justified. In the proof above, we have "projected" and talked about the "interior" and "exterior". In a more general situation, the last two concepts are not easy to define. However, in the present context all the geometric concepts can be replaced by simple definitions in terms of functions and inequalities. We leave this to the reader.

30.6 In this example we denote our triangular coordinates by x, y, w, where $w = 1 - x - y$. The polynomial is

$$xw + yw + x^2 + \sqrt{2}xy + y^2 = x + y - (2 - \sqrt{2})xy.$$

A way to see that this is not a convex function in the domain triangle is to look at it along the line $x = y$. There the function is given by $B^2 f(x, x) = 2x - (2 - \sqrt{2})x^2$, whose graph is a parabola opening downwards.

30.7 In our discussion of the definition of barycentric coordinates of a point P, we explained that for any choice of the point O we have $\overrightarrow{OP} = u\overrightarrow{OA} + v\overrightarrow{OB} + w\overrightarrow{OC}$. Taking $O = C$ we get (30.16). The Cosine Law now gives (30.17) and using $u + v + w = 1$ this can be transformed into the symmetric form (30.18).

30.9 If h_A is the altitude from the point A, the distance between the line $u = 0$ and the line $u = 1$ is h_A. Hence the distance between the lines $u = u_1$ and $u = u_2$ is $h_A|u_2 - u_1|$. Now it is easy to see the truth of the first inequality (30.20). To derive the second inequality, assume without loss of generality that $|u' - u| \le |v' - v| \le |w' - w|$. By (30.16),

$$|PP'| = |(u' - u)\,\overrightarrow{CA} + (v' - v)\,\overrightarrow{CB}| \le |PP'|_t \max(b, c)$$

from which the second inequality (30.20) follows. The relations between the Lipschitz constants based on straight line and triangular distance are obtained by applying the inequalities (30.20) to the definitions of these Lipschitz constants.

31.1 Consider the numbers to be z-coordinates of points above the nodes of $\mathcal{T}^n$. Let PQR and PRS be two triangles lying above adjacent triangles of $\mathcal{T}^n$ of this polyhedron. The condition then says that the midpoint of QS is the midpoint of PR. Thus any two adjacent triangles are coplanar and hence the points all lie in one plane. Thus the values assigned to the nodes are the values of a linear function $f_1(P)$. Let $f_2(P)$, $f_3(P)$ be the linear functions with values b, c, a and c, a, b at the vertices A, B, C. The sum of the values at the nodes is the same for f_1, f_2, f_3 and for $f = f_1 + f_2 + f_3$. The function f is linear and has the value $(a+b+c)/3$ at A, B, and C, hence it is constant. Hence the required sum is $(a+b+c)/3 \times$ (the number of nodes of $\mathcal{T}^n$) $= (a+b+c)(n+1)(n+2)/6$).

31.2 $B^3(1/2, 1/4, 1/4) = 7/4$.

31.3 We do the computation for b); a) is done similarly. An inverted triangle $T^n_{i,j,k}$ of $\mathcal{T}^n$ has vertices

$$\left(\frac{i-1}{n}, \frac{j}{n}, \frac{k}{n}\right), \quad \left(\frac{i}{n}, \frac{j-1}{n}, \frac{k}{n}\right), \quad \left(\frac{i}{n}, \frac{j}{n}, \frac{k-1}{n}\right),$$
$$i + j + k = n + 1, \quad i, j, k \geq 1.$$

The center C of similarity of $T^n_{i,j,k}$ and the domain triangle divides each of the three line segments connecting corresponding vertices of $T^n_{i,j,k}$ and the domain triangle in the ratio of the sizes of the triangles, which is $1 : n$. Hence

$$C = \frac{n}{n+1}\left(\frac{i-1}{n}, \frac{j}{n}, \frac{k}{n}\right) + \frac{1}{n+1}(1,0,0) = \left(\frac{i}{n+1}, \frac{j}{n+1}, \frac{k}{n+1}\right)$$

which is indeed the point $P^{n+1}_{i,j,k}$.

The relation between the 6-triangulation and the 7-triangulation is shown in Fig. 31.7. Here we add a figure showing just an interval subdivided into 6 and 7 equal parts, which may make it easier to see the reason for the relations we established.

Figure 31.7. Division of an interval into 6 and 7 congruent parts.

In the leftmost interval of the 6-division the subdivision point of the 7-division is 6/7th of the way to the right, etc. In the last interval it is only 1/7th of the way to the right.

Looking now at the 7-intervals, we see the leftmost one has no 6-subdivision point in its interior. The next one has one 1/6th of the way to the right, the one after that has it 2/6th of the way to the right, etc. The centers of similarity relation is somewhat easier to visualize here than in two dimensions.

32.1 The hypothesis is that the three-index array f is such that the equation $(E_i - E_k)(E_j - E_k)f \equiv 0$ holds for any permutation (i, j, k) of $(1, 2, 3)$. We abbreviate this as

(S.15) $$(E_j - E_i)(E_k - E_i) = 0.$$

Add the equation (S.15) with i and j interchanged to (S.15). The result can written in the form

$$E_i E_j = \tfrac{1}{2}E_i^2 + \tfrac{1}{2}E_j^2. \tag{S.16}$$

These equations imply that (S.15) holds even when i, j, and k are not distinct.

It will be easier to follow the rest of the computation if we use the operators $E_1 E_3^{-1} = \widetilde{E}_1$ and $E_2 E_3^{-1} = \widetilde{E}_2$ we introduced in (31.12). Then the equations (S.15) with $k = 3$ become

$$(\widetilde{E}_1 - I)(\widetilde{E}_1 - I) = 0,\ (\widetilde{E}_1 - I)(\widetilde{E}_2 - I) = 0,\ (\widetilde{E}_2 - I)(\widetilde{E}_2 - I) = 0. \tag{S.17}$$

Let $\widetilde{E}_1 - I = A,\ \widetilde{E}_2 - I = B$. Then

$$B^n f = (uE_1 + vE_2 + wE_3)^n f_{0,0,0} = (uA + vB + I)^n f_{0,0,n}.$$

Since $A^2 = AB = B^2 = 0$, the nth power of the operator reduces to just the terms that are linear in A and B:

$$B^n f = (nuA + nvB + I) f_{0,0,n} = (nuA + nvB + (u + v + w)I) f_{0,0,n}$$

This is a linear function of u, v, and w. Since the Bernstein polynomial has the given values at the three vertices, $B^n f$ must be the linear function given in the exercise.

Note that the above argument also furnishes an alternative proof of the important fact that a linear function is unchanged by the Bernstein operator, see Exercise 26.3).

Index